THE EMERGENCE OF COSMIC STRUCTURE

Previous Proceedings
in the Series of Annual Astrophysics Conferences
in College Park, Maryland

	Year	Short Title	Publisher	ISBN
Twelfth	2001		*"unpublished"*	
Eleventh	2000	Young Supernova	AIP Conference Proceedings 565	0-7354-0001-6
Tenth	1999	Cosmic Explosions	AIP Conference Proceedings 522	1-56396-943-2
Ninth	1998	After the Dark Ages	AIP Conference Proceedings 470	1-56396-855-X
Eighth	1997	Accretion Processes	AIP Conference Proceedings 431	1-56396-767-7
Seventh	1996	Star Formation	AIP Conference Proceedings 393	1-56396-678-6

Other Related Titles from AIP Conference Proceedings

655 Particle Physics and Cosmology: Third Tropical Workshop on Particle Physics and Cosmology – Neutrinos, Branes, and Cosmology
Edited by José F. Nieves and Chung N. Leung, February 2003, 0-7354-0112-8

624 Cosmology and Elementary Particle Physics: Coral Gables Conference on Cosmology and Elementary Particle Physics
Edited by B. N. Kursunoglu, S. L. Mintz, and A. Perlmutter, July 2002, 0-7354-0073-3

616 Experimental Cosmology at Millimetre Wavelengths: 2K1BC Workshop
Edited by Marco DePetris and Massimo Gervasi, May 2002, 0-7354-0062-8

609 Astrophysical Polarized Backgrounds: Wkshp on Astrophysical Polarized Backgrounds
Edited by Stefano Cecchini, Stefano Cortiglioni, Robert Sault, and Carla Sbarra,
March 2002, 0-7354-0055-5

586 Relativistic Astrophysics: 20th Texas Symposium
Edited by J. Craig Wheeler and Hugo Martel, October 2001, 0-7354-0026-1

579 Radio Detection of High Energy Particles: First International Wkshp; RADHEP 2000
Edited by David Saltzberg and Peter Gorham, July 2001, 0-7354-0018-0

575 Astrophysical Sources for Ground-Based Gravitational Wave Detectors
Edited by Joan M. Centrella, July 2001, 0-7354-0014-8

555 Cosmology and Particle Physics: CAPP 2000
Edited by Ruth Durrer, Juan Garcia-Bellido, and Mikhail Shaposhnikov, March 2001,
1-56396-986-6

523 Gravitational Waves: Third Edoardo Amaldi Conference
Edited by Sydney Meshkov, June 2000, 1-56396-944-0

To learn more about these titles, or the AIP Conference Proceedings Series, please visit the webpage
http://proceedings/aip.org/proceedings

THE EMERGENCE OF COSMIC STRUCTURE

Thirteenth Astrophysics Conference

College Park, MD   7-9 October 2002

EDITORS
Stephen S. Holt
Franklin W. Olin College of Engineering
Needham, Massachusetts

Christopher S. Reynolds
University of Maryland
College Park, Maryland

SPONSORING ORGANIZATIONS
NASA
  Goddard Space Flight Center
University of Maryland

Melville, New York, 2003
AIP CONFERENCE PROCEEDINGS ■ VOLUME 666

Editors:

Stephen S. Holt
Franklin W. Olin College of Engineering
1735 Great Plain Avenue
Needham, MA 02492
USA

E-mail: steve.holt@olin.edu

Christopher S. Reynolds
Department of Astronomy
University of Maryland
College Park, MD 20742
USA

E-mail: chris@astro.umd.edu

The articles on pp. 103-112, 171-181, 245-248, 337-346, and 347-354 were authored by U.S. Government employees and are not covered by the below mentioned copyright.

Authorization to photocopy items for internal or personal use, beyond the free copying permitted under the 1978 U.S. Copyright Law (see statement below), is granted by the American Institute of Physics for users registered with the Copyright Clearance Center (CCC) Transactional Reporting Service, provided that the base fee of $20.00 per copy is paid directly to CCC, 222 Rosewood Drive, Danvers, MA 01923. For those organizations that have been granted a photocopy license by CCC, a separate system of payment has been arranged. The fee code for users of the Transactional Reporting Service is: 0-7354-0128-4/03/$20.00.

© 2003 American Institute of Physics

Individual readers of this volume and nonprofit libraries, acting for them, are permitted to make fair use of the material in it, such as copying an article for use in teaching or research. Permission is granted to quote from this volume in scientific work with the customary acknowledgment of the source. To reprint a figure, table, or other excerpt requires the consent of one of the original authors and notification to AIP. Republication or systematic or multiple reproduction of any material in this volume is permitted only under license from AIP. Address inquiries to Office of Rights and Permissions, Suite 1NO1, 2 Huntington Quadrangle, Melville, N.Y. 11747-4502; phone: 516-576-2268; fax: 516-576-2450; e-mail: rights@aip.org.

L.C. Catalog Card No. 2003105090
ISBN 0-7354-0128-4
ISSN 0094-243X
Printed in the United States of America

CONTENTS

Preface ix

INTRODUCTORY OVERVIEW

Emergent Structure: The First Two Centuries of the First Two Eons 3
   V. Trimble

THE ORIGINS OF STRUCTURE

Beyond Cosmological Parameters 19
   M. Tegmark
The Origin of Density Fluctuations in a Cyclic Universe 33
   P. J. Steinhardt and N. Turok

STRUCTURE IN THE COSMIC MICROWAVE BACKGROUND

CMB Observables and Their Cosmological Implications 2002 45
   W. Hu
Repercussions of Structure Emergence on the CMB Polarization 59
   S. T. Staggs
The Absence of the Integrated Sachs-Wolfe Effect: Constraints on a Cosmological Constant 67
   S. P. Boughn and R. G. Crittenden

THE FIRST STARS, REIONIZATION, AND INDUCED STRUCTURE

The First Sources of Light 73
   V. Bromm and A. Loeb
On the Detectability of the Cosmic Dark Ages: 21-cm Lines from Minihalos 85
   H. Martel, P. R. Shapiro, I. T. Iliev, E. Scannapieco, and A. Ferrara
Photoevaporation of Minihalos during Reionization 89
   P. Shapiro, I. T. Iliev, A. C. Raga, and H. Martel
Mass Limits to Primordial Star Formation from Protostellar Feedback 93
   J. C. Tan and C. F. McKee
The Number of Supernovae from Primordial Stars in the Universe 97
   J. H. Wise and T. Abel

THE NATURE OF DARK MATTER

CDM Substructure in Gravitational Lenses: Tests and Results.............. 103
    C. S. Kochanek and N. Dalal
Probing the Distribution of Mass via Gravitational Lensing 113
    P. Natarajan
Galaxies and Halos in the Sloan Digital Sky Survey....................... 123
    T. A. McKay (for the SDSS Collaboration)
The Dark Matter Distribution in Galaxy Cluster Cores..................... 135
    J. S. Arabadjis, M. W. Bautz, and G. Arabadjis
X-Ray Measurement of Dark Matter "Temperature" in Abell 1795 139
    Y. Ikebe, H. Böhringer, and T. Kitayama
Dark Matter, AGNs, and Relativistic Jets................................. 143
    W. K. Rose
Cosmological Simulations of the Formation of Primordial Gas Clouds........ 147
    N. Yoshida
Halo Substructure and the Power Spectrum 151
    A. R. Zentner and J. S. Bullock

THE INTERGALACTIC AND INTRACLUSTER MEDIUM

The Lyman-α Forest as a Cosmological Tool 157
    D. H. Weinberg, R. Davé, N. Katz, and J. A. Kollmeier
The Entropy in Groups—A Clue to Galaxy Formation 171
    R. Mushotzky
Nonthermal Particles and Radiation Produced by Cluster
Merger Shocks... 183
    R. C. Berrington and C. D. Dermer
Intermittency and Large-Scale Structure in the Universe: A New
Window for the Study of the Nonlinear Regime of
Structure Formation.. 187
    P. Jamkhedkar and L.-Z. Fang
The Galaxy Proximity Effect in the Ly-α Forest 191
    J. A. Kollmeier, D. H. Weinberg, R. Davé, and N. Katz
Keck Absorption-Line Spectroscopy of Galactic Winds in Massive
Infrared-Luminous Galaxies 195
    D. S. Rupke, S. Veilleux, and D. B. Sanders
A Filament between Galaxy Clusters A3391 and A3395..................... 199
    E. Tittley and M. Henriksen

HIGH-REDSHIFT STRUCTURE

Active Galaxies from Z=0 to Z=6....................................... 205
    A. J. Barger
The Star Formation History of Galaxies from Infrared Observations 215
    A. Franceschini and G. Rodighiero

Formation and Evolution of Supermassive Black Holes in Galactic Centers: Observational Constraints........................227
  G. Hasinger (for the CDF-S Team)
Formation of Supermassive Black Holes................................237
  S. Kainer and W. K. Rose
Discovery of Galaxies in the z=1.5-2.5 "Bright Ages".....................241
  J. W. Colbert, M. Malkan, and M. Rich
Multi-wavelength Luminosity Functions of Galaxies.......................245
  J. P. Gardner
Galaxy Clusters to $z \leq 1$ from the Oxford Dartmouth Thirty Degree Survey...249
  M. Hammell, G. Wegner, L. Moustakas, P. Allen, G. Dalton, and E. Olding
Rest Frame Optical Spectra of Lyman Break Galaxies: Other Lensing Arcs around MS1512-cB58...253
  M. A. Malkan, H. I. Teplitz, and I. S. McLean
The Clustering of Massive Galaxies at $z \sim 1$.............................257
  R. A. Overzier, H. J. A. Röttgering, R. J. Wilman, and R. B. Rengelink
Faint AGN and the Ionizing Background................................261
  M. Schirber and J. S. Bullock
On the Nature of Lyman-α Emitters.....................................265
  J. Wang, S. Malhotra, J. Rhoads, M. Brown, T. Heckman, and C. Norman
Some Results from the Oxford-Dartmouth Thirty Degree Survey............269
  G. Wegner, M. Hammell, G. B. Dalton, P. D. Allen, E. J. Olding, and L. Moustakas

LOW-REDSHIFT STRUCTURE

Implications of 2d FGRS Results on Cosmic Structure.....................275
  J. A. Peacock (for the 2dF Galaxy Redshift Survey Team)
Testing Cosmological Models with Clusters of Galaxies....................291
  H. Böhringer and P. Schuecker
SHUFFLE: A New Statistical Bootstrap Method: Applied to Cosmological Filaments...303
  S. P. Bhavsar, S. Bharadwaj, and J. V. Sheth
A National Virtual Observatory Science Case: Properties of Very Luminous IR Galaxies...307
  K. D. Borne, S. Arribas, H. Bushouse, L. Colina, and R. Lucas
X-Ray Analysis of the MS0302 Supercluster.............................311
  D. J. Horner and M. Donahue
Recent Results from the NOAO Fundamental Plane Survey.................315
  J. E. Nelan, G. A. Wegner, R. J. Smith, M. J. Hudson, J. J. Malecki, J. R. Lucey, S. A. W. Moore, S. J. Quinney, D. Schade, N. B. Suntzeff, and R. L. Davies
Cosmic Structure Traced by Precision Measurements of X-Ray Brightest Galaxy Clusters in the Sky......................................319
  T. H. Reiprich, C. L. Sarazin, J. C. Kempner, M. F. Skrutskie, G. R. Sivakoff, H. Böhringer, and J. Retzlaff

FUTURE PROSPECTS

The Future of Microwave Background Physics 325
    A. Kosowsky

What Can We Learn about Cosmic Structure from Gravitational Waves? 337
    J. M. Centrella

Prospects for Future Observations in the Mid/Far IR 347
    J. C. Mather

OTHER TOPICS

Prevalence of Magnetic Field Forces and Universal Symmetry of Nature 357
    Y. Cinar

Time and Entropy 361
    V. I. Garaimov

The Joint Emergence of Large-Scale Structure and Large-Scale Magnetic Fields after Combination Time 365
    H. D. Greyber

List of Attendees 369
Author Index 373
Subject Index 375

Preface

This is the thirteenth installment of the current series of annual October Astrophysics Conferences in Maryland. These conferences are organized by astrophysicists at the Goddard Space Flight Center and the University of Maryland. The topic for each conference is selected with the help of an International Advisory Committee, the current membership of which is:

Marek Abramowicz, Göteborg   *Jim Peebles*, Princeton
Roger Blandford, Pasadena   *Sir Martin Rees*, Cambridge, UK
Claude Canizares, Cambridge   *Vera Rubin*, Washington
Arnon Dar, Haifa   *Joseph Silk*, Oxford
Alan Dressler, Pasadena   *David Spergel*, Princeton
Guenther Hasinger, Munich   *Rashid Sunyaev*, Moscow
Steve Holt, Needham   *Yasuo Tanaka*, Tokyo
Bob Kirshner, Cambridge   *Scott Tremaine*, Princeton
Dick McCray, Boulder   *Simon White*, Munich

In the spirit of this series, where we attempt to identify "hot" topics with broad appeal to both observers and theoreticians, *"The Emergence of Cosmic Structure"* clearly qualifies as an appropriate conference theme. Interestingly, this year's topic was close to that of the very first entry in this series twelve years before, which was entitled *"After the First Three Minutes."* The programme for this year's conference was developed by a Scientific Organizing Committee comprised of:

Chuck Bennett   *Bruce Margon*
Joan Centrealla   *Richard Mushotzky*
Guenther Hasinger   *Chris Reynolds*
Steve Holt   *Simon White*

After a brief welcoming address, the conference opened with what is now a traditional tandem of introductory invited presentations. The first of these was an review of the history of attempts to explain cosmic structure by *Virginia Trimble*, and the second, delivered by *Simon White*, defined the outstanding issues to which the conference attendees were obliged to address their attention. The programme of non-paralleled sessions then proceeded through the next two days. Each session was devoted to a specific topic with two or three invited talks and an extensive discussion period that included one-minute, one-viewgraph opportunities for each conference attendee to advertise his/her poster paper.

The banquet at the conclusion of the second day of each of these conferences always features a distinguished speaker. This year our speaker was *Neta Bahcall*, who shared some of her insights and experiences in a career devoted to the study of cosmic structure.

As usual, *John Trasco* and *Susan Lehr* made sure that all the logistics were handled flawlessly. Thanks also to *Chris Reynolds*, the co-editor of these Proceedings, and to all of the attendees who contributed to the success of the meeting.

Steve Holt
January 2003

INTRODUCTORY OVERVIEW

Emergent Structure: The First Two Centuries of the First Two Eons

Virginia Trimble[†]

*Astronomy Department, University of Maryland, College Park MD 20742
and Department of Physics, University of California, Irvine CA 92697*

Abstract. Scientific recognition of the existence, evolution, and significance of structure within the cosmos developed slowly. We follow the story here from the earliest times to the first systematic redshift surveys and the "Rubin-Ford effect," emphasizing the period beginning with William Herschel and ending about lunch time on Wednesday. The scientific issues cannot be put in any one linear order, because, for instance, some people were studying clusters of galaxies and measuring the mass of M31 while others still denied the existence of external galaxies.

INTRODUCTION

The images shown at the conference and the ideas presented there and here have been drawn from a very large number of secondary and primary sources. Important general ones include Berendzen (1976), Jaki (1972), Whitney (1971), Hoskin (1997), the introductory chapters of Peebles (1993), Harrison (1981, 1987), Smith (1982), and Martinez et al. (2002). Sources with primarily 20th century content include Bok and Bok (1945), Shapley (1943), Lundmark (1956), the chapters in Sandage et al. (1975) by D. Layzer, E. Holmberg, G.B. Field, C.D. Shane, G. de Vancouleurs, G.O. Abell, and A. Sandage, McVittie (1962), especially the articles by E L. Scott and G.B. van Albada, Neyman et al. (1961), and Trimble (1995, 1996, 1997, 1999). References that can be found in these last and in introductions to other "October" proceedings have often not been repeated here.

THE ANCIENTS TO NEWTON

The cosmologies of the Mesopotamians (one version of which you can find in Genesis) and of the Egyptians had a flat earth with square corners and one or more deities carrying out various tasks. The Egyptian Shu (air), for instance, held Nut (sky) up above Geb (earth), so that the sun (Aten) could sail his day and night boats around the configuration. The "stick man" of the Early harvard slides might be thought of as a modern Shu, though he inhabits a universe that is isotropic on larger scales. The Chinese Pan-ku (who also separated heavens from earth) served a similar function.

Assorted Greeks noticed that the earth is closer to a sphere than a plane and measured its size and the distance to the moon with reasonable precision. They also had a strong predilection for circles, spheres, and circular motion on all scales in the cosmos.

The merging of Greek philosophy with medieval church doctrine that was largely the work of Thomas Aquinas imposed this spherical symmetry on European thought after about 1260, along with immutability of the heavens, the four terrestrial elements + quintessence, and much else. Indeed many of the early readers of Copernicus were inclined to think that the most important thing he was saying was the primacy of uniform circular motion, whatever is at the center. But, before this, the 12th century universe of Hildegaard of Bingen had a spherical earth, but a pineapple-shaped lucidus ignis outside it. She also placed the fixed stars close to us, hail and lightening further out beyond the moon, followed by the sun and outer planets (as the stem of the pineapple). Thus she had not yet fully accepted immutability of the heavens as requiring hail, lightning, comets, meteors, and guest stars all to fall within the atmosphere. A Chinese model universe from about the same time is spherical and held up by dragons (Needham 1953). It also has an equatorial mount, rather than an ecliptic one like the European armillary spheres of the time, and the equatorial concept may well be a Chinese invention

Then the spheres and circles close in, first with earth at the center a la Ptolemy (and Martin Luther), then, increasingly, with the sun at the center, a la Copernicus.

Next is the question of where to put the stars. Outside the orbit of Saturn, clearly, but how far outside, how many of them, and how far should they extend? Early on, these questions tended to get stirred in with old philosophical considerations of whether voids and infinities were conceivable and hence possible. For some people, they probably still do.

Thomas Digges in 1576 extended his "orbe of starres fixed" to infinity (though not in the drawing) and allowed them to have planets and life. He said that the total assemblage though infinite was somehow vaguely spherical, but that it could have neither edges nor a center. Infinite, multiply inhabited, and more or less spherical universes are also to be found in the writings of Nicolas of Cusa (c. 1450), Giordano Bruno (before 1600!), and William Gilbert (d. 1602). Gilbert explicitly allowed the stars in his infinite distribution to be of intrinsically different size (meaning brightness).

Kepler (d. 1630) considered the possibility of an infinite, fairly uniform distribution of stars, of which the sun would be merely an undistinguished one, but rejected both this and a "stoic" universe (with infinite empty space beyond a single sphere of stars) in favor of a bounded cosmos, with a single spherical shell of stars not far outside the orbit of Saturn. Indeed he knew exactly where to put it by requiring that all the Platonic solids nest neatly within one another.

Supporters of finite vs. infinite distributions of stars co-existed right down to the 20th century. Descartes's 1636 world had an infinite number of vortices, with central stars illuminated by rotational motions (and planets forming in the whirls), and regions of influence that were neither spherical nor circular but look like Voronoi tessellation. Otto von Guericke, who in some sense discovered the vacuum (with the Magdeburg evacuated metal sphere and horses experiment) was perfectly happy to have a vacuum outside his sphere of stars. This was his solution to Olbers' paradox (see Harrison 1987 for more on how this consideration influenced early modern cosmologies). Newton in the 1660s seems to have been a finite "coelum stellatum" man, with some combination of chaos and void outside, but he later worried about the gravitational equivalent of Olbers in an infinite universe and evolved a sort of solution.

WILLIAM HERSCHEL AND THE SHAPE OF THE MILKY WAY

No one who actually looked at the sky would suppose that the distribution of stars around us is spherically symmetric, whether finite or infinite. The first disk galaxy does not, however, appear until the writings of Emmanuel Swedenborg (1734). He envisioned a sort of magnetic dipole structure (analogous to that found for the earth by Gilbert in 1600) and an infinite number of other Milky Ways filling space in a hierarchical arrangement.

Thomas Wright of Durham, who appears in the history paragraphs of many elementary astronomy texts, switched in about 1750 from a uniform ("promiscuous") distribution of stars to a slab, and his edge-on view of the Galaxy, though sun-centered, looks modern for its time, with stars of different intrinsic brightnesses and a reasonable diameter to thickness ratio. But his "face on" view is much more like a Greek central fire universe or an Eye of God. He also packed them hierarchically, again for partly Olbersian reasons.

William Herschel was an active contributor to astronomical thought from the 1760s until very close to his death in 1822. Thus he had plenty of time to change his mind and, eventually, rode all possible horses in the race involving nebulae all resolving into stars vs. some being truly diffuse, a single island universe (Galaxy) vs. other nebulae being other, comparable galaxies, and finite vs. infinite The picture with which he is most often associated dates from 1785 and charts the Milky Way via what he called star gauging. That is, assume all stars are the same real luminosities and figure out their distances from apparent brightness. Then distribute them more or less uniformly in space out to an edge set by the numbers you see in each direction. Inevitably, this puts the solar system in the plane of the Milky Way and very close to the center of a disk that extends a thousand light years in radius and a couple hundred in thickness., with slices in the edges corresponding to the Cygnus rift and such.

The elder Herschel's confidence in all stars being the same brightness was strong enough that he compiled lists of close pairs of unequal apparent brightness and followed their relative separation carefully) in the hopes of measuring parallax. In fact, he found the first few partial orbits of binary stars.

William Huggins settled the "resolvability" issue in 1868 in favor of truly diffuse, gaseous nebulae by finding emission lines in the specta of the Orion nebula and a couple of Herschel's planetaries. Majority opinion then settled in around a picture in which a disk-like galaxy or region of stars (with the sun near the center and filamented edges) had a region of nebulae on either side. The pictures, of course, had to have edges, so you could not be quite sure how far the zone of the nebulae extended!

Simon Newcomb (1878) favored such a picture for many years. He was, as director of the US Naval Observatory, the inevitable first president of the American Astronomical Society (called something else initially). He has had rather bad press within the community, largely for his opposition to astrophysics (meaning the introduction of spectroscopy into astronomy). It is, therefore, perhaps worth recalling that he also asked some very prescient questions at the turn of the previous century (Newcomb 1906). Among these were; (1) what is the size of the universe, and is there a boundary? (2) are the volume and duration finite or infinite? (3) what is the form and extent of the Milky Way, and is it conceivable that our apparent centrality means that we are the victims of some fallacy, like that afflicting Ptolemy? (Way to go, Simon!), and (4) where do the stars of large proper motion come from and go to? He had in mind that they would leave the visible galaxy in only millions of years, and the answer of course eventually involves one in identifying stellar energy sources, recognizing that star formation is a continuous process, and even invoking dark matter.

Some remarkable reconstructions of the Milky Way as it might appear from outside appear in the first decades of the 20th century, Easton (1900) knew that we were supposed to be at the center of the whole system but were clearly not at the center of the distribution of bright star clusters and nebulae. His face-on galaxy has a system of spiral arms whose center has been pushed off to one side in the direction of Cygnus, while we reside in a sparser region at $r = 0$ of a somewhat arbitrary circle. Eddington's edge on galaxy had a central cloud of stars and a ring around it (Eddington 1912).

Harlow Shapley counts as an official culture hero for getting us permanently out of the center of the Milky Way (Shapley 1919). He used pulsating variable stars to estimate distances to the globular clusters, which he recognized were heavily concentrated on one side of the sky, and so put us 20 kpc from the center of the whole Galaxy (and the spiral nebulae inside). Kapteyn, close to death in 1922, held by a much smaller disk, with the sun very near the center, and today is not much remembered, except for this "Kapteyn universe" and his selected areas.

The missing ingredient was, of course, interstellar obscuration (absorption and scattering) by dust. Kapteyn himself had worried about this in the early 1900s. Sanford (1917) and Curtis (1917, 1918) had assembled data suggesting to them that the "zone of avoidance" in the plane of the Milky Way, where few nebulae could be seen, was the result of obscuration in the plane, analogous to the dark lanes to be seen in many edge-on spirals. Indeed the whole Curtis-Shapley debate belongs to this era, and has as one of its subtexts the issue of interstellar dimming of light (as well as the existence of external galaxies and the official topic of the size of the galaxy).

Credit for the discovery of the effects of systematic obscuration goes, however, to Trumpler (1930) and his assemblage of star clusters, whose apparent angular diameters fell less steeply with apparent magnitude than implied by $1/r^2$ alone. Trumpler's (1941) own image of the galaxy at first looks like a wrong-headed attempt to please all parties by putting a Kapteyn universe off on one edge of a Shapley galaxy. What he meant, however, is that there is a small, off-center circle of the disk that we can survey, while the inner disk, center, and far side are largely obscured, except at high latitude. Plaskett's (1939) drawing is very much like the one we all put on the (black, marker, or virtual) boards in classrooms today. The sun is 10 kpc from the center (which was right from Oort's work in the 1930s until Baade shrank the galaxy in 1953, and again from a 1965 IAU resolution in favor of 10 to a successor in 1988, which moved us back in to 8.5 kpc).

Plaskett, Bok and Bok (1945, who drew just the outline of their Milky Way), and we all know that the galaxy is rotating, and that you must be sure to subtract off the rotation velocity if you want to do cosmology with redshifts. The rotation hypothesis was put forward by Lindblad (1924-26), and additional supporting data assembled by Oort (1927), who frequently gets credit for the discovery, though he titled his paper "Observational evidence confirming Lindblad's hypothesis of a rotation for the galactic system."

THE NUMBERS AND NATURE OF NEBULAE

At this point, we have to back up, because some astronomers were already working on this section while others were still arguing about the size of the Milky Way. Lundmark (1956) notes that 21 fuzzy things (including the Hyades and Pleiades, as well as the Orion nebula and Andromeda) were known to the ancients. Some of these must have been lost, since Halley was aware of only six in 1715. LaCaille catalogued 42 southern fuzzies in 1755, and Messier's better known, but less systematic, catalogue has 107. He apparently found their nature of no interest, except that they were non-comets.

William Herschel observed about 1000 nebulae and left drawings of many. Some are (with 20-20th century hindsight) obviously planetaries (which is what he called them); some are irregular HII regions; and others look like edge-on spirals. His telescopes did not, however, reveal any real structure within face-on disks, and he drew no spirals.

Meanwhile, however, Michell (1767) had "pre-discovered" star gauging and estimated the size of the Milky Way (1000 LY or so) . He then calculated that, if some of the more conspicuous nebulae (like M31 and M33) were about the same size, then their brightest individual stars should have V = 13.8. The modern value is 15.6 (but both our distance scales and maximum stellar luminosities have grown a good deal in the interim). Another way to see and say this (though apparently Michell did not) is that if M31 is 4 ° across in the sky and is rather like the Milky Way, then the distance between us is about 15 times the individual sizes.

The 19th was, altogether, a messy century for nebular astronomy. At midpoint, Rosse (1851) published a drawing of M51 that clearly shows the arms. He described it as a spiral (his word) with stars. Messier had called it a double nebula with no stars, W. Herschel a bright nebula with a halo and a companion, and J. Herschel a divided ring.

This brings us back to the beginning of the 20th century again, with the community still pretty firmly divided on the reality of other galaxies. Agnes Clerke is the person perhaps most often quoted to the effect that no astronomer in full possession of the data (and his faculties) believed in external galaxies. But others, whose names are now also largely forgotten (like Wolfe and Very) felt the same way in various languages.

Meanwhile, Lundmark (1919) had put M31 at 220,000 pc from the novae in it. Curtis's (1917) number was 6 Mpc for several nebulae with novae. And Opik (1922) had estimated a distance and a mass on the reasonable assumption that M/L for the disk ought to be like that of the solar neighborhood. He also had a rotation curve to work with obtained by Pease at Mt, Wilson.

Vesto Melvin Slipher is another of the undersung heroes, who belongs in this section because he took the first nebular spectrogram from which a radial velocity (though not rotation) could be measured (M31 in 1912). He got, rightly, about -300 km|sec. He had been set to the task of getting nebular spectra by Percival Lowell, who expected to find evidence that the spirals were solar systems in formation. Slipher initially concurred,, but by the time he had a dozen examples (most with large positive velocities) he began to suspect that he was seeing something else. He did not attempt a plot of those velocities vs. anything, at least in published form, though others did, by including a distance-dependent K term in their solutions for solar motion. You may, if you wish, regard one or more of these as the "discovery" of Hubble's law. Lemaitre and Wirtz (separately) have perhaps the best claims.

THE REALITY OF LARGER STRUCTURES

We will have to back up again, because many people (including both the Herschels) had noticed that the nebulae were not randomly scattered on the sky outside the zone of avoidance, long before the nature of those nebulae had been settled. In fact, if you happen to know how to distinguish a proper extragalactic nebula from a star cluster, HII region, or planetary nebula, you can discover Virgo, the Local Group, and, marginally, the Leo, CVn, and UMa clusters in Messier's catalog. If, however, you plot all his objects, then the largest concentration is actually globular clusters in the general vicinity of RH = 16-20 hours, declination = 0 to -40 (the galactic bulge), and M1 is, of course, the Crab Nebula, because the expected path of Halley's comet in its 1758 return came nearby.

In the same sort of way, Virgo, Coma, and one or two others can be seen in plots of the objects in J. Herschel's General Catalogue (e.g, Proctor 1869)and Dreyer's 1888 New General Catalogue (e.g. Waters 1894), though there are also concentrations that represent the Large and Small Magellanic Clouds and a certain number of "selected areas" where the cataloguers had looked harder than average. The last person to make such a plot without community agreement about what he was plotting was Charlier (1908, 1922). He was a strong supporter of a fractal or hierarchical universe and so expected to find structure on a range of scales, as indeed he did.

To make further progress, one needs some way of ruling out alternatives to real structure. Differential obscuration lives in Sect. 9. Let's look here at statistical flukes (such as the one that puts the brightest quasar, 3C 273, in the field of the Virgo cluster).

Michell (1767) tackled the statistics problem for binary stars, using the same sort of "one-minus-the-probability-of-the-opposite" that one uses to figure out how many people have to be in a room before two will share a birthday. He got the details somewhat wrong (S. Stigler, pr. comm. 2001), but was completely correct in concluding that (contrary to what Wm. Herschel thought) most pairs of stars in the sky, even of very unequal brightness, are physically associated, as are the star clusters. Polya (1919) redid the calculation, by thinking of putting n points on a sphere and not wanting any one to fall within an angle less than S of another. His answer was

$$P(n/S) = \cos^{2n}(S/2) \approx exp(-nS^24).$$

And Bok (1934) did it again, this time having galaxies in mind

As in the stellar case, galaxy pairs were the first structure statistically established, initially by Lundmark (1932), who found about 100 close pairs in NGC and the Shapley-Ames (1932) Catalogue, and then, more carefully, by Holmberg (1937), who was Lundmark's student and who had a very large (55,000) but very non-homogeneous sample of galaxies. He found 695 pairs less than 0.1° apart, where chance predicted 42. Holmberg was, in this same paper, the first to estimate galaxy masses from the hypothesis that the pairs were mostly bound by gravitation. His answer was an M/L ratio larger than that found by Hubble for single galaxies, but smaller than that found by Zwicky (1933) for the Coma cluster, but this is part of a different story (one version of which appears in Trimble 1987).

Next after binaries come small groups. Indeed they might well be the same thing, observed to different limiting magnitudes. The Local Group is a binary if you see down only to 10^{10} solar luminosities in V (and a triple a couple of magnitudes below that). The LG was recognized and named by Hubble (1936), who remarked that these (then) six galaxies were grouped much more closely than the general run of extragalactic nebulae. This is a three-dimensional remark. LG members are scattered across much of the sky.

Notice that we have skipped right over the "miracle decade" during which Hubble established, to nearly everyone's satisfaction, that nebulae contain Cepheid variables, leading to distances for them well outside the Milky Way and that these nebulae are moving apart from one another on average. Incidentally, while the distances in the linear velocity-distance-relation were Hubble's own, the velocities came first, from Slipher and, later from Milton Lassell Humason. Again that is part of a slightly different story (Trimble 1995, 1996)

Hubble's Local Group (to which another 30 or so members have been added since 1936) had a radius of about 1 MLY on his distance scale, and so was too small by a factor 3-4. Thus he concluded that the Milky Way outshone Andromeda by about 10 times, while Andromeda, in turn, outshone not just the other LG members but nearly all the other nebulae by another factor 10. In other words, his distance scale outside the LG was too small by a somewhat larger factor 5-7.

THE PREVALENCE OF CLUSTERING

The recognition that most or all galaxies are grouped and clustered took several decades to pervade the community, though not so long as the time required for all to regard spirals as other galaxies. As late as the 1960s, mainstream astronomers were still sometimes writing or saying that individual galaxies were "the basic building blocks" of the universe (G.O. Abell lectures in 1963, McCrea 1968, Chiu 1968)

Hubble, though he had identified the Local Group, retained a life-long slant in favor of most galaxies being randomly distributed on the sky (e.g. Hubble 1934, 1936) Shapley (1932, 1934), who was an early exponent of "mostly clustered" thought this was because Hubble insisted on using large telescopes, whose fields of view could take in only about one galaxy at a time. Hubble's response was that Shapley would have used large telescopes if he had had access to them. There was indeed little love lost between them (Christianson 1997), at least partly because Hubble had volunteered for active duty as soon as he had defended his 1917 thesis, rather than taking up a position at Mt. Wilson Observatory. By the time he returned from France, Shapley (offered a position at nearly the same time, which he accepted) had been observing for a couple of years and had done some of the things that Hubble later said he had planned for. Shapley moved on to be director of Harvard College Observatory in 1921, while Hubble remained at Mt. Wilson until his death in 1953. Both were, however, members of the IAU Commission (28: Nebulae and Star Clusters) that, at the 1925 General Assembly, emphasized the need for a modern catalogue of nebulae to replace or complement NGC. Shapley's observatory was already at work in the southern hemisphere, and the Shapley-Ames catalog was one of the products. It is characteristic of the men that Hubble belonged only to Comm. 28, while Shapley belonged to 6 others and was president of 27, Variable Stars. Shortly before Hubble's death, he was not even an IAU member, while Shapley was president of

Comm 39 (International Observatories), belonged to 28 and three others, and had family members covering three more .

Specific clusters, like Coma and Virgo, recognizable in 19th century data, had been pointed out by Hinks, Wolf, Wirtz, and others before the settling of the, "island universe" issue and were generally accepted. The next was Hercules (Shapley 1934). Hubble (1934, 1936a) added a couple more, Number 7 was identified by Carpenter (1931) who used it as a point on a density-size plot; Lundmark (1931) picked out a couple, and Tombaugh (1937) found one while he was engaged in another search, which was also successful.

But the time had come again for statistics. Bok (1934), with his reformulation of the "birthday method" said clustering was the norm. He was, of course, at Harvard, where Shapley was still director. Still more, the time had come for a uniform data base. This was provided by Shane and Wirtanen (1954), who covered the northern sky with $6 \times 6°$ plates taken at Lick and counted galaxies in small boxes. This yielded a map in which dense regions seem to connect up over wide angles, sometimes in clumps, sometimes in filaments or shells around relative empty regions. And the heaviest possible statistical guns were brought to bear on these counts by Neyman, Scott, and Shane (1953). Their conclusions were consistent with essentially all galaxies being in groups or clusters, with a continuous distribution in membership from a few to a few thousand. They also found evidence for residual correlations on still larger scales, which they described as possible second-order clustering.

At this higher-order level, binaries were once again first, Shapley (1933) having reported a close cluster pair.

THE EXISTENCE OF SUPERCLUSTERING AND BEYOND

Meanwhile, the 48-inch Schmidt telescope had come into operation in about 1948, with a field of view as large as those of the Lick 36-inches and a much fainter limiting magnitude. Most of the 4000-plus plates (for which the plate holder weighed just over 40 pounds) were taken by George Abell and Albert Wilson (both of whom later said that the process had destroyed their first marriages). This was the Palomar Observatory Sky Survey (POSS).

Wilson (1964, 1969) then drifted off into industry and some rather peculiar cosmological speculations, while Abell stayed on station and compiled an extensive catalog of rich clusters (Abell 1958). Zwicky, et al. (1961-68) independently examined the POSS plates and compiled a six-volume catalog of galaxies and clusters. The Zwicky et al. definition of a cluster was very different from Abell's, and thereby hangs much of the ensuing squabble. Zwicky's idea, both then, earlier, and later (e.g. Zwicky 1938, 1963) was that the universe was completely divided into cluster cells, each inhabited by precisely one cluster, of characteristic size 20-40 Mpc (Zwicky and Rudnicki 1963). Zwicky used his cell size to estimate the mass of the graviton as $10^{16} M_\odot$. A Zwicky cluster could have considerable substructure within that cell volume.

In other words, Zwicky's clusters were not so very different from the superclusters that Abell (1961) was advocating – and determining masses near 10^{16}M. for – by the time of the Berkeley IAU and associated symposia. Incidentally, in case you might be feeling that Zwicky took unfair advantage of the hard work of Abell and Wilson, remember that he had been the first to advocate the use of Schmidt telescopes for serious professional astronomy and had worked hard to bring both the 18-inch and 48-inch to Palomar Mountain. He and Wilson remained lifelong friends (through the second marriage of each).

It is, of course, possible to pick out tracers that are not obviously superclustered or even clustered, for instance the 3C and Parkes bright radio sources (Holden 1966, Payne 1967, with a few hundred radio galaxies and quasars in each catalog). For that matter, the 31,000 brightest northern 6-cm sources still look remarkably smooth on the sky to the eye or the Michell test. Nevertheless, it seems a bit odd in retrospect that de Vaucouleurs (1956, 1958, 1970 and many other places) thought himself a lone warrior in the fight for superclustering. He could easily have declared victory in the mid 1960s and gone on to other things.

Instead, de Vaucouleurs (1970 e.g.), who, like McVittie, habitually included a cosmological constant in his mental map of the universe, became a strong advocate of still larger, hierarchical or fractal structure, in the tradition of Swedenborg, Wright, Lambert, and Charlier. He started the argument by drawing an analogy with estimates for the age of the universe. His graph showed 6000 years at the time of Newton and Archbishop Ussher, 30 million in the mid 19th century (the Kelvin-Helmholtz timescale, so called because it was discovered by Meyer and Waterston), and closely spaced, rising points from the 2 Gyr associated with Hubble's value of H to the 10-20 Gyr associated with his and Sandage's H's. The historical data cannot be gainsaid. Indeed a few years earlier, a bunch of us Caltech grad students had plotted H vs. time and concluded that it would go negative in about 1970 and the universe start to recontract. This did not happen. Instead, H has leveled off in a band that was 50-100 in 1970 and perhaps 55-80 today. De Vaucouleurs, however, asked rhetorically, "Would it not be remarkable if our knowledge of the time scale of the universe, after this long period of growth, should suddenly come to a halt just as I am writing this paper?" Remarkable, perhaps, but true,

In a similar way, he plotted the sizes and densities of the largest structures known, pointing out that size had gone up and density down systematically over the century with the recognition of galaxies, clusters, and superclusters. Again he asked, "Would it not be remarkable...?" Again perhaps remarkable, but true. The volumes of the universe surveyed have expanded enormously since 1970, most recently with the 2dF and SDSS results presented at this meeting. But the largest structures seen are no longer the sizes of the survey volumes (the fact which aroused much of de Vaucouleurs's suspicion). Instead, observers now find the same sorts of clusters, sheets, filaments, voids, bubbles, and sponges over and over again.

The largest structures in the universe have, finally, been identified, and we can concentrate, as this conference did, on figuring out how they arose. Indeed one speaker specifically advised that, "If you have a graduate student who wants to work on fractal universes, don't let him." So also wrote Field (1975, but written in about 1970). Nevertheless, you can't quite go home yet. We still have about 1.5 more issues to examine.

LARGE SCALE STRUCTURE IN VELOCITY SPACE

All those lumps of matter, light and dark. are going to pull on each other. The most obvious manifestation is the large velocity dispersion of clusters (Zwicky 1933), but one might also expect net relative motions on larger scales. So said Gamow (1946), and I do not know that he was first. He suggested that, since everything from moons to the Milky Way is rotating, so might be the universe as a whole, or large regions of it.

The master's thesis of Vera Cooper Rubin (1951, 1954) looked for such rotation and other possible systematic motions in a set of about 100 published redshifts. Now, if you live on the outskirts of some structure of poorly known size and location then, with redshift data alone, it is not very easy to distinguish between rotation around some point in a given direction and net motion toward and away from the directions 90° away. Rubin's (1951) analysis suggested both possibilities. The eye of 2002 naturally slides over the "rotation" numbers and focuses on the "expansion/contraction" numbers between 180 and 370 km/sec in roughly the direction of the north polar cap. Oh, one says, she saw what we now call Virgocentric infall, that is, the fact that the expansion of the universe looks slower in the general direction of the center of the supercluster to which we belong.

The first, and for some years almost the only, citations of the work came from de Vaucouleurs (1953, 1958, 1959/60), who invoked the evidence for rotation or retarded expansion or both as support for the existence of his supercluster. His later papers (de Vaucouleurs 1958, 1959/60, 1964) say that the same asymmetry, most likely to be interpreted as inhibited expansion, is to be seen in the data of Humason, Mayall, and Sandage (1956) a classic compilation, whose appendix dropped the Hubble constant further from about 250 to 180 km/sec/Mpc. Not all authors signed all subsections of the paper, and de Vaucouleurs carefully distinguishes the redshifts of Humason and Mayall from the interpretations of Sandage (who later became one of the great, all-time measurers of redshifts). His last sally is the remark that such anisotropy means that the velocities determined from galaxies in the direction of Virgo will be too small by 30% or so and the Hubble constant therefore also too small (and he will fight that battle for the rest of his life, holding on to $H = 90$-100 to the end).

Close the curtain for about 20 years and bring it up again on a now fully established Vera C. Rubin and her colleagues associated with the Department of Terrestrial Magnetism. She and they embarked on what they described as an unconventional investigation, designed not to measure H or even q but to look for deviations from isotropy of the Hubble law. The first hint of a "yes" answer appears in Rubin et al. (1973). The full-blown result was presented at the 1975 Bloomington AAS meeting and promptly dubbed "the Rubin-Ford effect." This means that you aren't exactly questioning the accuracy of the data, but you don't much like the implications and don't want to be identified with them. The data appear in Rubin et al. (1976).

The problem, as in all such studies, is that, while redshifts require only a large telescope, good detectors (Rubin was one of the first converts to image intensifiers), and dogged determination, distances require ingenuity or divine inspiration. They chose to look only at Sc I galaxies and assume that all were the same real brightness (or rather would be if corrected to face-on), leading to individual luminosity distances for the galaxies and so to velocity residuals relative to smooth Hubble expansion. And yes, there was systematic variation across the sky, and yes, the error bars are large. Their examination of E and S0 velocities reported about the same time by Sandage found residuals that varied across the sky in much the same way. Both implied that the contents of a largish volume of space (including the Local Group) were moving at about 500 km/sec relative to the general expansion.

Another decade passes, and, in January 1986, 68 astronomers gather in Hawaii to discuss Galaxy Distances and Deviations from Universal Expansion (Madore and Tully 1986. According to the editors, "scarcely a single active worker in the field of the distance scale missed the event." Both Sandage and de Vaucouleurs were there and stand adjacent, though not obviously together, in the front row of the conference photo. De Vaucouleurs spoke on H (which was, of course, large), not on superclustering. Sandage is not represented in the proceedings. Nor, more strangely, is Tammann, who was also a participant. Neither Rubin nor Ford of "the effect" participated (and only one or two papers cited them). And Abell was by then dead. But the proceedings include (and the participants heard) two key first reports. One was from the group later called the Seven Samurai, who collected redshifts and distance estimates for elliptical galaxies and searched for deviations from Hubble flow. Their result was consistent, to within the rather large error bars, with that of Rubin et al, (1976) for both amplitude and direction of the large scale peculiar motion we share. Alphabetically (which was not perhaps the actual order of command) they are D. Burstein, R. Davies, A. Dressler, S. Faber, D. Lynden-Bell, R. Terlevich, and G. Wegner.

Also reported in Kona were the first convincing detections of a dipole anisotropy in the 3K background radiation, found almost simultaneously by groups at Princeton (the best-known name was D. Wilkinson) and UC Berkeley (P. Lubin). Their detectors flew on balloons. The amplitude (about 600 km/sec) and direction (galactic longitude = 272 ± 5, galactic latitude = $30 \pm 5°$) for the motion of the Local Group are described by the authors (Lubin and Villela, in Madore and Tully 1986) as completely consistent with the Rubin et al. (1976) numbers, though later discussions have been less confident about this point. The CMB numbers represent the motion of the Local Group relative to the surface of last scattering, at a redshift near 1000. Lubin and Villela (p 169) quote a "private communication" number from Sandage for the motion of the sun in the Local Group.

TIDYING UP LOOSE ENDS

Can we now say that all astronomers now take large scale structure and streaming seriously, as real phenomena that models of galaxy formation must retrodict or hindcast? Perhaps. Maxwell, Pauli, and others have been credited with remarks along the general lines that "everyone now believes in the wave theory of light, because all the people who believed in the particle theory of light are dead." Hidden under the carpet throughout the last section or two has been a small group of astronomers who have persisted in attributing apparently large scale structure on the sky (in both two and three dimensions) to differential obscuration in the Milky Way. The suggestion appears in a paper by Viktor A. Ambartsumian (1940), who eventually focussed on a different idea, that large scale clusters existed but were not gravitationally bound and somehow represented matter newly appearing or newly changing state in the universe. Tadeos Agekyan (1957), a student of Ambartumyan, carried the differential absorption point of view to Leningrad/St. Petersburg, where he passed it on to B. I. Fasenko. Fasenko (1996) carried the banner for a number of years, but has not published (on that or anything else) since 1996 in journals that are covered by A&A Abstracts. Agekyan still appears in the IAU directory at St. Petersburg University (Fasenko was never a member) but also has not published on this subject in a number of years.

Thus it seems that all the people who favored differential obscuration over very large scale structure are, if not dead, anyhow no longer on the front lines. This is, of course, not quite the same as saying that we have hold of all parts of the right answer. Conference participants perhaps gradually became aware that all the overheads that had words near the top had reproductions of well-known optical illusions and impossible objects (about half from Escher) at the bottom, ending with the profile of a fairly scholarly-looking chap, drawn by Paul Agule, who rotated 90 ° to the hand-written word "liar."

ACKNOWLEDGMENTS

Thanks, as always, to Stephen Holt and the SOC for the invitation to prepare and present this retrospective and to John Trasco, Susan Lehr, and Mary Ann Phillips for making everything run so smoothly. If there are conferences in heaven, I hope they are in charge. Perhaps the best way to thank one's colleagues in these beancounting days is to cite their papers. Of the roughly 80 individuals in the references 32 are, or more often, I fear, were, friends or at least acquaintances. To them a special thank you. The friends of friends group is another 21 (for instance Eddington via McCrea).

REFERENCES

Abell, G.O. 1958 ApJ5 3, 211,
1961. AJ 66, 607
Agekyan, T.A. 1957. A. Zh. 11, 366
Ambartsumian, V.A. 1940. Bull. Abast. Obs. 4, 17
Berendzen, R. et al. 1976. Man Discovers the Galaxies, (NY: Science History Publications)
Bok, B.J. 1943. Harv. Obs. Bull. 895
Bok, B.J. & P.F. Bok. 1945. The Milky Way, 2nd Ed. (Blakiston Co.)
Carpenter, E.F. 1931. PASP 43, 247
Charlier, C.V.L. 1908. Ark. f. Mat. Astr. Fis. 4, 1
..... 1922. Lund Med. Ser I. No. 48
Chiu, H.Y. 1968. Science. 4, 33
Christianson, G.E, 1997. Edwin Hubble, Mariner of the Febulae (Inst. of Physics)
Curtis, H.D. 1917. PASP 29, 145 and Lick Obs. Bull. 300
.....1918. Lick Obs. Publ. 13, 45
de Vaucouleurs, G 1953. ApJ 158, 30
.... 1956. Vistas in Astron. 2, 1584
.....1958. AJ 63, 253
.....1959/60. Sov. Astron. 3, 897

.... 1964. AJ 69, 737
.....1970. Science 167, 1203
Easton, C. 1900. ApJ 12, 136
Eddington, A,S. 1912. Stellar Movements (London)
Fesenko, B.I. 1996. Sov. Astron. (Astron. Rep., A . Zh.) 40, 616
Field, G.B. 1975. in A.Sandage et al. Eds. Galaxies and the Universe (U. Chicago Press). p. 359
Gamow G. 1946. Nature 158, 549
Harrison, E.R. 1981, Cosmoloty (Cambridge Univ, Press)
.....1987. Darkness at Night (Harvard Univ. Press)
Holden, D.J. 1966. MURAS 113, 225
Holmberg, E. 1937 Lund Annals 6, and ApJ 92, 2000
Hoskin, M,A. 1997. Cambridge Illustrated History of Astronomy (CUP) Hubble, E.P 1934. ApJ 79, 8
.... 1936. Realm of the Nebulae (Oxford Univ. Press)
.....1936a. ApJ 84, 517
Humason, M.L., N.U. Mayall & A.R. Sandage 1956. AJ 61, 97
Jaki. S. 1972. The Milky Way (NY: Science History Publications)
Lindblad, B. 1924-26, Uppsala Medd. nos. 3,4,6,13
Lundmark, K. 1919. AN 209, 369
.....1931. Lund Obs. Cir. 1, 30
.....1932. Harvard Ann. 88, 4, 43
.....1956. Vistas in Astr. 2, 1607
Madore, B.F. and R.B. Tully eds. 1986. Galaxy Distances and Deviations from Universal Expansion (Reidel, NASO ASI series)
Martinez, V., V. Trimble, & M. Pons Bordeira eds. 2002. Historical Development of Modern Cosmology (ASP Conf. Ser. 257)
McCrea, W.H. 1968. Science 160, 1295
McVittie, G.C ed. 1962. IAU Symp. 15, Problems in Extragalactic Research (Macmillan)
Michell, J. 1767. Phil. Trans. Roy. Soc. 57, 234
Needham, J. 1955. Vistas in Astron. 1, 67
Newcomb, S. 1878. Popular Astronomy
.....1906. Sidelights on Astronomy and Kindred Fields of Popular Science (NY: Harper)
Neyman, J., T. Page & E. Scott eds. 1961. Conference on Instability of Systems of Galaxies, AJ 66, 533-636
Neyman, J. E. Scott, & C.D. Shane 1953. ApJ 117, 92
Oort, J.H. 1927. BAN 3, 275
Opik E.J. 1922. ApJ 55, 406
Payne, A.0, 1967. Aust. J. Phys. 20, 291
Peebles, P.J.E. 1993. Principles of Physical Cosmology (Princeton Univ. Press)
Plaskett. 3.S, 1939. Popular. Astron. 47, 255
Polya, G. 1919. AN 208, 175
Proctor, R,A. 1869. MNRAS 29, 337
Reynolds, J.H. & V.M, Slipher 1925. in A. Fowler, ed. Trans IAU II (London Imperial College) p. 116 & 132
Rosse, Wm. Parsons, Lord 1850. Phil. Trans. Roy Soc.
Rubin, V.C, 1951. AJ 56, 47
.....1954. PNAS 40, 541
Rubin, V.C. et al. 1973. ApJ 181, L111 .....1976. AJ 81, 769
Sandage, A. et al. eds. 1975. Galaxies and the Universe (U Chicago Press)
Sanford, R.F. 1917. Lick Obs. Bull 9,80
Shane, C.D & C.A. Wirtanen 1954. AJ 58, 285
Shapley, H. 1919. ApJ 48, 754; 49, 311; 50, 107
..... 1932. Bull. HCO No. 880
..... 1933. PNAS 19, 591 & 1011
.....1934. MNRAS 94, 815
.... .1943. Galaxies (Harvard U Press)

Shapley, H & A. Ames 1932. Ann. HCO 88 No, 2
Tombaugh, C.W. 1937 PASP 49, 259
Trimble, V 1987. ARA&A 25, 425
.....1995. PASP 107, 1133
.....1996. PASP 108, 1073
.....1997. Space Sci. Rev 79, 793
.....1999. in S S. Holt & E M. Smith eds. After the Dark Ages, AIP Conf. Ser 470, 3
Trumpler, R.3. 1930. Lick Obs. Bull. 14, 154
.....1941. PASP 53, 155
van Albada, G B. 1962. in McVittie (1962) op. cit
Waters, S. 1894. MNRAS 54, 426
Whitney, C.A 1971. The Discovery of Our Galaxy (NY, Knopf)
Wilson, A.G. 1964. PNAS 52, 849 .....1969. Science 165, 202
Zwicky, F. 1933. Helv. Phys, Acta 6O 110 v .....1937. ApJ 86, 218
.....1938. PASP 50, 218
.....1963. Ap3 137, 707
Zwicky, F, & K Rudnicki 1963. ApJ 137, 718
Zwicky, F, et al. 1961-68. Catalogue of Galaxies and Clusters of Galaxies (6 vols, Calif. Inst. of Technology)

THE ORIGINS OF STRUCTURE

Beyond Cosmological Parameters

Max Tegmark

Dept. of Physics, Univ. of Pennsylvania, Philadelphia, PA 19104; max@physics.upenn.edu

Abstract. Observational constraints on spacetime from supernovae Ia, CMB, lensing, Lyman α Forest and galaxy clustering are reviewed, focusing on how the underlying physics (dark matter, dark energy, gravity) can be tested rather than assumed. This is possible because we are simply measuring spacetime, and to 1st order, all observations probe just three cosmological functions, not a dozen dubious cosmological parameters: the cosmic expansion history $H(z)$, the primordial power spectrum $P*(k)$ and the fluctuation growth $g(z,k)$. Measuring these functions will clarify the nature of dark energy (from how its density depends on time), the nature of dark matter (from its power spectrum growth) and the nature of early universe (from the primordial power spectrum).

INTRODUCTION

Traditionally, space was merely a three-dimensional static stage where the cosmic drama played out over time. Einstein's theory of general relativity [1] replaced this by four-dimensional spacetime, a dynamic geometric entity with a life of its own, capable of expanding, fluctuating and even curving into black holes. Now the focus of research is increasingly shifting from the cosmic actors to the stage itself. Triggered by progress in detector, space and computer technology, an avalanche of astronomical data is revolutionizing our ability to measure the spacetime we inhabit on scales ranging from the cosmic horizon down to the event horizons of suspected black holes, using photons and astronomical objects as test particles. The goal of this article is to review these measurements and future prospects, focusing on three key issues:

1. The global topology and curvature of space
2. The expansion history of spacetime and evidence for dark energy
3. The fluctuation history of spacetime and evidence for dark matter

In the process, I will combine constraints from the cosmic microwave background [2], gravitational lensing, supernovae Ia, large-scale structure (LSS), and the hydrogen Lyman α forest[3]. Although it is fashionable to use cosmological data to measure a dozen free "cosmological parameters", I will argue that improved data allow raising the ambition level beyond this, testing rather than assuming the underlying physics. I will discuss how with a minimum of assumptions, one can measure key properties of spacetime itself in terms of a few cosmological functions: the expansion history of the universe, the spacetime fluctuation spectrum and its growth.

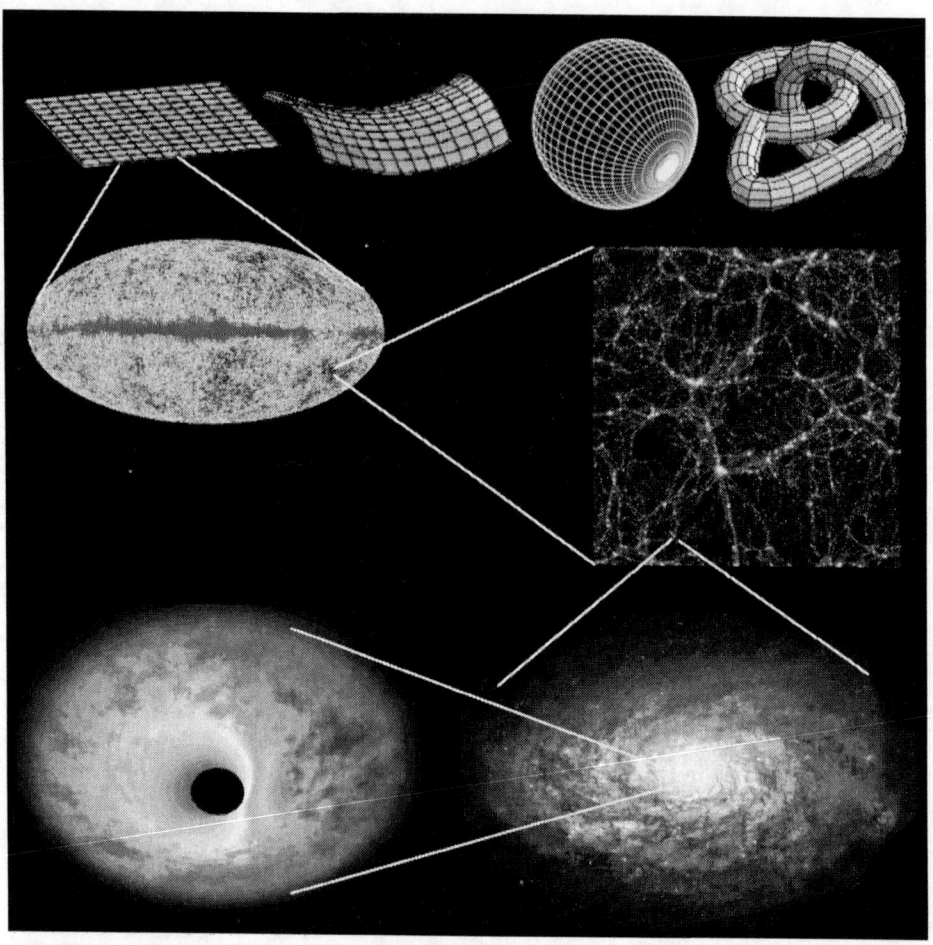

FIGURE 1. Summary of the spacetime issues discussed in this article. One can use photons and astronomical objects as test particles to measure spacetime over 22 orders of magnitude in scale, ranging from the cosmic horizon (probing the global topology of and curvature of space — top) down to galaxies (giving evidence for dark matter), galactic nuclei and binary stellar systems (giving evidence for black holes). The figure illustrates how spacetime ripples at the 10^{-5} level will be imaged by the cosmic microwave background satellite MAP [4] and has grown via gravitational instability into cosmic large-scale structure [5], galaxies and, it seems, black holes [6].

Goals and tools

Before embarking on a survey of spacetime, a brief review of what it is we want to measure and the basic tools at our disposal [7, 8] is in order, as well as a general picture of how spacetime related to the structure of the universe. According to general relativity, spacetime is what mathematicians call a manifold, characterized by a topology and a metric. The topology gives the global structure (Fig. 1, top): is space infinite in all directions like in high-school geometry or multiply connected like say a hypersphere or doughnut so that traveling in a straight line could in principle bring you back home — from the other direction? The metric determines the local shape of spacetime, *i.e.*, the distances and time intervals we measure, and is mathematically specified by a 4×4 matrix at each point in spacetime.

General relativity theory (GR) consists of two parts, each providing a tool for measuring the metric. The first part of GR states that in the absence of non-gravitational forces, test particles (objects not heavy enough to have a noticeable effect on the metric) move along geodesics in spacetime, generalized straight lines, so the observed motions of photons and astronomical objects allow the metric to be reconstructed. I will refer to this as *geometric* measurements of the metric. The second part of GR states that the curvature of spacetime (expressions involving the metric and its first two derivatives) is related to its matter content — in most cosmological situations simply the density and pressure, but sometimes also bulk motions and stress energy. I will refer to such measurements of the metric as *indirect*, because they assume the validity of the Einstein field equations (EFE) of GR.

The broad brush picture

The current consensus in the cosmological community is that spacetime is extremely smooth, homogeneous and isotropic (translationally and rotationally invariant) on large ($\sim 10^{23}$m-10^{26}m) scales, with small fluctuations that have grown over time to form objects like galaxies and stars on smaller scales. Cosmic microwave Background (CMB) observations [2] have shown that space is almost isotropic on the scale of our cosmic horizon ($\sim 10^{26}$m), with the metric fluctuating by only about one part in 10^5 from one direction to another, and combining this with the so-called cosmological principle, the assumption that there is nothing special about our vantage point, implies that space is homogeneous as well. Three-dimensional maps of the galaxy and quasar distribution give more direct evidence for large-scale homogeneity [9].

The fact that the CMB fluctuations are so small is useful, because it allows the intimidating nonlinear partial differential equations governing spacetime and its matter content to be accurately solved using a perturbation expansion. To zeroth order (ignoring the fluctuations), this fixes the global metric to be of the so-called Friedman-Robertson-Walker (FRW) form, which is completely specified except for a curvature parameter and a free function giving its expansion history. To first order, density perturbations grow due to gravitational instability and gravitational waves propagate through the FRW background spacetime, all governed by *linear* equations. Only on smaller scales ($\lesssim 10^{23}$m) do the fluctuations get large enough that nonlinear dynamics becomes important — in the realm of galaxies, stars and, perhaps, black holes. We will limit our discussion to what can be learned from the 0th and 1st order measurements, covered in turn in the following two sections.

OVERALL SHAPE OF SPACETIME

Curvature of space

The question of whether space is infinite was answered last year with a resounding *maybe*. For a FRW metric, answering this question is equivalent to measuring the curvature of space Fig. 1, specifically the *radius of curvature R*. R is the radius of the hypersphere if space is finite, $R = \infty$ if space is flat, and R is an imaginary number($R^2 < 0$) for saddle-like curvature. Because the three angles of a triangle will add up to $180°$ in flat space, more if $R^2 > 0$ (like on a sphere) and less if $R^2 < 0$ (like on a saddle) cosmologists have measured R using the largest triangle available: one with us at one corner and the other two corners on the hot opaque surface of ionized hydrogen that delimits the visible universe and emits the CMB, merely 400,000 years after the big bang. Photographs of this surface reveal hot and cold spots of a characteristic angular size that can be predicted theoretically. This characteristic spot size (or, more rigorously, the first peak in the CMB power spectrum [10]) subtends about $0.5°$ — like the Moon — if space is flat. Sphere-like curvature would make all angles appear larger, so characteristic spots much larger than the Moon would indicate a finite universe curving back on itself, whereas smaller spots would indicate infinite space with negative curvature. Recent experiments have observed the first peak and hints of additional smaller scale peaks [2].

The universe may be infinite, because the measured characteristic spot size is so close to $0.5°$ that we still cannot tell whether space is perfectly flat or very slightly curved either way. The sharpest current limits on R, obtained by combining all CMB experiments with galaxy clustering data [12, 13] to constrain other parameters, are $|R| > 20 h^{-1} \mathrm{Gpc} \approx 10^{27}$m. This is in sharp contrast to a few years ago, when the most popular models had negatively curved space with $|R| \approx 4 h^{-1}$ Gpc. In other words, space now seems to be either infinite or much larger than the observable universe, whose radius is about $9 h^{-1}$ Gpc. In 1900, Karl Schwarzschild discussed the possibility that space was curved and published a paper with a lower limit $R > 2500$ light-years $\approx 2 \times 10^{19}$m [14]. A century later, we thus know that the universe is at least another 40 million times larger!

Topology of space

Even if space turns out to be negatively curved or perfectly flat, it might be finite. General relativity does not prescribe the global topology, so various possibilities are possible (Fig. 1, top). The simplest non-trivial model has flat space and the topology of a three-dimensional torus, where opposing faces of a cube of size $L \times L \times L$ are identified to be one and the same. Living in such a universe would be indistinguishable from living in a perfectly periodic one: if $L = 10$m, you could see the back of your own head 10m away, and additional copies at 20 m, 30 m, and so on — searches for multiple images of cosmological objects have constrained such models. Also, just as a finite guitar string has a fundamental tone and overtones, linear spacetime fluctuations in such a toroidal universe could have only certain discrete wavenumbers. As a result, its CMB power

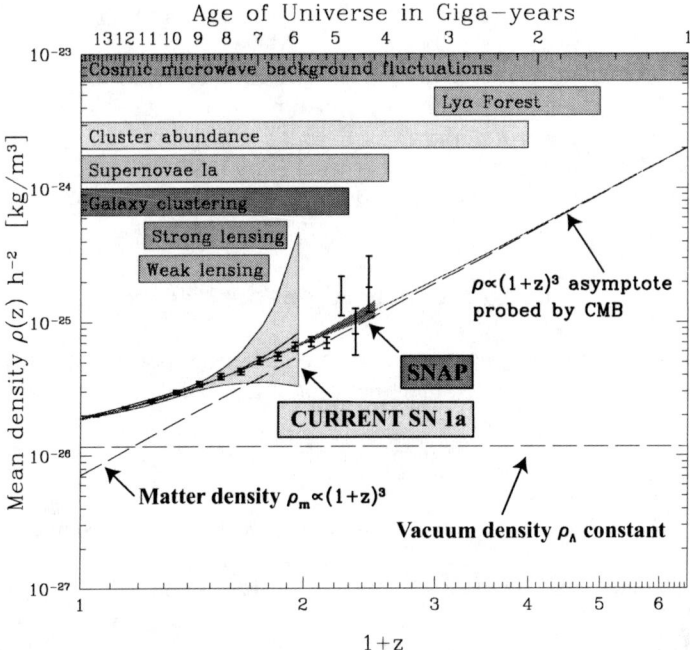

FIGURE 2. Solid curve shows the concordance model [12] for the evolution of the cosmic mean density $\rho(z) \propto H(z)^2$. This curve uniquely characterizes the spacetime expansion history. The horizontal bars indicate the rough redshift ranges over which the various cosmological probes discussed are expected to constrain this function. Because the redshift scalings of all density contributions except that of dark energy are believed to be straight lines with known slopes in this plot (power laws), combining into a simple quartic polynomial, an estimate of the dark energy density $\rho_X(z)$ can be readily extracted from this curve. Specifically, $\rho \propto (1+z)^4$ for the cosmic microwave background (CMB), $\rho \propto (1+z)^3$ for baryons and cold dark matter, $\rho \propto (1+z)^2$ for spatial curvature, $\rho \propto (1+z)^0$ for a cosmological constant and $\rho \propto (1+z)^{3(1+w)}$ for dark energy with a constant equation of state w. Measurement errors are for current SN Ia constraints and a forecast for what the SNAP satellite [19] can do, assuming flat space as favored by the CMB. Error bars are for a non-parametric reconstruction with SNAP.

spectrum would differ on large scales, and it was shown that if the universe were such a torus, then L must be at least of the order of the cosmic horizon [15, 16]. Indeed, it was shown that all three dimensions of the torus must at least about this large to explain the absence of a type of approximate reflection symmetry in the CMB sky [17]. Cosmic topology is now a burgeoning field of study[18], but all available data so far is still consistent with the simplest possible three-dimensional space, the infinite flat Euclidean space that we learned about in high school. The global structure of our four-dimensional spacetime also depends on the beginning and end of time, to which we now turn.

SPACETIME EXPANSION HISTORY

One of the key quantities that cosmologists yearn to measure is the function $a(t)$, describing the expansion of the universe over time — if space is curved, a is simply the magnitude of the radius of curvature, $a = |R|$. A mathematically equivalent function more closely related to observations is the Hubble parameter as a function of redshift, $H(z)$, giving the cosmic expansion rate and defined by $H \equiv \frac{d}{dt}\ln a$, $1+z \equiv a(t_{\text{now}})/a(t)$.

What $\rho(z)$ tells us about dark energy

Squaring our curve $H(z)$ gives us the cosmic matter density (Fig. 2). If the Einstein Field equations of GR are correct, then the mean density of the universe is given by the Friedmann equation

$$\rho(z) = \frac{3H(z)^2}{8\pi G}. \qquad (1)$$

Here G is Newton's gravitational constant and, if space is curved, the density ρ is defined to include an optional curvature contribution $\rho_{\text{curv}} \equiv -3c^2/8\pi G R^2$, where c is the speed of light. Conveniently, all standard components of the cosmic matter budget contribute simple straight lines to this plot, because their densities drop as various power laws as the universe expands. For instance, the densities of both ordinary and cold dark matter particles are inversely proportional to the volume of space, scaling as $\rho \propto (1+z)^3$.

The cosmic density $\rho(z)$ measured from SN Ia and CMB (yellow band in Fig. 2) was higher in the past, but rises slower than $(1+z)^3$ towards higher z, with a shallower slope than 3 at recent times. Fig. 2 shows that although the cosmic density $\rho(z)$ measured from SN Ia and CMB was indeed higher in the past, the curve rises slower than this towards higher redshift, with a shallower slope than 3 at recent times. This is evidence for the existence of *dark energy*, a substance whose density does not rise rapidly with z. Adding a cosmological constant contribution $\rho_\Lambda \approx 4 \times 10^{-26}$ kg/m^3 (about 2/3 of the current matter budget) whose density is, by definition, constant, provides a good fit to the measurements (Fig. 2). This 1998 discovery [20, 21] stunned the scientific community and triggered a worldwide effort to determine the nature of the dark energy. A model-independent approach involves measuring the curve $\rho(z)$ more accurately with (Fig. 2), to determine whether independent measurements of $\rho(z)$ agree, so that we can rule out problems with observations, and to determine the time-dependence of dark energy density $\rho_X(z)$. If it is constant, we may have measured vacuum energy/Einstein's cosmological constant, and if not, we should learn interesting physics about a new scalar quintessence field, or whatever is responsible. A less ambitious approach that is currently popular is assuming that the equation of state (pressure-to-density ratio) w of the dark energy is constant [22, 23], which is equivalent to assuming that $\rho_X(z)$ is a straight line in Fig. 2 with a free amplitude and slope.

What $\rho(z)$ tells us about our origin and destiny

If we can understand the different components of the cosmic matter budget well enough to extrapolate the curve $\rho(z)$ from Fig. 2 to the distant past and future, we can use the Friedmann equation to solve for $a(t)$ and obtain information about the origin and ultimate fate of spacetime. $a(t) = 0$ in the past or future would correspond to a singular big bang or big crunch, respectively, with infinite density $\rho(z)$. As to the past, such extrapolation seems justified at least back to the first second after the big bang, given the success of big bang nucleosynthesis in accounting for the primordial light element abundances [24]. Regarding the very beginning, the jury is still out. Extrapolation back to the very beginning is more speculative. According to the currently most popular scenario, a large and nearly constant value of ρ at $t \lesssim 10^{-34}$ seconds caused exponential expansion $a(t) \propto e^{Ht}$ during a period known as inflation [25], successfully predicting both negligible spatial curvature and a nearly scale-invariant adiabatic scalar power spectrum[10] with subdominant gravitational waves. A rival ekpyrotic model inspired by string theory and a related eternally oscillating model have attracted recent attention[26, 27, 28]. If the density approaches the Planck density (10^{97} kg/m^3) as $t \to 0$, quantum gravity effects for which we lack a fundamental theory should be important, and a host of speculative scenarios have been put forward for what happened at $t \sim 10^{-43}$ seconds. A very incomplete sample includes the Hawking-Hartle no-boundary condition [29], God creating the universe, the universe creating itself [30], and so-called pre-big-bang models [31]. Another possibility is that the Planck density was never attained and that there was no beginning, just an eternal fractal mess of replicating inflating bubbles, with our observed spacetime being merely one in an infinite ensemble of regions where inflation has stopped [32, 33].

As to the future, the expansion can only stop ($H = 0$) if the effective density $\rho(z)$ drops to zero. The only two density contributions that can in principle be negative are those of curvature (measured to be negligible) and dark energy (measured to be positive), suggesting that the universe will keep expanding forever. Indeed, if the dark energy density stays constant, we are now entering another inflationary phase of exponential expansion ($a(t) \propto e^{Ht}$), and in about 10^{11} years, our observable universe will be dark and lonely with almost all extragalactic objects having disappeared across our cosmic horizon [34]. However, such conclusions must clearly be taken with a grain of salt until the nature of dark energy is understood.

How to measure $\rho(z)$

In conjunction with the R, the curve $H(z)$ can be measured geometrically, using photons as test particles. Objects of known luminosity (called standard candles) or known physical size (called standard yardsticks) at different redshifts can be used to determine $H(z)$ by comparing their measured brightness or angular size with theoretical predictions, which follow from computing the trajectories of nearly parallel light rays and depend only on $H(z)$ and the (apparently negligible) curvature of space [35, 36]. The best standard candles to date are supernovae of type Ia, and 92 SN Ia [20, 21] were used [36] to measure $H(z)$ and thereby $\rho(z)$ (Fig. 2). The best standard yardstick so far is the characteristic CMB spot size, suggesting that space is flat.

$H(z)$ can also be measured indirectly. As discussed below, $H(z)$ affects the growth of density fluctuations and can therefore be probed by galaxy clustering and other techniques (Fig. 2). Such fluctuation measures have constrained matter to make up no more than about a third of the critical density needed to explain why space is flat. This Enron-like accounting situation provides supernova-independent evidence for dark energy [12, 13, 39].

GROWTH OF COSMIC STRUCTURE

While SN Ia and CMB peak locations have recently revolutionized our knowledge of the metric to 0th order (curvature, topology and expansion history), other observations are probing its 1st order fluctuations with unprecedented accuracy. These perturbations come in two important types. The first are gravitational waves, hitherto undetected ripples in spacetime which propagate at the speed of light without growing in amplitude. The second are density fluctuations, which can get amplified by gravitational instability (Fig. 1) and are being measured by CMB, gravitational lensing and the clustering of extragalactic objects, notably galaxies and gas clouds absorbing quasar light (the so-called Lyman α forest, LyαF) over a range of scales and redshifts (Fig. 3).

Plane wave perturbations of different wavenumber evolve independently by linearity, and are so far consistent with having uncorrelated Gaussian-distributed amplitudes as predicted by most inflation models [25]. The 1st order density perturbations are therefore characterized by a single function $P(k,z)$, the *power spectrum* [10], which gives the variance of the fluctuations as a function of z and wavenumber k. $P(k,z)$ depends on and can therefore teach us about three things: The cosmic matter budget, the seed fluctuations created in the early universe and Galaxy formation. A challenge is to robustly disentangle the three. We are not there yet, but new data is making this increasingly feasible because each of the probes in Fig. 3 involve different physics and is affected by the three in different ways as outlined below.

Given the profusion of recent measurements of $H(z)$ and $P(k,z)$, it is striking that there is a fairly simple model that currently seems to fit everything (Fig. 2 and Fig. 4). In this so-called concordance model [12, 13, 39, 40], the cosmic matter budget consists of about 5% ordinary matter (baryons), about 30% cold dark matter, about 0.1% hot dark matter (neutrinos) and about 65% dark energy based on CMB and LSS observations, in consistent with LyαF [41], lensing [42, 43] and SN Ia [20, 21]. The seed fluctuations created in the early universe are consistent with the inflation prediction of a simple power law $P(k,z) \propto k^n$ early on, with $n = 0.9 \pm 0.1$ [12, 13]. Galaxy formation appears to have heated and reionized the universe not too long before $z = 6$ based on the LyαF [44].

Although inferences about things like the expansion history, the matter budget and the early universe involve many assumptions (about the nature of dark energy, dark matter, gravity, galaxy formation, and so on), the avalanche of new cosmology data allows raising the ambition level and testing these assumptions about the underlying physics. Given the matter budget and the expansion history $H(z)$, theory predicts the complete

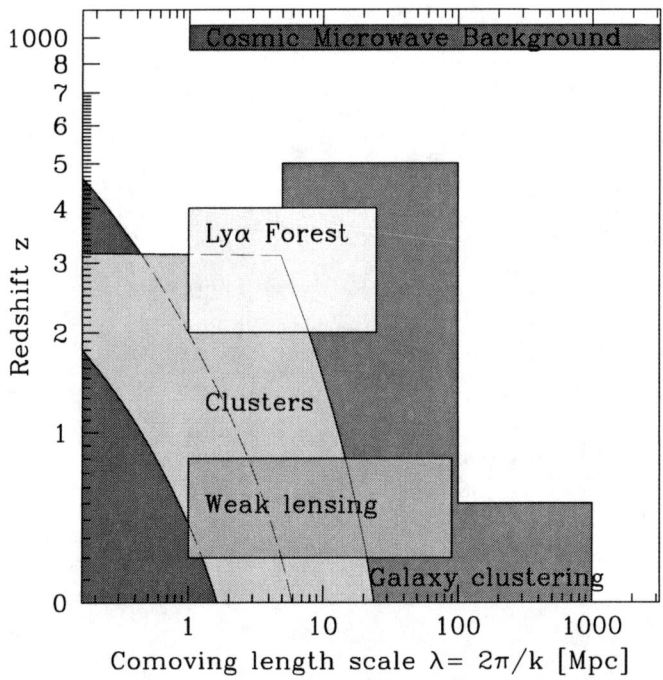

FIGURE 3. Shaded regions show ranges of scale and redshift over which various observations are likely to measure spacetime fluctuations over the next few years. The lower left region, delimited by the dashed line, is the non-linear regime where rms density fluctuations exceed unity in the "concordance" model from [12].

time-evolution of linear clustering, so measuring its redshift dependence (Fig. 3) offers redundancy and powerful cross-checks.

Many power spectrum probes are improving rapidly (Fig. 3). **Gravitational lensing** uses photons from distant galaxies as test particles to measure the metric fluctuations caused by intervening matter, as manifested by distorted images of distant objects, and first measured $P(k,z)$ in 2000. Three-dimensional mapping of the universe with **galaxy redshift surveys** offers another window on the cosmic matter distribution, through its gravitational effects on galaxy clustering. This field is currently being transformed by the 2 degree Field (2dF) survey and the Sloan Digital Sky Survey, and complementary surveys will map high redshifts and the evolution of clustering. The abundance of **galaxy clusters** at different epochs, as probed by optical, x-ray, CMB or gravitational lensing surveys, is a sensitive probe of $P(k,z)$ on smaller scales and the **LyαF** offers a new and exciting probe of matter clustering on still smaller scales when the universe was merely 10-20% of its present age. **CMB** experiments probe $P(k,z)$ through a variety of effects as far back as to $z > 10^3$, with exciting new fronts being increased sensitivity [4], increased angular resolution and CMB polarization (still undetected).

These complementary probes can be combined to break each others' degeneracies and

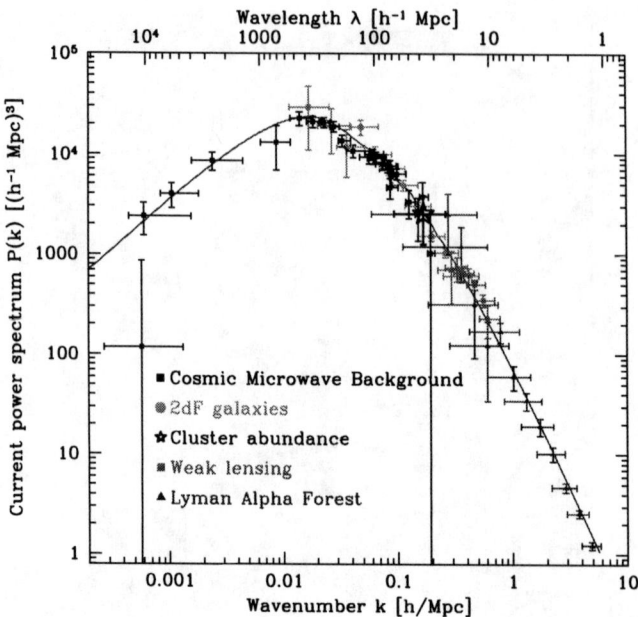

FIGURE 4. Measurements of the current ($z = 0$) power spectrum of density fluctuations computed as described in [48], assuming the matter budget of [13] and reionization at $z = 8$. The CMB measurements combine the information from all experiments to date as in [48]. LSS points are from a recent analysis [45] of the 3D distribution of 2dF galaxies [9], and correcting them for bias shifts them vertically ($b = 1.3$ assumed here) and should perhaps blue-tilt them slightly. The cluster error bars reflect the spread in the literature. The lensing points are based on [49]. The LyαF points are from a reanalysis [50] of [41] and have an overall calibration uncertainty around 17%. The curve shows the concordance model of [13].

independently measure the matter budget, the primordial power spectrum and galaxy formation details [47]. The power spectra measured by CMB, LSS, lensing and LyαF are the product of the three terms: the primordial power spectrum, a so-called transfer function quantifying the subsequent fluctuation growth, and, for LSS and LyαF only, a so called bias factor accounting for the fact that the measured galaxies and gas clouds may cluster differently than the underlying matter [11].

Disentangling bias and systematic errors: Galaxy bias has now been measured from data and found to be of order unity for typical 2dF galaxies [40, 46], and LyαF bias may be computable with hydrodynamics simulations [41, 50]. Although CMB, LSS, lensing and LyαF each comes with caveats of their own, their substantial overlap (Fig. 3) should allow disagreements between data sets to be distinguished from disagreements between data and theory.

Disentangling primordial power from the matter budget: The transfer function can be disentangled from the primordial power since it depends on the matter budget, and conveniently in rather opposite ways for CMB than for lower z $P(k)$ measurements (LSS, lensing, LyαF). For instance, increasing the cold dark matter density $h^2\Omega_c$ shifts the galaxy power spectrum up to the right and the CMB peaks down to the left if the primordial spectrum is held fixed. Adding more baryons boosts the odd-numbered CMB peaks but suppresses the galaxy power spectrum rightward of its peak and also makes it wigglier. Increasing the dark matter percentage that is hot (neutrinos) suppresses small-scale galaxy power while leaving the CMB almost unchanged. This means that combining CMB with other data allows unambiguous determination of the matter budget, and the primordial power spectrum can then be inferred. Combining CMB temperature and polarization measurements also helps in this regard, because the characteristic wiggles imprinted by the baryons and dark matter are out of phase for the two, whereas wiggles due to the primordial spectrum would of course line up for the two [48].

Although the best is still to come in this area, the basic conclusion that the universe is awash in nonbaryonic dark matter is already supported independently by CMB, LyαF, galaxy surveys, cluster counting and lensing. The agreement on the baryon density between fluctuation studies (CMB + galaxy surveys) and nucleosynthesis and on the dark energy density between fluctuation studies and SN Ia are both indications that spacetime fluctuation measurements are on the right track and will live up to their promise in this decade of precision cosmology.

Constraints on dark matter from nonlinear clustering

On small scales, the linear perturbation expansion eventually breaks down as density fluctuations grow to be of order unity, collapsing to form a variety of interesting astrophysical objects. Although the theoretical predictions are more difficult in this regime, the metric can still be accurately measured using photons and astrophysical objects as test particles. The gravitational potential well is probed by strong gravitational lensing of photons through its distorting effect on background objects [51] and also by the motions of massive objects like galaxies, stars or gas clouds. The orbital parameters in a binary system reveal the masses of the two objects, just as we once weighed the Sun exploiting Earth's orbit around it. In more complicated systems, the central mass distribution can be inferred statistically from velocity dispersions observed in the vicinity. These basic tools have revealed surprises on three vastly different scales: dark matter in galaxies and clusters ($\sim 10^{20-23}$m), supermassive black holes in galactic bulges ($\sim 10^{10} - 10^{13}$m) and stellar-mass black holes ($\sim 10^4 - 10^5$m). Recent black hole reviews include [52, 53, 54].

As noted by Zwicky in 1933 [55], the amount of mass in galaxies and galaxy clusters inferred from rotation curves or velocity dispersions exceeds the mass of luminous matter by a large factor. Precision measurements with a variety of techniques have confirmed this finding, providing evidence that both galaxies and clusters are accompanied by roughly spherical halos of cold dark matter. This dark matter evidence is independent of that from linear perturbation theory described above, yet produces roughly consistent estimates of the total cosmic dark matter density [56, 57].

New measurements such as mapping tidal streamers, stripy remnants of galaxies cannibalized by the Milky Way in the past, may allow 3D reconstruction of our own dark matter halo, and early results suggest that it may be elliptical rather than perfectly spherical [58]. Measurements of the shape and substructure of dark matter halos can probe the detailed nature of the dark matter. Indeed, computer simulations with cold dark matter composed of weakly interacting particles appear to predict overly dense cores in the centers of galaxies and clusters, and that there should be about 10^3 discrete dark matter halos in our Galactic neighborhood (the Local Group), in contrast to the less than 10^2 galaxies actually observed. These halo profile and substructure problems have triggered talk of a cold dark matter crisis and much recent interest in self-interacting dark matter [59], warm dark matter [60] and other more complicated dark matter models which suppress cores and substructure. It is not obvious that there really is a crisis, because baryonic feedback properties may be able to reconcile vanilla cold dark matter with observations and substantial halo substructure has recently been detected with gravitational lensing [61], but this active research area should teach us more about dark matter properties whatever they turn out to be.

OUTLOOK

I have surveyed recent measurements of spacetime over a factor of 10^{22} in scale, ranging from the cosmic horizon down to the event horizon of black holes. On the largest scales, evidence supports a flat infinite space and eternal future time. The growth of spacetime fluctuations has suggested that about 30% of the cosmic matter budget is made up of (mostly cold) dark matter, about 5% ordinary matter and the remainder dark energy. There is further evidence for the same dark matter in the halos of galaxies and clusters. Finally, spacetime seems to be full of black holes, both supermassive ones in the centers of most galaxies and stellar mass ones wherever high mass stars have died [52, 53, 54].

The Devil's advocate requires that we clarify the assumptions underlying these conclusions. The geometric test particle observations have measured the spacetime metric, but all inferences about dark energy, dark matter and the inner parts of black holes assume that the Einstein Field Equations (EFE) of GR are valid. Attempts have been made to explain away all three by modifying the EFE. So-called scalar-tensor gravity has been found capable of giving accelerated cosmic expansion without dark energy [63]. Although not an ab initio theory, the approach known as modified Newtonian dynamics (MOND) attempts to explain galaxy rotation curves without dark matter [64]. It is not inconceivable that the EFE can be modified to avoid black hole singularities [66].

So could dark energy, dark matter and black holes be merely a modern form of epicycles, which just like those of Ptolemy can be eliminated by modifying the laws of gravity [36, 65, 67, 68]? The way to answer this question is to test the EFE observationally, by embedding them in a larger class of equations and quantifying the observational constraints. So far, the true theory of gravity has been shown to extremely close to GR in the regime probed by solar system dynamics and binary pulsars [7, 8] and the MOND-loophole has been at least partially closed [69, 70, 71]. However, this does *not* imply that the true theory of gravity must be indistinguishable from GR in all contexts, in particular for very compact objects [62] or for cosmology [7, 8], so testing gravity remains a fruitful area of research. Such tests continue even in the laboratory [72], testing the gravitational inverse square law down to millimeter scales to probe possible extra dimensions [73].

In conclusion, the coming decade will be exciting: an avalanche of astrophysical observations are measuring spacetime with unprecedented accuracy, allowing us to test whether it obeys Einstein's field equations, and consequently whether dark energy, dark matter and black holes are for real.

ACKNOWLEDGMENTS

The author wishes to thank Roger Blandford, Angélica de Oliveira-Costa Harold Shapiro and Matias Zaldarriaga for helpful comments. Support for this work was provided by NSF grants AST-0071213 & AST-0134999, NASA grants NAG5-9194 & NAG5-11099, and a David and Lucile Packard Fellowship.

REFERENCES

1. E. Einstein, *Relativity: the Special and the General Theory* (New York, Random House, 1920).
2. The cosmic microwave background (CMB) is the oldest light around, emanating from the hot opaque hydrogen plasma that filled the universe during its first 400,000 years. An up-to-date review is M. White & J. Cohn, astro-ph/0203120 (2002).
3. The Lyman α forest (LyαF) is the plethora of absorption lines in the spectra of distant quasars caused by neutral hydrogen in overdense intergalactic gas along the line of sight. It allows us to map the cosmic gas distribution out to great distances.
4. http://map.gsfc.nasa.gov
5. http://www.mpa-garching.mpg.de/Virgo/
6. B. C. Bromley, W. A. Miller & V. I. Pariev, *Nature* **391**, 54 (1998).
7. C. M. Will, *Theory & Experiment in Gravitational Physics* (Cambridge, Cambridge U. P., 1993).
8. C. M. Will, gr-qc/9811036 (1998).
9. M. Colless *et al.*, *MNRAS* **328**, 1039 (2001).
10. The CMB power spectrum is the level of temperature fluctuations as a function of angular scale, defined as the variance of the spherical harmonic coefficients of a sky map. A matter power spectrum is the level of 3D density fluctuations as a function of spatial scale, defined as the variance of the Fourier coefficients of the density field.
11. The measured power spectrum of galaxies or LyAF clouds may differ from the power spectrum that we really care about — that for the underlying dark matter. Bias is defined as the square root of the ratio of the two power spectra.
12. X. Wang, M. Tegmark & M. Zaldarriaga, *PRD* **65**, 123001 (2002).
13. G. Efstathiou *et al.*, *MNRAS* **330L**, 29 (2002).
14. K. Schwarzschild, *Vier. d. Astr. Ges.* **35**, 337 (1900).
15. D. Stevens, D. Scott & J. Silk, *PRL* **71**, 20 (1993).
16. A. de Oliveira-Costa & G. F. Smoot, *ApJ* **448**, 477 (1995).
17. A, de Oliveira-Costa, G. F. Smoot & A. A. Starobinski, *ApJ* **468**, 457 (1996).
18. J. Levin, gr-qc/0108043 (2001).
19. http://snap.lbl.gov
20. S. Perlmutter *et al.*, *Nature* **391**, 51 (1998).
21. A. G. Riess *et al.*, *Astron. J.* **116**, 1009 (1998).
22. I. Maor, R. Brustein & P. J. Steinhardt, *PRL* **86**, 6 (2002).
23. D. Huterer & M. S. Turner, *Phys. Rev. D* **64**, 123527 (2001).
24. S. M. Carroll & M. Kaplinghat, *PRD* **65**, 063507 (2002).

25. A. R. Liddle & D. H. Lyth, *Cosmological Inflation and Large-Scale Structure* (Cambridge, UK, Cambridge U. P., 2000).
26. J. Khoury et al., *PRD* **64**, 123522 (2001).
27. R. Kallosh, L. Kofman & A. D. Linde, *PRD* **64**, 123523 (2001).
28. P. J. Steinhardt & N. Turok, *Science* **??**, ?? (2002).
29. J. B. Hartle & S. W. Hawking, *PRD* **28**, 2960 (1983).
30. R. J. Gott & L. X. Li, *PRD* **58**, 023501 (1998).
31. G. Veneziano, hep-th/0002094 (2000).
32. A. D. Linde, *Particle Physics and Inflationary Cosmology* (Chur, Switzerland, Harwood, 1990).
33. A. Vilenkin, *PRL* **74**, 846 (1995).
34. A. Loeb, *PRD* **65**, 047301 (2002).
35. Y. Wang & P. M. Garnavich, *ApJ* **552**, 445 (2001).
36. M. Tegmark, *PRD* **66**, 103507 (2002).
37. M. Rowan-Robinson, *The Cosmological Distance Ladder* (New York, Freeman, 1985).
38. W. L. Freedman et al., *ApJ* **553**, 47 (2001).
39. N. Bahcall et al., *Science* **284**, 1481 (1999).
40. O. Lahav et al., astro-ph/0112162 (2001).
41. R. A. C. Croft et al., astro-ph/0012324 (2000).
42. M. Bartelmann & P. Schneider, *Phys. Rep.* **340**, 291 (2001).
43. L. Van Waerbeke et al., astro-ph/0101511 (2001).
44. N. Y. Gnedin, astro-ph/0110290 (2001).
45. M. Tegmark, A. J. S. Hamilton & Y. Xu, *MNRAS* **335**, 887 (2002).
46. L. Verde et al., astro-ph/0112161 (2001).
47. D. J. Eisenstein, W. Hu & M. Tegmark, *ApJ* **518**, 2 (1999).
48. M. Tegmark & M. Zaldarriaga, *PRD* **66**, 103508 (2002).
49. H. Hoekstra, H. Yee & M. Gladders, *ApJ* **577**, 595 (2002).
50. N. Y. Gnedin & A. Hamilton J S, *MNRAS* **334**, 107 (2002).
51. J. Wambsganss, astro-ph/001242 (2000).
52. A. Celotti, J. C. Miller & D. W. Sciama, *Class. Quant. Grav.* **16**, A3 (2000).
53. A. Marconi, astro-ph/0201504 (2002).
54. J. Frank, King A & D. Raine, *Accretion power in astrophysics, 3rd ed.* (Cambridge, Cambridge U. P., 2002).
55. F. Zwicky, *Helv. Phys. Acta* **6**, 110 (1933).
56. H. Hoekstra et al., *ApJL* **548**, L5 (2001).
57. N. Bahcall et al., *ApJ* **541**, 1 (2000).
58. H. J. Newberg, astro-ph/0111095 (2001).
59. D. N. Spergel & P. J. Steinhardt, *PRL* **84**, 3760 (2000).
60. P. Bode, J. P. Ostriker & N. Turok, astro-ph/0010389 (2000).
61. N. Dalal & C. S. Kochanek, astro-ph/0111456 (2002).
62. S. A. et al.. Hughes, astro-ph/0110349 (2001).
63. B. Boisseau et al., *PRL* **85**, 2236 (2000).
64. M. Milgrom, *ApJ* **270**, 365 (1983).
65. S. S. McGaugh, *ApJL* **541**, L33 (2000).
66. P. O. Mazur & E. Mottol, astro-ph/0103466 (2001).
67. P. J. E. Peebles, astro-ph/9910234 (1999).
68. J. A. Sellwood & A. Kosowsky, astro-ph/0009074 (2000).
69. L. M. Griffiths, A. Melchiorri & J. Silk, *ApJ* **553**, L5 (2001).
70. M. White, D. Scott & E. Pierpaoli, astro-ph/0004385 (2000).
71. M. White & C. S. Kochanek, *ApJ* **560**, 539 (2001).
72. C. D. Hoyle et al., *PRL* **86**, 1418 (2001).
73. L. Randall, *Science* **296**, 1422 (2002).

The Origin of Density Fluctuations in a Cyclic Universe

Paul J. Steinhardt* and Neil Turok[†]

*Department of Physics, Princeton University, Princeton, NJ 08544, USA
[†]DAMTP, Centre for Mathematical Sciences, Wilberforce Road, Cambridge, CB3 0WA, UK

Abstract. This paper is a brief introduction to the "cyclic model of the universe," a radical alternative to standard big bang/inflationary theory in which space and time exist indefinitely, inflation is avoided, and the universe undergoes periodic epochs of expansion and conclusion. This introduction explains the novel way in which density perturbations are generated which seed large scale structure formation and produce spatial variations in the cosmic microwave background temperature.

The cyclic model of the universe [1, 2] is a radical alternative to the standard inflationary/big bang model [3, 4]. One key difference is the general flow of the cosmic evolution. In the standard model, the universe expands from the big bang and proceeds monotonically, transforming the cosmos from hot to cold and from dense to dilute. In the cyclic model, the evolution is periodic. The universe undergoes epochs of expansion during which its temperature and density decrease, but, after the density becomes negligibly small, a sequence of events occurs which creates new matter and radiation that reheats the universe to high temperature and density and triggers a new period of expansion and cooling.

A second key difference is the events that shape the large scale structure of the universe. In the standard picture, a brief period of hyper-rapid expansion (inflation) shortly following the big bang makes the universe homogeneous and isotropic, flattens the spatial curvature, and creates a nearly scale-invariant spectrum of density fluctuations. In the cyclic model, inflation is replaced by physical processes that occurred a cycle ago, before the last bang. These processes entail energies and timescales that are one hundred orders of magnitude different from inflation. Furthermore, the physics that creates the density fluctuations that is fundamentally different.

Despite these extraordinary differences, the cyclic model appears capable of reproducing all of the successful predictions of the standard model with fewer ingredients. Not all predictions are identical, either. The cyclic and inflationary models have exponentially different predictions for the spectrum of primordial gravitational waves, for example. So, there are future tests which can discriminate the two models. At present, though, both models are in equally good agreement with the data.

THE CYCLIC MODEL: THE BRANE PICTURE

The cyclic model is inspired by recent concepts in superstring theory, particularly M-theory, the Horava-Witten model, branes, orbifolds and extra dimensions [6, 1], as well as a precursor cosmological model known as the "ekpyrotic scenario" [7]. The cosmological model does not *require* these features. We can pose the theory in a field theoretic language that is more easily compared to inflation. However, the M-theory description provides a simple and compelling geometrical picture that provides a natural intuition about how the model works. So, we will present here both the M-theory and field theory descriptions here.

In heterotic M-theory models, our three-dimensional universe is a hypersurface embedded in a spacetime with an extra spatial dimension. (Actually, in the Horava-Witten model, there are 6 additional spatial dimensions compactified on a Calabi-Yau manifold, but the manifold is so small that the six dimensions can be neglected for our purposes.) The hypersurface is a boundary or orbifold plane of the extra dimension separated by a finite distance from a second boundary/orbifold. The orbifolds have energy and momentum. They can interact through gravity and exchange virtual membranes. In the cyclic model, the orbifolds are drawn together by these interactions, and they collide and bounce at regular intervals.

The model goes through the following stages. Each cycle begins with a "bang," a collision between branes that creates matter and radiation. The universe proceeds directly to the radiation-dominated epoch without encountering any inflation. The model must introduce a mechanism for making the universe homogeneous, isotropic and flat, and for creating a nearly scale-invariant spectrum of density perturbations. However, this will be accomplished by a sequence of events that occurs at a different point in the cycle. Hence, the universe proceeds directly after the bang to radiation domination to matter domination and, finally, to dark energy domination.

In the big bang/ inflationary model, dark energy comes as a complete surprise. It is not predicted or required. Rather, dark energy is added *ad hoc* to make the model consistent with the recent observations of cosmic acceleration [8, 9, 10].

In the cyclic model, dark energy moves to center stage as an essential element of the cyclic model. Its source is the potential energy associated with the interaction between branes. When the branes are far apart, the energy is presumed to be small and positive, acting as a form of quintessence that causes the branes to stretch at an accelerating rate, expanding by a factor of two every 15 billion years. Continued for 100 doublings or a trillion years, the dark energy thins out the matter and radiation in the universe to a point where the universe approaches a homogeneous vacuum. Furthermore, any warps or curvature in the branes are stretched out. Hence, two of the roles of inflation, making the universe homogeneous and flat, are replaced by dark energy in the cyclic model. Dark energy also is important in making the cyclic solution a classical attractor. That is, if the branes are kicked away from the ideal cyclic orbit, the period of dark energy domination causes the evolution to converge after a cycle or two to the ideal evolution.

After the matter and radiation have been thinned out, the universe begins a period of "contraction." But, unlike earlier cyclic models discussed in the 1920's and 1930's [11] our three dimensions do not contract and the temperature and density do not diverge. Rather, the extra dimension between the orbifolds contracts as the two branes approach one another and head towards collision. The contraction ends in a "crunch" at which matter and radiation are created. The two branes bounce apart, but now filled with the newly created hot matter-radiation whose density dominates the older, thinned out matter-radiation from the previous cycle. Due to gravity, the new matter and radiation causes the branes to begin to stretch again and damp the motion of branes. The universe has returned to the same state as it was after the last bang and the cycle begins anew.

During the contraction phase, the branes undergo quantum fluctuations that cause them to wrinkle. For simple, exponentially decreasing interbrane potentials, the wrinkles form a scale-invariant spectrum. As a result of the wrinkles, the branes collide, bounce and reheat at different times at different locations. The collision thereby imprints a scale invariant spectrum of spatial variations in the temperature on the branes after the collision.

THE CYCLIC MODEL: THE FIELD THEORETIC PICTURE

As we have noted, the cyclic story can be described in terms of an ordinary four-dimensional field theory, which can be obtained by taking the long wavelength limit of the brane picture [1, 2]. The distance between branes becomes a moduli (scalar) field ϕ. The interbrane interaction is replaced by a scalar field potential, $V(\phi)$. The different stages in the cyclic model in the brane picture are in one-to-one correspondence to the motion of the scalar field along the potential. See Fig. 1.

Then, the action S describing describing gravity, the scalar field ϕ, and the matter-radiation fluid is:

$$S = \int d^4x \sqrt{-g} \left(\frac{1}{16\pi G} \mathcal{R} - \frac{1}{2}(\partial \phi)^2 - V(\phi) + \beta^4(\phi) \rho_R \right), \tag{1}$$

where g is the determinant of the Friedmann-Robertson-Walker metric $g_{\mu\nu}$, G is Newton's constant and $\mathcal{R}$ is the Ricci scalar.

Particularly notable is the coupling $\beta(\phi)$ between ϕ and the matter-radiation (ρ_R) density. This coupling is crucial because it accounts for the fact that the temperature and density do not diverge at the crunch. If β were set to unity, the solution to the equation of motion would be $\rho_R \propto 1/a^4$, where a is the scale factor. Then, at the crunch where $a \to 0$, the density would diverge. (This describes the case of the older cyclic models.) However, the β factor has the property that $\beta \to \infty$ as $a \to 0$ such that $a\beta \to$ constant. The revised solution to the equation of motion is $\rho_R \propto 1/(a\beta)^4$ which approaches a constant as $a \to 0$. The energy, once thinned out during the dark energy dominated phase, remains thinned out at the bounce.

If we had begun with the field theory and simply introduced the β-factor by hand, it would have seemed incredibly fine-tuned to choose a form which diverges as $1/a$. However, we now understand that the form arises automatically if ϕ is the modulus field that describes the size of the extra dimension. The β-factor simply reflects the fact that the extra-dimension collapses but our three-dimensions do not. As a result, entropy produced during one cycle is not concentrated at the crunch and does not contribute significantly to the entropy density at the beginning of the next cycle. Hence, cycles can

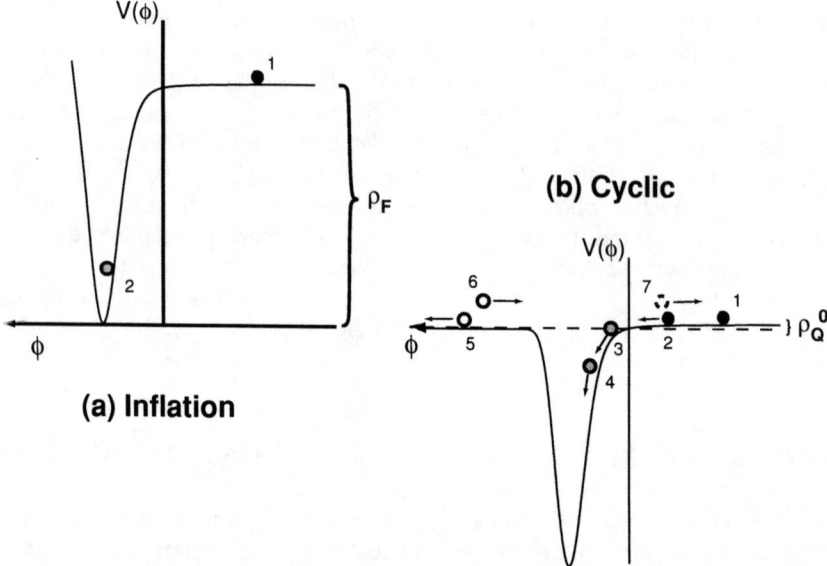

FIGURE 1. Schematic plot of the potential $V(\phi)$ as a function of the field ϕ for (a) inflationary cosmology and (b) cyclic models. For inflation, accelerated expansion and production of density perturbations occur in stage 1; reheating occurs at stage 2. For the cyclic model, present-day accelerated expansion occurs in stages 1 and 2; deceleration in stage 3; contraction and production of density perturbations begins in stage 4; contraction (dominated by scalar field kinetic energy) in stage 5; bounce, production of matter-radiation, and re-expansion in stage 6; matter-radiation dominated epoch begins in stage 7, and the cycle begins anew.

continue for an arbitrarily long time and there is no practical way of distinguishing one cycle from the next.

ORIGIN OF DENSITY PERTURBATIONS

If the cyclic model can be described in terms of ordinary field theory, then it may seem surprising that it is possible to generate a nearly scale invariant spectrum density perturbations. Perturbations in theories with scalar fields were investigated in the 1980's [12, 13], and it was found that a nearly scale-invariant arises if the scalar field has a nearly de Sitter equation of state. How does the cyclic model fit in? The perturbations cannot be generated during the dark energy dominated phase because the dark energy density is exponentially too small to generate a spectrum with the right amplitude.

The explanation, as has been recently discovered [14], is that there are actually three distinct ways of producing a nearly scale-invariant spectrum, and that inflation represents only one of them. The three ways can be characterized by $w \equiv (\frac{1}{2}\dot{\phi}^2 - V)/\frac{1}{2}\dot{\phi}^2 + V)$, the effective equation of state of the scalar field. Case I is where $w \approx 1$ and the universe is expanding, the example of inflation. Case II is a contracting universe with $w \approx 0$, an example which has not been used in cosmological models to date. Case III is a contracting universe with $w \gg 1$ – the situation that applies in the cyclic model. (Here I am only considering cases where w is nearly constant, cases which can be obtained with simple potentials; contrived examples can also be constructed in which w is time-dependent.)

We know how design a scalar field potential so that $w \approx -1$. If the potential is sufficiently flat (V'/V and V''/V are very small), then the field ϕ rolls slowly down the potential, V is nearly constant for an extended period, and w approaches -1. It is in regime where the potential is flat that the perturbations are produced and, hence, where there are tight constraints on the form of $V(\phi)$. After the flat portion of the potential, there is a great deal of freedom in choosing the potential shape.

What is required to obtain $w \gg 1$? From the expression for w, it is apparent that this is only possible if the potential is negative. In particular, for a negative exponentially *steep* potential $V \approx -\exp(c\phi)$, the solutions to the equation of motion have a scaling solution in which $\dot{\phi}^2/2V$ is constant and approximately -1. Consequently, w is much greater than unity and nearly constant. Curiously, a potential which rolls from positive to negative is just what is needed to go from an accelerating universe to a contracting universe, so the requirements for a scale-invariant spectrum dovetail with overall scenario. In analogy with inflation, this steep regime of the potential corresponding to the generation of fluctuations is where there are tight constraints on the form of the potential. Although the standard example of a potential has a flat positive plateau on one side and rise of V on the other [1, 2], there is actually tremendous flexibility in choosing the shape of $V(\phi)$ way from the steep portion [15]. See Fig. 2. For example, the positive plateau may be replaced by an increasing function or even a locally stable (positive energy) minimum. Similarly, the potential need approach zero or even have a minimum. (These features are put in the standard example motivated by M-theory [1, 16], but they are not required for cyclic cosmology.)

The generation of fluctuations for $w \gg 1$ can be understood *heuristically* by examining the perturbed Klein-Gordan equation [17]:

$$\delta\phi_k'' = -\left(k^2 + \frac{a''}{a} + V_{,\phi\phi}\right)\delta\phi_k \qquad (2)$$

where $\phi(\mathbf{x},t)$ has been expanded in fourier components $\delta\phi_k(t)$ with wavenumber k and prime is derivative with respect to conformal time η. The a''/a term is due to gravitational expansion, and the last term is due to the self-interaction of the scalar field. This equation applies equally to inflation and to cyclic models. The well-known result[] from inflation is that, in order to obtain a scale invariant spectrum, the combination of the last two terms on the right hand side must behave as $(2/\eta^2)\delta\phi_k$. In the case of inflation, $a(\eta) = -1/\eta$ and $V_{,\phi\phi}$ is negligible. So, the scale invariant fluctuations are due entirely to the gravity term. A second solution exists where $a''/a = 2/\eta^2$ and the gravity term dominates: namely, $a = \eta^2$, the dust-like $w = 0$ universe, Case II above.

The cyclic model corresponds to the limit where the gravity term is negligible and, instead, the perturbation equation is driven by the potential term. For the negative exponential potential, for example, the scaling solution corresponds to $V_{\phi,\phi} \approx 2/\eta^2$.

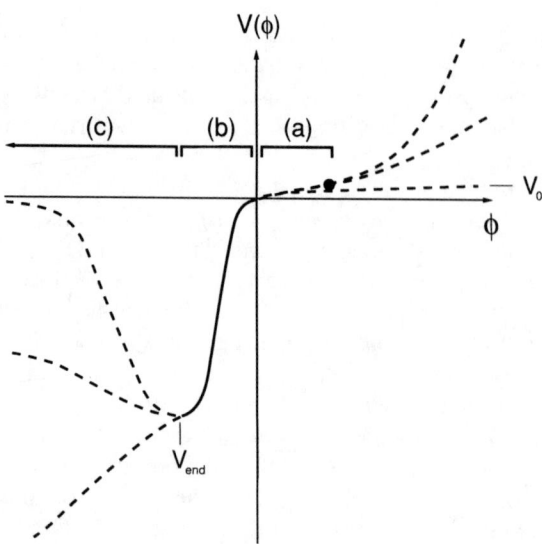

FIGURE 2. Plots of possible cyclic potentials showing how they can be viewed as having three separate parts: (a) positive potential energy density; (b) steep, negative potential; (c) less steep or increasing potential. The figure is intended to emphasize that the only tight constraints are for the negative, steep portion of the potential (b) where perturbations are generated.

In sum, we have seen a remarkable result: The perturbations are produced in inflationary models under conditions that are extremely different from cyclic models – hyper-rapid expansion as opposed to ultra-slow contraction – and yet the perturbation equations expressed in terms of η are isomorphic and the outcome is the same. Both produce a nearly scale-invariant, gaussian spectrum of adiabatic density perturbations.

We should note that the derivation of the density perturbation spectrum outlined here is not rigorous and it does not follow what happens to the perturbations at the bounce. At the bounce, fluctuations in the scalar field (or branes) lead to variations in the time of collision and of reheating to high temperature [1, 17]. In this way, the fluctuations are converted directly into a spectrum of density perturbations after the bounce. This argument is similar to the "time-delay" approach [13] to deriving perturbations in inflation.

Formal methods should use gauge invariant variables and include gravitational backreaction. The formal analysis reveals the same conclusions as the heuristic derivation above; the gravitational backreaction is negligible [17, 14]. Some other authors attempting their own methods have claimed the opposite result. A common error has been to choose a gauge invariant variable that does not include the growing mode perturbations to leading order. This mistake is easy to make since the appropriate gauge invariant variables for inflationary cosmology fail to include the growing mode perturbation in a contracting cyclic phase. Failure to take proper account of this difference results in accidentally projecting out the scale-invariant spectrum of fluctuations. A second common error has been to ignore the radiation produced at the bounce. In our computation, this leads to elimination of the scale-invariant fluctuations [17]. The reason is clear: in the no-radiation limit, the evolution is time-reversal invariant. Any perturbations generated during contraction translate into purely decay perturbations after the bounce. The radiation breaks the symmetry between contracting and expanding. Now the scale-invariant perturbations produced during contraction match to a combination of growing and decaying modes after the bounce [17]. The growing mode provides the needed scale-invariant spectrum of density perturbations. However, this spectrum only exists if the bounce is not time-reversible.

GRAVITATIONAL WAVES

If both produce the same outcome, what is the difference? The outcome is only the same when it comes to density perturbations. The difference shows up in the spectrum of gravitational waves [1, 2, 17, 18]. To see this, consider that the perturbation equation above applies to all fields. For inflation, the gravity term dominates over the potential term, which is proportional to the mass of the inflaton during inflation. Gravity will, therefore, dominate in the equation describing any light mass fields, including the two massless polarization modes of the metric fluctuations. All such fields will have scale-invariant fluctuations. For most fields, this is irrelevant because the fluctuations leave no distinctive cosmic signature. Fluctuations of the massless metric perturbations, though, propagate as gravitational waves, leaving a distinctive signature in the cosmic microwave background anisotropy as well as in gravitational wave detectors. Hence, inflation predicts that there should be a nearly scale-invariant spectrum of gravitational waves in addition to density fluctuations.

For the cyclic model, the potential term dominates and the gravitational term is irrelevant. Hence, cyclic models only produce scale invariant fluctuations in the scalar field. For gravitational waves, there is only the a''/a term and the spectrum is very blue. That is, the amplitude drops from some small value at very short wavelengths to exponentially smaller values at cosmic scales.

This dramatic difference from inflationary cosmology (see Fig. 3 from [18]) is the most promising approach for distinguishing the two scenarios. In the near-term, measuring the polarization of the cosmic microwave background is most sensitive method for searching for stochastic gravitational waves. Gravitational waves leave a characteristic "B-mode" polarization pattern that is smaller in amplitude than the "E-mode" created by density perturbations, but distinctive [19, 20]. Detector sensitivities anticipated in the coming decade should be sufficient to cover a wide span of the most likely inflationary models [21]. As suggested in Fig. 3, direct detection of gravitational waves from inflation is not feasible in the next decade. Detection of the cosmic gravitational waves would clearly support inflation and disprove the cyclic scenario. The absence of detection would not disprove inflation, but it would force us towards more arcane inflationary models versus the comparatively simpler cyclic model.

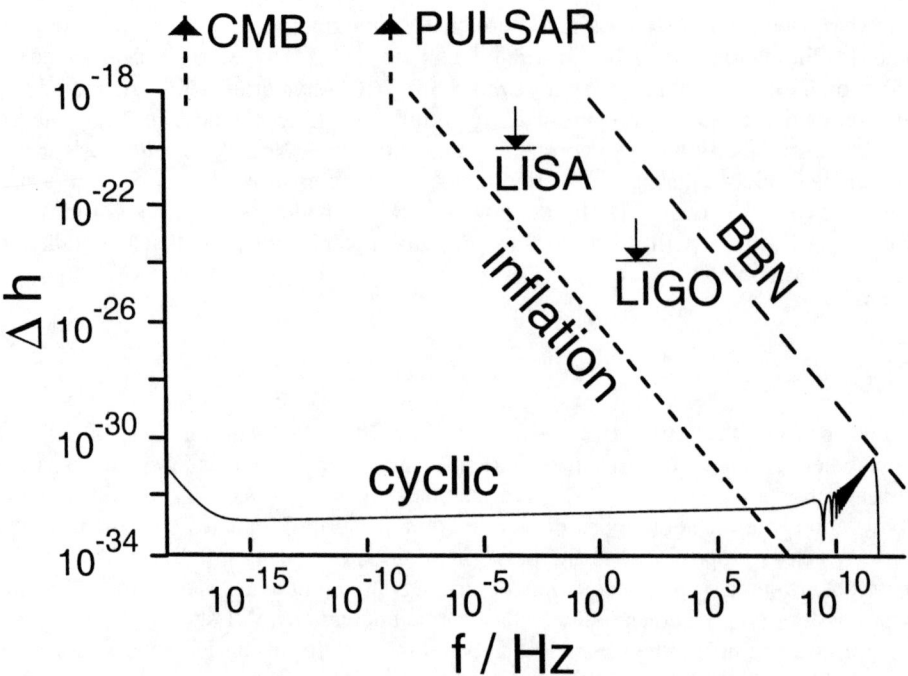

FIGURE 3. Plots of the dimensional strain Δh vs. frequency for the gravitational waves anticipated in cyclic (solid) and inflationary models (short-dashed) Included are constraints from big bang nucleosynthesis (BBN, excludes range to upper right of long-dashed line) and best limits anticipated from the proposed LISA and advanced LIGO gravitational wave detectors. The cosmic microwave background (CMB) and pulsar constraints are shown dashed to indicate that they lie beyond the top of the figure. Both models satisfy the BBN constraints for typical parameters. CMB is currently the most promising approach for observing inflationary gravitational waves. The cyclic prediction is orders of magnitude below current and projected tests.

CONSTRUCTING THE CYCLIC POTENTIAL

We have summarized the basic ingredients needed to compute the spectrum of density perturbations, and the reader can look to Refs. [14, 17] for the technical details. But how does this consideration constrain the effective potential for the cyclic model? For inflation, the most stringent constraints are on the flat part of the potential, the range of the inflaton field where the density perturbations are generated. The constraints are commonly expressed as bounds on two "slow-roll" parameters:

$$\varepsilon = \left(\frac{V'}{V}\right)^2 \ll 1 \text{ and} \tag{3}$$

$$\eta = \frac{V''}{V} \ll 1. \tag{4}$$

For the cyclic model, the analogous constraints are on the steep portion of the potential where perturbations are generated. See Fig. 2. The constraints can be expressed in terms of two "fast-roll" parameters [14, 15]:

$$\bar{\varepsilon} = \left(\frac{V}{V'}\right)^2 \ll 1 \text{ and} \tag{5}$$

$$\bar{\eta} = 1 - \frac{V''V}{(V')^2} \ll 1. \tag{6}$$

The first constraint forces the slope to be steep and the second fixes the curvature, where each applies to the range of ϕ where the fluctuations are generated that are within the horizon today.

The result is that the constraints in the two models are remarkably similar. The bad news is that the degree of tuning in cyclic models is no better than in inflationary models. The good news is that it is no worse. Also, recall, the cyclic model avoids the need for an inflaton, the inflation potential, and two episodes of accelerated expansion. So, there is actually some net gain for cyclic models, even if not in terms of degree of tuning. A further discussion of designing potentials for cyclic models can be found in [15].

In sum, the cyclic model has rapidly developed into a promising and provocative alternative to the standard big bang inflationary picture. There remain open issues, most especially a rigorous demonstration that the bounce can occur when quantum fluctuations are included. (A related, unproven issue for inflationary cosmology is a rigorous demonstration that the universe emerges from the big bang with the right conditions to have inflation.) But, other than this uncertainty, the other key aspects of the model are well-developed in technical detail. So, it appears for the next few years, there will be continued development of the models vigorous debates about which is theoretically preferable. Ultimately, though the answer must be determined by observation, either the detection of primordial gravitational waves or of other distinguishing features yet to be determined. What is at stake is nothing less than our understanding of the past history and future fate of the universe.

This work was supported in part by US Department of Energy grant DE-FG02-91ER40671 (PJS) and by PPARC-UK (NT).

REFERENCES

1. Steinhardt, P.J. and Turok, N., *Science* **296**, 1436 (2002).
2. Steinhardt, P.J. and Turok, N., *Phys. Rev.* **D65** 126003 (2002).
3. Guth, A.H., *Phys. Rev.* **D23** 347 (1981).
4. Linde, A.D., *Phys. Lett.* **B108** 389 (1982).
5. Albrecht, A. and Steinhardt, P.J., *Phys. Rev. Lett.* **48** (1982) 1220.
6. Hořava, P. and Witten, E., *Nucl. Phys.* **B460** (1996) 506; **B475** (1996) 94.
7. Khoury, J., Ovrut, B.A., Steinhardt, P.J. and Turok, N., *Phys. Rev.* **D64** 123522 (2001).
8. Perlmutter, S. *et al*, *Ap. J.* **517** 565 (1999).
9. Riess, A.G. *et al*, *Astron. J.* **116** 1009 (1998).
10. Garnavich, P.M. *et al*, *Ap. J.* **509** (1998) 74.
11. Tolman, R.C., *Relativity, Thermodynamics and Cosmology*, (Oxford U. Press, Clarendon Press, 1934).

12. Bardeen, J., Steinhardt, P.J., and Turner, M.S., *Phys. Rev. D***28**, 679 (1983).
13. Guth, A.H. and Pi, S.Y., *Phys. Rev. Lett.* **49** (1982) 1110.
14. Gratton, S., Khoury, J., Steinhardt, P.J. and Turok, N., in preparation.
15. Khoury, J., Steinhardt, P.J., and Turok, N., in preparation.
16. Khoury, J., Ovrut, B.A., Seiberg, N., Steinhardt, P.J., and Turok, N., *Phys. Rev. D***65**, 086007 (2002).
17. Khoury, J., Ovrut, B.A., Steinhardt, P.J. and Turok, N., *Phys. Rev. D***66** 046005 (2002).
18. Boyle, L., Steinhardt, P.J., and Turok, N., in preparation.
19. Seljak, U. and Zaldarriaga, M., *Phys. Rev. Lett.* **78**, 2054 (1997).
20. Kamionkowski, M. and Kosowsky, A., *Phys. Rev. D***57**, 685 (1998).
21. Lange, A., Proceedings of the NAS Symposium on Challenges to the Standard Paradigm, in preparation.

STRUCTURE IN THE COSMIC MICROWAVE BACKGROUND

CMB Observables and Their Cosmological Implications 2002

Wayne Hu

Center for Cosmological Physics and Department of Astronomy and Astrophysics, University of Chicago, Chicago IL 60637

Abstract. The tremendous experimental progress in cosmic microwave background (CMB) temperature and polarization anisotropy studies over the last few years has helped establish a standard paradigm for cosmology at intermediate epochs and has simultaneously raised questions regarding the physical processes at the two opposite ends of time. We review the acoustic phenomenology that forms the cornerstone of the standard cosmological model and discuss internal consistency relations which lend credence to its interpretation. We touch on future milestones in the study of CMB anisotropy and their implications for inflationary and dark energy models.

INTRODUCTION

The pace of discovery in the field of Cosmic Microwave Background (CMB) anisotropy has been accelerating over the last few years. With it, the basic elements of the cosmological model have been falling into place: the nature of the initial seed fluctuations that through gravitational instability generated all of the structure in the universe, and the mixture of matter-energy constituents that drives its expansion.

In 1992, the COBE DMR experiment reported the first detection of cosmological anisotropy in the temperature of the CMB. The 10^{-5} variations in temperature detected on scales larger than the 7° resolution provided strong support for the gravitational instability paradigm. These variations represent the direct imprint of initial gravitational potential perturbations through their redshifting effect on the CMB photons, called the Sachs-Wolfe effect [1], and are of the right amplitude to explain the large-scale structure of the universe. From 1992-1998, a host of experiments detected a rise and fall in the level of anisotropy from degree scales to arcminute scales. In the following two years, the Toco, Boomerang, and Maxima experiments measured a clearly defined first peak in the spectrum and provided empirical evidence that the small scale anisotropy is dominated by coherent acoustic phenomena, long predicted to exist [2, 3]. This first peak measurement has subsequently been confirmed by several groups, with the best measurement to date from the Archeops experiment. These measurements firmly establish initial gravitational potential or curvature fluctuations as the primary source of structure in the universe (c.f. cosmological defect models), provide clear evidence that the universe is close to spatially flat and, in conjunction with other cosmological measurements of the dark matter, an indicate that there is a component of missing or dark energy. These findings provide support for the inference from distant supernovae that the expansion is currently accelerating under the influence of the dark energy.

TABLE 1. CMB anisotropy milestones

Phenomena	Experiments	Date
Sachs-Wolfe	COBE DMR	'92
Degree-scale	many	'93-99
First peak	Toco, Boom, Maxima	'99-'00
Secondary peaks	DASI, Boom	'01
Damping tail	CBI	'02
Polarization detection	DASI	'02
Secondary anisotropy	CBI?, BIMA?	'02?
Polarization peaks	–	future
non/Gaussianity	–	future
Reionization bump	–	future
Lensing of peaks	–	future
Dark energy ISW	–	future
Grav.-wave polarization	–	future

In 2001, the DASI and Boomerang experiments announced a detection of secondary acoustic peaks in the spectrum which was subsequently confirmed by the VSA experiment. These experiments provided a precise measurement of the baryon-photon ratio that is in excellent agreement with inferences from big bang nucleosynthesis, lending further confidence in the underlying cosmological model. Moreover they provided the first direct evidence that dark, non-baryonic, matter exists in the early universe. In the past year, two more predicted phenomena have been discovered: the dissipation of the acoustic peaks at small scales by the CBI experiment and, following a series of increasingly tighter upper limits, the polarization of the anisotropy by the DASI experiment. These observed phenomena provide the best internal evidence that the physical assumptions underlying the interpretation of the peaks are justified.

The past few years have been an era of milestones for experiments and millstones for theoretical speculation. The standard cosmological paradigm of structure formation by gravitational instability of cold dark matter in a nearly homogeneous and isotropic universe has been so thoroughly tested that it is extremely difficult to find viable contenders that differ in any fundamental way. The acoustic peaks are furthermore strikingly consistent with two fundamental predictions of simple inflationary models of the early universe: a flat spatial geometry and a nearly scale-invariant spectrum of initial curvature fluctuations. Alternatives to inflation, must now be nearly indistinguishable phenomenologically to be consistent with the data (e.g. [4]).

Key past and future milestones are summarized in Table 1. In this review we shall try to place this somewhat bewildering array of phenomena in its cosmological context. We begin with a description of CMB observables in §, proceed through the physical basis of their generation in § to the phenomenology of the acoustic peaks in §. Finally in § we touch on future milestones of CMB temperature and polarization anisotropy.

CMB OBSERVABLES

The fundamental observable in the CMB is the intensity of radiation per unit frequency per polarization at each point in the sky. The polarization state of the radiation in a direction of sky denoted $\hat{\mathbf{n}}$ is described by the intensity matrix $\langle E_i(\hat{\mathbf{n}})E_j^*(\hat{\mathbf{n}})\rangle$ where $\mathbf{E}$ is the electric field vector and the brackets denote time averaging.

The radiation is measured to be a blackbody of $T = 2.728 \pm 0.004\text{K}$ (95% CL) [5] with fractional variations across the sky at the 10^{-5} level and fractional polarization at the 10^{-6} level. It is then convenient to describe the observables by a temperature fluctuation matrix decomposed in the Pauli basis (e.g. [6, 7])

$$\mathbf{P} = C\langle\mathbf{E}(\hat{\mathbf{n}})\mathbf{E}^\dagger(\hat{\mathbf{n}})\rangle = \Theta(\hat{\mathbf{n}})\mathbf{I} + Q(\hat{\mathbf{n}})\sigma_3 + U(\hat{\mathbf{n}})\sigma_1 + V(\hat{\mathbf{n}})\sigma_2, \tag{1}$$

where we have chosen the constant of proportionality so that the Stokes parameters (Θ,Q,U,V) are dimensionless, e.g. $\Theta(\hat{\mathbf{n}}) \equiv \Delta T(\hat{\mathbf{n}})/T$ averaged over polarization states. Linear polarization is generated in the CMB by Thomson scattering of anisotropic radiation much like polarization by reflection; circular polarization V is absent cosmologically. Note that under a counterclockwise rotation of the coordinate axes by ψ, $Q \pm iU \to e^{\mp 2i\psi}(Q \pm iU)$.

The temperature and polarization fields are decomposed as [8, 9]

$$\Theta_{lm} = \int d\hat{\mathbf{n}} Y_{lm}^*(\hat{\mathbf{n}})\Theta(\hat{\mathbf{n}}), \quad E_{lm} \pm iB_{lm} = -\int d\hat{\mathbf{n}}\,_{\pm 2}Y_{lm}^*(\hat{\mathbf{n}})[Q(\hat{\mathbf{n}}) \pm iU(\hat{\mathbf{n}})], \tag{2}$$

in terms of the complete and orthogonal set of spin harmonic functions, $_sY_{lm}$, which are eigenfunctions of the Laplace operator on a rank s tensor [10, 11]. $Y_{lm} = {}_0Y_{lm}$ is the ordinary spherical harmonic. For small sections of sky, the spin-harmonic expansion becomes a Fourier expansion with $Y_{lm} \to e^{i\mathbf{l}\cdot\hat{\mathbf{n}}}$ and $_{\pm 2}Y_{lm} \to -e^{\pm 2i\phi_l}e^{i\mathbf{l}\cdot\hat{\mathbf{n}}}$, where ϕ_l is the azimuthal angle of the Fourier wavevector $\mathbf{l}$. Note that the E and B modes are then simply the Q and U states in the coordinate system defined by $\mathbf{l}$ [12], i.e. B-modes have a polarization orientation at $45°$ to the wavevector.

The primary observable is the two point correlation between fields $X, X' \in \{\Theta, E, B\}$

$$\langle X_{lm}^* X'_{l'm'}\rangle = \delta_{ll'}\delta_{mm'} C_l^{XX'}, \tag{3}$$

and are described by power spectra C_l as long as the fields are statistically isotropic. If parity is also conserved then B has no correlation with Θ or E. If in addition the fluctuations are Gaussian distributed, the power spectra contain all of the statistical information about the fields. The measurements of these power spectra to date are shown in Fig. 1.

PHOTON-BARYON PLASMA

Fluid Dynamics

Before recombination at $z_* \approx 10^3$, the photons and baryons form a plasma in which Thomson scattering is so rapid that the system can be considered a nearly perfect

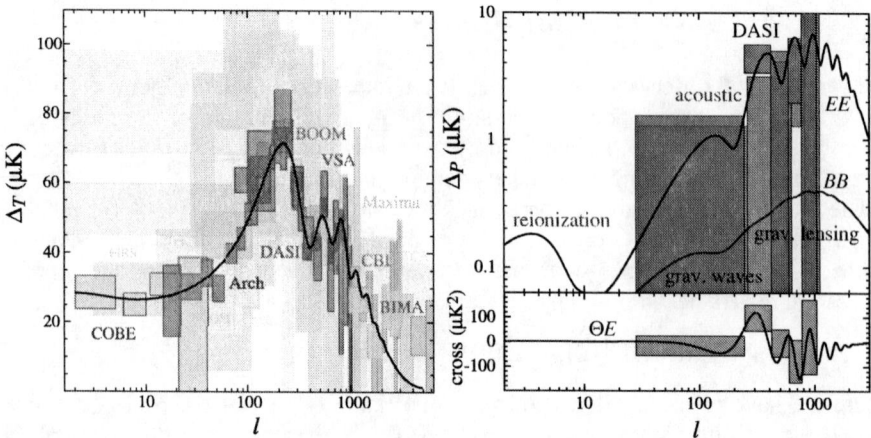

FIGURE 1. Power spectra data plotted as the rms contribution per logarithmic interval $[l(l+1)C_l/2\pi]^{1/2}$ with error boxes representing 1σ error bars and approximate multipole bandwidth. Overplotted is a scale-invariant, flat cosmological model with $\Omega_m = 1/3$, $\Omega_\Lambda = 2/3$, $h = 0.7$, $\Omega_b h^2 = 0.02$, reionization redshift $z_i = 7$, and an inflationary energy scale of $E_i = 2.2 \times 10^{16}$ GeV.

fluid. Fluctuations in the CMB are then described in Fourier space by the temperature $\Theta(\mathbf{k})$, bulk velocity $v_\gamma(k)$ and a small residual radiative viscosity or anisotropic stress $\pi_\gamma(\mathbf{k})$. These generate the temperature and polarization anisotropy by simple projection [34]. The analogous quantities for the baryons are the density perturbation δ_b and bulk velocity v_b. For gravity, we choose a conformal Newtonian representation where the perturbations are defined by the Newtonian potential Ψ (time-time metric fluctuation) and the curvature fluctuation Φ (space-space metric fluctuation $\approx -\Psi$).

Covariant conservation of energy and momentum requires that the photons and baryons satisfy separate continuity equations

$$\dot{\Theta} = -\frac{k}{3}v_\gamma - \dot{\Phi}, \qquad \dot{\delta}_b = -kv_b - 3\dot{\Phi}, \qquad (4)$$

where overdots represent conformal time $\eta = \int dt/a$ derivatives, and coupled Euler equations

$$\dot{v}_\gamma = k(\Theta + \Psi) - \frac{k}{6}\pi_\gamma - \dot{\tau}(v_\gamma - v_b),$$
$$\dot{v}_b = -\frac{\dot{a}}{a}v_b + k\Psi + \dot{\tau}(v_\gamma - v_b)/R, \qquad (5)$$

where $R = (p_b + \rho_b)/(p_\gamma + \rho_\gamma) \approx 3\rho_b/4\rho_\gamma$ is the photon-baryon momentum density ratio. Here $\dot{\tau} = n_e \sigma_T a$ is the opacity to Thomson scattering.

The continuity equations represent particle number conservation. For the baryons, $\rho_b \propto n_b$. For the photons, $T \propto n_\gamma^{1/3}$, which explains the 1/3 in the velocity divergence term. The $\dot{\Phi}$ terms come from the fact that Φ is a perturbation to the scale factor and so they are the perturbative analogues of the cosmological redshift and density dilution from the expansion. The Euler equations have similar interpretations. The expansion makes particle momenta decay as a^{-1}. The cosmological redshift of T accounts for this effect in the photons. For the baryons, it becomes the expansion drag on v_b ($\dot{a}/a$ term). Potential gradients $k\Psi$ generate potential flow. For the photons, stress gradients in the fluid, both isotropic ($k\delta p_\gamma/(p_\gamma+\rho_\gamma) = k\Theta$) and anisotropic ($k\pi_\gamma$) counter infall. Thomson scattering exchanges momentum between the two fluids ($\dot{\tau}$ terms).

For scales much larger than the mean free path $\dot{\tau}^{-1}$, the Euler equation may be expanded to leading order in $k/\dot{\tau}$, such that the photons are isotropic in the baryon rest frame $v_\gamma = v_b$ and so the joint Euler equation becomes

$$\frac{d}{d\eta}[(1+R)v_\gamma] = k[\Theta + (1+R)\Psi]. \tag{6}$$

Combining this with the continuity equation leads to the oscillator equation (e.g. [20])

$$\frac{d}{d\eta}[(1+R)\dot{\Theta}] + \frac{k^2}{3}\Theta = -\frac{k^2}{3}(1+R)\Psi - \frac{d}{d\eta}[(1+R)\dot{\Phi}], \tag{7}$$

and a small residual anisotropic stress or quadrupole that tracks the evolution of the fluid velocity [22], $\pi_\gamma = (32/15)(k/\dot{\tau})v_\gamma$. This dependence reflects the fact that a local quadrupole can arise from a gradient in the velocity field, for example as photons from two hot crests of a plane wave fluctuation meet at the trough in between, but is suppressed by scattering.

Equation (7) is the fundamental relation for acoustic oscillations. The change in the momentum of the photon-baryon fluid is determined by a competition between the pressure restoring and gravitational driving forces. Note that the frequency is determined by the sound speed which is given by $c_s^2 \equiv \dot{p}/\dot{\rho} = 1/3(1+R)$.

The residual quadrupole moment π_γ, the source of polarization, tracks the motion of the fluid. Polarization is generally small, at most $\sim 10\%$ of the anisotropy itself, since it requires scattering for its generation but its quadrupole source is suppressed if the scattering is too rapid. Given the initial conditions and gravitational potentials, these equations predict the phenomenology of the acoustic oscillations in the temperature and polarization fields.

Initial Conditions

The simplest inflationary models make a set of definite predictions for the initial conditions of the acoustic oscillations and hence their successful observation provides strong support for the inflationary paradigm. Quantum fluctuations in the scalar field that drives inflation imprints a nearly spectrum of Gaussian curvature (potential) fluctuations $k^3 P_\Phi/2\pi^2 \propto k^{n-1}$ where $n \approx 1$ on a spatially flat background metric. Gravitational infall into these initial potentials eventually generates all of the structure in the universe. Quantum fluctuations in the gravitational wave degrees of freedom also produce a nearly scale-invariant spectrum of fluctuations whose power depends on E_i^4 where E_i is the energy scale of inflation.

A Newtonian gravitational potential $\Psi \approx -\Phi$ necessarily imparts an initial temperature perturbation since Ψ represents a spatially varying time-time perturbation to the metric away from coordinates where the temperature is homogeneous. The perturbation is equivalent to a change in the scale factor since $a \propto t^{2/3(1+p/\rho)}$. This change produces a temperature perturbation from the cosmological redshift $T \propto a^{-1}$ of

$$\Theta = -\frac{2}{3(1+p/\rho)}\Psi \qquad (8)$$

or $-\Psi/2$ in the radiation dominated era. We call $\Theta + \Psi$ the effective temperature since it also accounts for the redshift a photon experiences when climbing out of a potential well, also known as the Sachs-Wolfe effect [1]. In the matter dominated epoch, $\Theta + \Psi = \Psi/3$.

There are three important aspects of these results. First, inflation sets the *temporal* phase of all wavemodes by starting them all at the initial epoch. We shall see that this predicts a coherent set of peaks in the CMB spectrum with a definite phase. Cosmological defect models predict a more random distribution of acoustic phases which produces incoherent acoustic phenomena. Defects can be now be ruled out as a primary mechanism for structure formation in the universe. More generally, without inflation or some other modification to the matter-radiation dominated universe, curvature perturbations cannot be generated outside the apparent horizon and so build up only by the causal motion of matter. This generally at least entails a delay in temporal phase of the oscillations which is not observed.

Secondly, since the power spectrum of the effective temperature is directly related to scale-invariant curvature fluctuations from inflation, the acoustic oscillations should also be approximately scale invariant in amplitude in the tight coupling regime. Observations are in excellent agreement with this fundamental prediction with tight constraints on the index $n = 0.94^{+0.11}_{-0.04}$ ([23], see also [24, 25, 26]). We will therefore base the discussion of the acoustic phenomenology on models with nearly scale-invariant initial curvature fluctuations as predicted by inflation. Finally, the scale-invariant gravitational wave background leads to a quadrupolar distortion in the CMB temperature field just like its effect on a ring of test masses. Thomson scattering of the quadrupole anisotropy then leaves a signature in the B-mode polarization.

ACOUSTIC PHENOMENOLOGY

Acoustic Scale

It is instructive to first consider a simplified model where the universe is always matter-dominated in its expansion and the dynamical effects of the baryons are negligible. This amounts to holding Φ and Ψ constant in the oscillator equation (7) and setting $R = 0$. The solution for an initial curvature fluctuation is then simple

$$[\Theta + \Psi](k,\eta) = \frac{1}{3}\Psi(k,0)\cos(ks), \qquad v_\gamma(k,\eta) = \frac{\sqrt{3}}{3}\Psi(k,0)\sin(ks), \qquad (9)$$

where $s = \int_0^\eta c_s d\eta'$ is the distance sound can travel by η or the *sound horizon*.

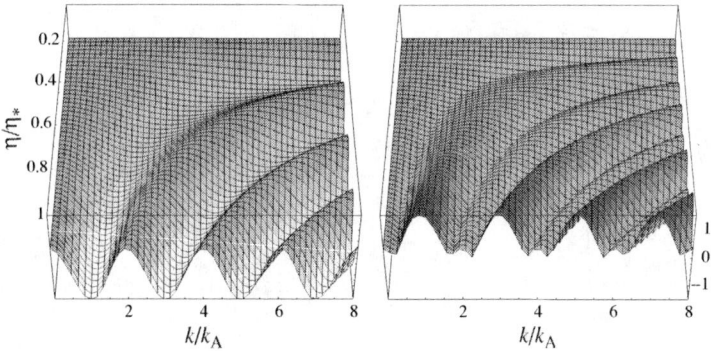

FIGURE 2. Temporal evolution of the effective temperature as a function of wavenumber predicted by the simplified model of Eqn. (11). Left: amplitude; right: rms fluctuation. All modes begin with the same phase and are frozen in at the same time $\eta = \eta_*$. The result is a harmonic series of extrema with a fundamental scale of $k_A = \pi/s_*$. Baryons, here shown with an $R = 0.6(\eta/\eta_*)^2$, displace the zero point of the oscillation. In the rms, this leads to enhanced odd numbered peaks.

In the limit of scales large compared with the sound horizon $ks \ll 1$, the perturbation is frozen into its initial conditions. This is the gist of the statement that the large-scale anisotropy measured by COBE directly measure the initial conditions. On small scales, the amplitude of the Fourier modes exhibits temporal oscillations corresponding to compression and rarefaction of the plasma inside gravitational potential wells. All modes are frozen in at recombination η_* (see Fig. 2 left). Modes that are caught at either maxima *or* minima of their oscillation at recombination correspond to peaks in the rms fluctuation or power (see Fig. 2 right). Given the solution in Eqn. (9), these modes are harmonics $k_n = nk_A$ of a fundamental acoustic scale $k_A = \pi/s_*$, where n is an integer (see Figure 2a). These become angular peaks in the anisotropy power with a characteristic angular scale of $l_A = k_A D_*$. The Doppler effect from the line-of-sight velocity has an rms $v_\gamma/\sqrt{3}$ and so contributes an equal amplitude fluctuation. It is $\pi/2$ out of phase with the temperature. This is because extrema represent turning points in the oscillation where the velocity vanishes. Nonetheless, because of the geometry of the projection, these oscillations do not contribute peaks in the angular power spectrum [20]. We shall see that they do however predict the peaks in the polarization spectrum through the residual quadrupole.

Under the flat, matter-dominated assumption the horizon distance scales as $\eta \propto a^{1/2}$. Thus $s_* \approx D_*/\sqrt{3000}$ and so $l_A \approx 200$. In a spatially curved universe, the distance used in the conversion from k to l no longer equals the comoving distance $D_* \neq \eta(a = 1) - \eta_*$. Consider first a closed universe with radius of curvature $R = H_0^{-1}|\Omega_{\text{tot}} - 1|^{-1/2}$. Suppressing one spatial coordinate yields a 2-sphere geometry with the observer situated at the pole (see Fig. 3 left). Light travels on lines of longitude. A physical scale λ at fixed latitude given by the polar angle θ subtends an angle $\alpha = \lambda/R\sin\theta$. For $\alpha \ll 1$, a Euclidean analysis would infer a distance $D = R\sin\theta$, even though the *coordinate distance* along the arc is $d = \theta R$; thus $D = R\sin(d/R)$. For open universes, simply replace sin with sinh. The peak locations $l_n = nk_A D_*$ making them extremely sensitive

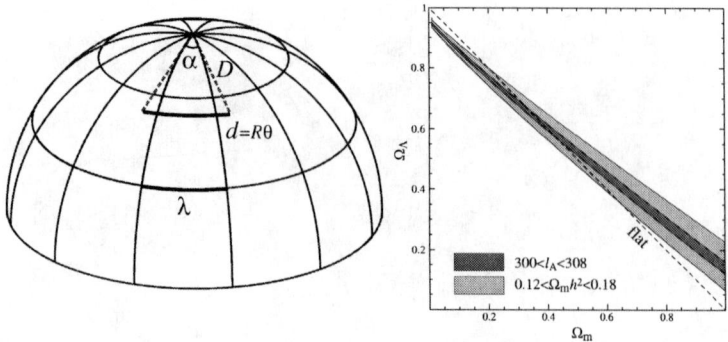

FIGURE 3. The acoustic scale. Left: geometrical effect in a closed universe. Objects in a closed universe are further than they appear ($D < d$). Consequently the measured angular scale of the peaks is extremely sensitive to the curvature. Right: translation of a constraint on the acoustic scale $300 < l_A < 308$ [24] (68% CL) onto the (Ω_m, Ω_Λ) plane with all other parameters fixed versus constraints with uncertainties in the physical matter density $\Omega_m h^2$ from CMB determinations [23] folded in.

to the spatial curvature [3, 28]; measurements constrain the geometry to be close to flat. Since a flat universe is at the critical density, local measurements of a low matter density indicate that there is a missing or "dark" component of the energy density, in good agreement with indications of an accelerating expansion from distant supernovae.

These simple scalings must be modified for precise predictions. The main effect is from the radiation density which changes the expansion rate and so shifts the acoustic scale to higher multipoles. For a flat universe with $\Omega_m h^2 \approx 0.15$, $l_A \approx 300$ – a large effect. The baryons lower the sound speed and so have a similar but smaller effect. The dark energy also provides a smaller effect through a decrease in the distance to recombination D_* and hence a lowering of l_A, with an increase in Ω_e or w. The sensitivity of the acoustic scale to cosmological parameters is approximately (see also Fig. 4)

$$\frac{\Delta l_A}{l_A} \approx -1.1 \frac{\Delta \Omega_{\text{tot}}}{\Omega_{\text{tot}}} - 0.24 \frac{\Delta \Omega_m h^2}{\Omega_m h^2} - 0.17 \frac{\Delta \Omega_e}{\Omega_e} - 0.11 \Delta w_e + 0.07 \frac{\Delta \Omega_b h^2}{\Omega_b h^2}, \quad (10)$$

around a model of $\Omega_{\text{tot}} = 1$, $\Omega_m h^2 = 0.15$, $\Omega_e = 0.65$, $w_e = -1$ and $\Omega_b h^2 = 0.02$ [29]. A joint analysis of the data yields $l_A = 304 \pm 4$ [24] and with these errors, the uncertainty in the cosmological interpretation in the dark energy and curvature domain is already dominated by uncertainty in $\Omega_m h^2$ not peak measurement error (see Fig. 3 right). Fortunately as we shall see, this parameter can be internally determined from the CMB and serves as an example where future increased precision will have important implications for a quantity of fundamental interest, the dark energy.

Finally, there is a separate effect on the first peak. We shall see that the decay of the gravitational potential in a radiation dominated universe generates anisotropy from gravitational redshifts after last scattering, filling in the rise to the first peak and shifting its location downwards off of the harmonic series to $l_1 \approx 3l_A/4$, placing $l_1 \approx 220$ in agreement with the observed location.

Baryon loading

Baryons add to the mass of the photon-baryon plasma without adding to the pressure. An examination of Eqn. (7) shows that their effect comes solely through the baryon-photon momentum density ratio $R \approx 0.6(\Omega_b h^2/0.02)(a/10^{-3})$. It is instructive to look at the solution to the oscillator equation (9) in the approximation that R is constant. Again under the assumption of a matter-dominated universe, the solution is that of a simple harmonic oscillator in a constant gravitational field but with an increased mass term

$$[\Theta + \Psi](k, \eta_*) = [\Theta + (1+R)\Psi](k, 0)\cos(ks)$$
$$= [1 + 3R]\frac{1}{3}\Psi(0)\cos(ks) - R\Psi(0), \qquad (11)$$

where the sound speed entering into the sound horizon calculation is reduced since $c_s^2 = 1/3(1+R)$. Aside from this reduction, baryons have two distinguishing effects: they enhance the amplitude of the oscillations by $1 + 3R$ fractionally and shift the zero point of the oscillation by $-R\Psi$. The latter modulates the amplitude of neighboring peaks: odd numbered peaks will be enhanced over the zero-baryon case by $1 + 6R$; even numbered peaks remain the same (see Fig. 2, right). Physically, the baryon mass enhances compression inside gravitational potential wells.

These qualitative results remain true in the presence of the real time-variable R. Measurement of these baryonic signatures currently limits $\Omega_b h^2 = 0.022^{+0.004}_{-0.002}$ [23]. This value is strikingly consistent with inferences of the baryon density from big bang nucleosynthesis. Consequently there have been no significant changes in the baryon-photon ratio from an energy scale of an MeV to an eV in the expansion history of the universe.

Dark Matter and Radiation

The matter-to-radiation ratio scales as $\rho_m/\rho_r \approx 3.6(\Omega_m h^2/0.15)(a/10^{-3})$ and so the universe is only barely matter-dominated at last scattering. Moreover fluctuations corresponding to the higher peaks entered the sound horizon at an earlier time, during radiation domination. As we have seen, including the radiation changes the expansion rate of the universe and hence the physical scale of the sound horizon at recombination. Fortunately, radiation has the more unique effect of driving the acoustic oscillations by making the gravitational force evolve with time [20].

The exact evolution of the potentials is determined by the relativistic Poisson equation. Its qualitative content is clear: since the background density is decreasing with time, the density fluctuations must grow unimpeded by pressure to maintain constant potentials. In particular, in the radiation dominated era once pressure begins to fight gravity at the first compressional extrema of the oscillation, the Newtonian gravitational potential and spatial curvature must decay.

This decay actually drives the oscillations: it is timed to leave the fluid maximally compressed with no gravitational potential to fight as it turns around. The net effect is

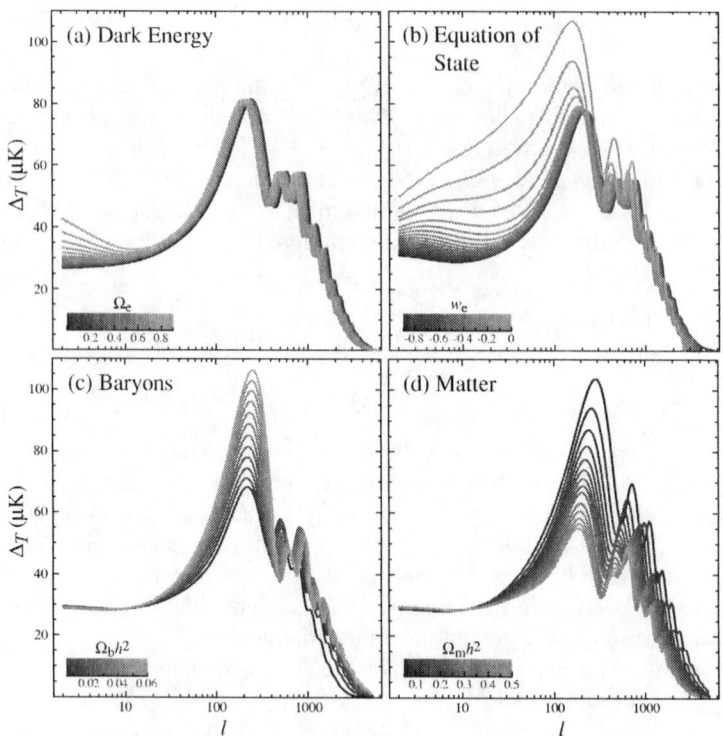

FIGURE 4. Sensitivity of the temperature power spectrum to: the energy density of the dark energy today Ω_e in units of the critical density, the equation of state parameter of the dark energy w_e, the physical baryon density $\Omega_b h^2$ and the physical matter density $\Omega_m h^2$. All are varied around a fiducial flat model of $\Omega_e = 0.65$, $w_e = -1$, $\Omega_b h^2 = 0.02$ and $\Omega_m h^2 = 0.15$ with $n = 1$.

doubled since the redshifting from the spatial metric fluctuation Φ also goes away at the same time. When the universe becomes matter dominated the gravitational potential is no longer determined by photon-baryon density perturbations but by the pressureless dark matter. Therefore, the amplitudes of the acoustic peaks increase as the matter-to-radiation ratio decreases [30, 20]. The net result is that across the horizon scale at matter-radiation equality the acoustic amplitude increases by a factor of $(\frac{1}{3}\Psi - 2\Psi)/\frac{1}{3}\Psi = 5$ for a pure photon and dark matter universe, and a factor of 4 when including the effect of neutrinos and baryons [31] (see also Fig. 5). By eliminating gravitational potentials, radiation also eliminates the alternating peak heights from baryon loading. The observed high third peak (see Fig. 1) is a good indication that matter dominates the energy density at recombination. Finally, the effect of the decaying potential after recombination leads to so-called integrated Sachs Wolfe contributions to the temperature fluctuations through the continuity equation (4) and shifts the first acoustic peak downwards off of the acoustic series [20].

Observations of these matter-radiation phenomena (see Fig. 4) currently constrain the total matter density to be $\Omega_m h^2 = 0.15 \pm 0.03$ [23] and are crucial in internally resolving

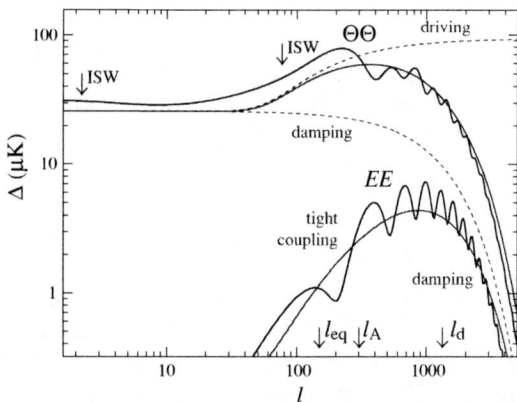

FIGURE 5. Phenomenological contributions to the anisotropy form. Radiation drives the amplitude of the acoustic oscillations up across the equality scale l_{eq} and down across the damping scale l_D (functional forms from [34]). The radiation ISW effect pushes $l_1 < l_A$ due to the radiation and raises the large angle anisotropy due to the dark energy. The polarization scales as l/l_D for $l < l_D$ due to the tight-coupling suppression of its quadrupole source and is also damped below l_D.

ambiguities in the interpretation of the peak scale. Since this number greatly exceeds the measured baryon density, these observations are also the first empirical indication that the dark matter exists at high redshift.

Damping

The photon-baryon fluid has slight imperfections corresponding to shear viscosity π_γ and heat conduction $(v_b - v_\gamma)$ in the Euler equation (5) [32, 33]. These imperfections damp acoustic oscillations. These are both associated with the diffusion of the photons which is especially important during recombination when the mean free path of the photons rises dramatically. As we have seen the viscosity $\pi_\gamma \sim k v_\gamma / \dot{\tau}$. With the continuity equation $k v_\gamma \approx -3\dot{\Theta}$, this leads to a Θ damping term in the oscillator equation. The heat conduction term can be shown to have a similar effect by expanding the Euler equations in $k/\dot{\tau}$. An examination of the resulting damped oscillator equation shows diffusion suppresses the oscillation amplitude by a factor of order $e^{-k^2 \eta / \dot{\tau}}$. The damping scale k_d is thus of order $\sqrt{\dot{\tau}/\eta}$, corresponding to the geometric mean of the horizon and the mean free path. Detailed numerical integration of the equations of motion are required to track the rapid growth of the mean free path and damping length through recombination itself. These calculations show that the damping scale is of order $k_d s_* \approx 10$ leading to a substantial suppression of the oscillations beyond the third peak (see Fig. 5). Observations of the damping phenomena provide a check on the fundamental assumptions underlying the interpretation of the acoustic peaks. With $\Omega_b h^2$ and $\Omega_m h^2$ and the acoustic scale l_A fixed by the first few peaks, the damping scale is uniquely predicted given the atomic physics of recombination. Any change in recombination, for example due to a variation in the fine structure constant, will be revealed as a discrepancy in the predictions. Likewise, any misinterpretation of l_A due to say initial conditions that were not purely curvature fluctuations resulting in a phase shift in the peaks would also show up as a discrepancy. The data in the damping tail to date are beautifully consistent with the predictions of the model (see Fig. 1).

Polarization

The dissipation of the acoustic oscillations leaves a signature in the polarization of the CMB in its wake. The fact that the polarization source is the quadrupole explains the shape and height of the polarization spectra in Fig. 1. Since the quadrupole is of order $k v_\gamma / \dot{\tau} \sim (k/k_d)(k_d \eta_*)^{-1} v_\gamma$, the polarization spectrum rises as l/l_D to peak at the damping scale with an amplitude of about 10% of the temperature fluctuations before falling due to the elimination of the acoustic source itself due to damping. Since v_γ is out of phase with the temperature, the polarization peaks are also out of phase with the temperature peaks. Furthermore, the phase relation also tells us that the polarization is correlated with the temperature perturbations. The correlation power $C_l^{\Theta E}$ being the product of the two, exhibits oscillations at twice the acoustic frequency. As in the case of the damping, the predicting the precise value requires numerical work [6] since $\dot{\tau}$ changes so rapidly near recombination. Nonetheless the detailed predictions shown in Fig. 5 bear these qualitative features.

The acoustic polarization and cross correlation has recently been detected by the DASI experiment. Like the damping scale, the acoustic polarization spectrum is uniquely predicted from the temperature spectrum once $\Omega_b h^2$, $\Omega_m h^2$ are specified. Polarization thus represents a sharp test on the assumptions of the recombination physics and power law curvature fluctuations in the initial conditions used in interpreting the temperature peaks.

These acoustic peaks in the polarization appear exclusively in the EE power spectrum due to the azimuthal symmetry of the plane wave fluctuations. During the break down of tight coupling that occurs at last scattering, any gravitational waves present will also imprint a local quadrupole anisotropy to the photons and hence a linear polarization to the CMB [35]. These contribute to the BB power and their detection would provide invaluable information on the origin of the fluctuations (see e.g. [4]). Specifically, in simple inflationary models their amplitude gives the energy scale of inflation. The gravitational wave amplitude h oscillates and decays once inside the horizon, so the associated polarization source scales as $\dot{h}/\dot{\tau}$ and so peaks at the $l \approx 100$ horizon scale and not the damping scale at recombination (see Fig. 1). This provides a useful scale separation of the various polarization effects.

DISCUSSION

The tremendous experimental progress in CMB anisotropy studies over the last few years has helped establish a standard paradigm for cosmology at intermediate epochs but has simultaneously raised questions about the physics at the two opposite ends of time. Simple inflationary models of the early universe have so far passed the test of the acoustic peaks. They predict the near scale-invariant initial curvature fluctuations in a spatially flat background that have now been observed. Moreover, spatial flatness appears to be maintained today by a mysterious form of missing or dark energy.

The first step into the future will come with the release of data from the MAP satellite in early 2003 which in the full course of the mission will provide the definitive measurement of temperature anisotropy across the whole sky down to a quarter of a degree. The MAP satellite will test the spectrum of initial conditions beyond the simple constraints on the power law index which can be extracted today (e.g. [40, 41]). These advances will come with the increased precision in the temperature measurements and determination of the acoustic peaks in the polarization and cross correlation. Measurement of the polarization peaks will not only eventually double the amount of statistical information that can be extracted from the CMB but can also provide a sharp distinction between small changes in the dynamics of the plasma, which affect temperature and polarization differently, and small deviations from scale invariance of in the initial conditions (e.g. [42]), which affect the two alike. Studies of the Gaussianity of the acoustic fluctuations will also provide a strong test of the inflationary model. The increased precision on the matter-radiation ratio will sharpen considerably the constraints on the dark energy. MAP should also be able to detect or place limits on the amount of reionization in the universe through a large angle bump in the polarization (see Fig. 1) and hence remove a central ambiguity for dark energy and dark matter studies from the peaks.

The next generation of experiments dedicated to fine-scale, secondary structure in the anisotropy will enable new tests of the dark energy from structures in the universe such as galaxy clusters. The effect from unresolved clusters may have already been detected as a small scale excess by the CBI and BIMA experiments. Confirmation must await determination of its frequency spectrum. Secondaries will also provide a new handle on reionization through the Doppler shift of CMB photons off of moving structures.

Further down the road lies the milestone of the gravitational lensing of the peaks. Gravitational lensing of the temperature and polarization fields distorts their properties, most visibly in the generation of B-mode polarization and in the higher order correlation of multipole moments. Detection will not only require more sensitive instruments and higher angular resolution but also subtraction of galactic and extragalactic foreground contaminants [43, 44, 45]. In principle higher order statistics can be used to reconstruct the lensing potential with a large gains in sensitivity available if the B-modes from lensing can not only be detected but accurately mapped [36]. From this reconstruction the dark matter and dark energy dependent growth of structure can be measured. For example gravitational lensing can enable tests of the scalar-field hypothesis for the nature of the dark energy through cross-correlation with the integrated Sachs-Wolfe effect from decaying potentials [37].

The ultimate future milestone for the CMB is the detection of gravitational waves from inflation through B-mode polarization. Detection would represent strong evidence for the inflationary model and pin down its energy scale. The window of detectability can be extended to energy scales above a few 10^{15} GeV but only if the B-modes from gravitational lensing are removed from a direct reconstruction.

These future milestones will require much experimental effort to achieve and much theoretical effort to interpret. Nevertheless they provide the hope that the next decade of studies of CMB temperature and polarization anisotropy will be as fruitful as the last.

REFERENCES

1. R.K. Sachs, A.M. Wolfe, *Astrophys. J* **147** (1967), 73.
2. P.J.E. Peebles, J.T. Yu, *Astrophys. J* **162** (1970), 815.
3. A.G. Doroshkevich, Y.B. Zel'dovich, R.A. Sunyaev, *Sov. Astron.* **22** (1978), 523.
4. P.J. Steinhardt, N. Turok, *Phys. Rev. D* **65** (2002), 126003.
5. D.J. Fixsen, et al. *Astrophys. J* **473** (1996), 576.
6. J.R. Bond, G. Efstathiou, *Mon. Not. R. Ast. Soc.* **226** (1987), 655.
7. A. Kosowsky, *Ann. Phys.* **246** (1996), 49.
8. M. Kamionkowski, A. Kosowsky, A. Stebbins, *Phys. Rev. D* **55** (1997), 7368.
9. M. Zaldarriaga, U. Seljak, *Phys. Rev. D* **55** (1997), 1830.
10. E. Newman, R. Penrose, *J. Math Phys.* **7** (1966), 863.
11. J.N. Goldberg, et al. *J. Math Phys.* **7** (1966), 863.
12. U. Seljak, *Astrophys. J* **482** (1997), 6.
13. P.J.E. Peebles, *Astrophys. J* **153** (1968), 1.
14. Y. Zel'dovich, V. Kurt, R. Sunyaev, *Sov. Phys.–JETP* **28** (1969), 146.
15. S. Seager, D.D. Sasselov, D. Scott, *ApJS* **128** (2000), 407.
16. U. Seljak, M. Zaldarriaga, *Astrophys. J* **469** (1996), 437.
17. A. Lewis, A. Challinor, A. Lasenby, *Astrophys. J* **538** (2000), 473.
18. J.R. Bond, G. Efstathiou, *Astrophys. J* **285** (1984), 45.
19. N. Vittorio, J. Silk, *Astrophys. J* **285** (1984), L39.
20. W. Hu, N. Sugiyama, *Astrophys. J* **444** (1995), 489.
21. W. Hu, M. White, *Phys. Rev. D* **56** (1997), 596.
22. N. Kaiser, *Mon. Not. R. Ast. Soc.* **202** (1983), 1169.
23. J.R. Bond, et al. *Theoretical Physics, MRST 2002* (2002), astro-ph/0210007.
24. L. Knox, N. Christensen, C. Skordis, *Astrophys. J Lett.* **563** (2001), 95.
25. X. Wang, M. Tegmark, M. Zaldarriaga, *Phys. Rev. D* **65** (2002), 123001.
26. W.J. Percival, et al. *Mon. Not. R. Ast. Soc.* in press (2002), astro-ph/0206256.
27. L. Knox, Y.S. Song, *Phys. Rev. Lett.* **89** (2002), 011303.
28. M. Kamionkowski, D.N. Spergel, N. Sugiyama, *Astrophys. J* **426** (1994), L57.
29. W. Hu, M. Fukugita, M. Zaldarriaga, M. Tegmark, *Astrophys. J* **549** (2001), 669.
30. U. Seljak, *Astrophys. J* **435** (1994), L87.
31. W. Hu, N. Sugiyama, *Astrophys. J* **471** (1996), 542.
32. J. Silk, *Astrophys. J* **151** (1968), 459.
33. S. Weinberg, *Astrophys. J* **168** (1971), 175.
34. W. Hu, M. White, *Astrophys. J* **479** (1997), 568.
35. A. Polnarev, *Sov. Astron.* **29** (1985), 607.
36. W. Hu, T. Okamoto, *Astrophys. J* **574** (2002), 566.
37. W. Hu, *Phys. Rev. D* **65** (2002), 023003.
38. http://map.nasa.gsfc.gov
39. http://astro.estec.esa.nl/Planck
40. Y. Wang, D.N. Spergel, M.A. Strauss, *Astrophys. J* **510** (1999), 20.
41. M. Tegmark, M. Zaldarriaga, *Phys. Rev. D* in press (2002), astro-ph/0207047.
42. R. Easther, B.R. Greene, W.H. Kinney, G. Shiu, *Phys. Rev. D* **66** (2002), 023518.
43. M. Tegmark, D.J. Eisenstein, W. Hu, A. de Oliviera Costa, *Astrophys. J* **530** (2000), 133.
44. S. Prunet, S.K. Sethi, F.R. Bouchet, *Mon. Not. R. Ast. Soc.* **314** (2000), 348.
45. C. Baccigalupi, et al. *Mon. Not. R. Ast. Soc.* submitted (2002), astro-ph/0209591.

Repercussions of Structure Emergence on the CMB Polarization

S. T. Staggs

Physics Department; Jadwin Hall; Princeton, NJ 08544

Abstract. Acoustic oscillations in the early universe, jumpstarted by superhorizon fluctuations left behind by inflation, give rise to a series of coherent peaks in the spatial power spectrum of the cosmic microwave background (CMB). An inevitable consequence of this scenario is that the CMB photons are left with small polarization anisotropies. Polarization patterns may be separated by parity into E-modes and B-modes; acoustic oscillations give rise only to E-modes. A recent detection of the E-modes agrees well with predictions. Many experiments are poised to make further measurements of E-modes, and to begin searching for B-modes, which should arise at subdegree scales because of weak lensing deflections of the original E-modes en route to the observer. At larger angular scales, gravitational waves from inflation could produce B-modes. Experimental prospects are outlined.

MOTIVATIONS FOR CMB POLARIMETRY

Most details of the early universe present themselves only obliquely, through shreds of evidence requiring clever detective work to tease out their meaning: one quarter of the baryonic universe is helium; gas at the edges of galaxies rotates too fast; axions are hard to find. By contrast, the cosmic microwave background comprises photons so old they bring to Earth an actual image of the early universe, when it was only a few hundred thousand years old. Of course, comprehension of why the universe looked like that requires clever detective work. Radiation is described fully by its wavelength distribution, number density, and polarization state. Full appreciation of the most ancient image in the universe depends on examination of all these properties.

The Inevitable Emergence of Polarization Anisotropies

Introduction. Peaks in the spatial power spectrum of the cosmic microwave background (CMB) intensity arise from coherent sloshing of the plasma contents of the early universe in and out of gravity wells. Figure 1 illustrates theoretical predictions for the power spectrum. Excellent elucidation of this acoustic oscillations picture may be found in several places, most recently in the article by Hu in this volume. We sketch the source of coherent peaks in the power spectrum, and then outline why polarization anisotropy results from the acoustic oscillation scenario. These ideas also have been detailed in a number of other sources (see [1], [2], [3], and [4], for a start). We summarize in the context of existing and proposed experiments.

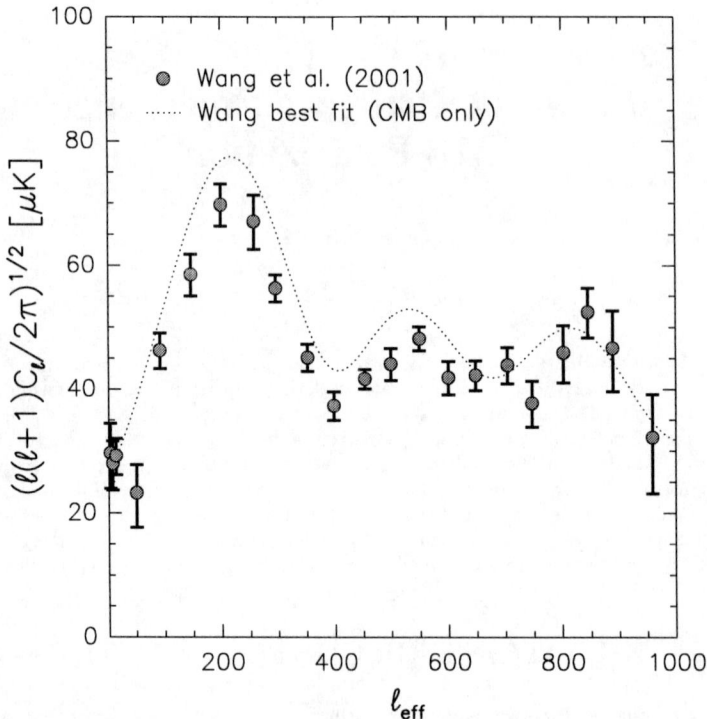

FIGURE 1. The power spectrum of the CMB temperature anisotropy: theoretical prediction for a typical model, and a subset of recent data based on the compilation of [5]. The authors binned 105 measurements of the power spectrum, including the effects of beam and calibration uncertainties. The individual measurements came from a variety of experiments, detailed by Wang, Tegmark & Zaldarriaga. Here, ℓ is the multipole moment of the spherical harmonics $Y_{\ell m}$. The temperature distribution on the celestial sphere is expanded in spherical harmonics with coefficients $a_{\ell m}$, and $C(\ell) = <|a_{\ell m}|^2>$, where the brackets denote averaging over m. Figure courtesy of M. Nolta.

Acoustic Oscillations. Some time after the inflationary epoch of the universe, and after nucleosynthesis, a proton-electron plasma suffused space, not only within a local Hubble volume (of radius ct, where t is the age of the Universe at the time), but well outside such volumes. Blackbody radiation supported the plasma in thermal equilibrium; electromagnetic interactions kept the baryons tightly coupled to the photons so that the amalgam may be considered a single fluid. The fluid evolved under gravitational and electromagnetic forces according to its initial conditions: its density and velocity as a function of position. The equations governing interactions of the fluid decouple to first order under Fourier transform (see, eg, [6]); thus it suffices to consider the evolution of individual Fourier modes of the fluid density. Fourier modes with wavelength larger than the horizon size (ct) had nonzero density amplitudes (having been generated causally and then stretched faster than the speed of light by inflation), but no velocity. The CMB data themselves indicate that each mode entered the horizon with nearly equal amplitude.[1] As the horizon grew, more and more Fourier modes "entered the horizon" and began oscillating. Each mode thus began oscillating with the same initial conditions (maximum density, zero velocity), at a time proportional to its wavelength. The epoch of decoupling arrived when the universe cooled to allow hydrogen to form from the plasma's electrons and protons, and electromagnetic coupling between the photons and baryons ended. The transition from a completely ionized universe to an essentially neutral one took place over a relatively short time interval, so the power spectrum of the CMB today records the square of the amplitude of each Fourier mode of the photon-baryon fluid at essentially that single instant.[2] The pattern which emerges looks like a sinusoidal oscillation because the abscissa coordinate (proportional to wavevector) traces out the phase of oscillation which has been reached by a given mode at the time of decoupling. The smallest modes (those at large multipole moment ℓ) have been inside the horizon for a long time, and thus have oscillated more than once.

Polarization from Anisotropy. An electron illuminated by an anisotropic distribution of unpolarized light may emit polarized radiation toward observers in certain locations. Consider for example an electron impinged upon by bright radiation from the $\hat{z}$ direction. The amplitudes of the incoming electric fields in the $\pm \hat{x}$ and $\pm \hat{y}$ directions are equal: the radiation is unpolarized. The electron emits dipole radiation in response to both E_x and E_y. However, an observer on the $\hat{x}$-axis only sees radiation polarized in the $\hat{y}$ direction since the dipole pattern from E_x has a null along the $\hat{x}$-axis. It turns out (see, e.g., [2]) that only a quadrupolar anisotropy around the electron leads to polarized emison.

At the epoch of decoupling, when photons underwent their last scatters before the plasma went neutral, the mean free path of the photons changed abruptly. Prior to the last scatters, the plasma was opaque enough that polarized emission could not arise from local density quadrupoles. Instead, electrons in the middle of convergent or divergent flows of the photon-baryon fluid emitted polarized radiation in response to the anisotropic pattern of Doppler shifts in the radiation impinging on those electrons. So, finally, we see that the power spectrum of the polarization of the CMB primarily reflects the velocity of the fluid.

Thus, measurement of the polarization anisotropy of the CMB constrains cosmological parameters affecting the acoustic oscillations scenario, allowing confirmation of effects seen in the temperature anisotropy, and providing additional constraints in some cases.

---

[1] Note that it is typical to plot $\ell^2 \sqrt{P(\ell)}$, where $P(\ell)$ has the usual power spectrum definition as the average of the squared amplitude of all modes with multipole moment ℓ. The first factor of ℓ effects plotting of the power spectrum *per logarhythmic interval* and the second ℓ undoes the effect of averaging since there are $2\ell+1$ values of m for each ℓ.

[2] More accurately, the angular power spectrum represents an average over three-dimensional Fourier modes projected onto the surface of last scattering.

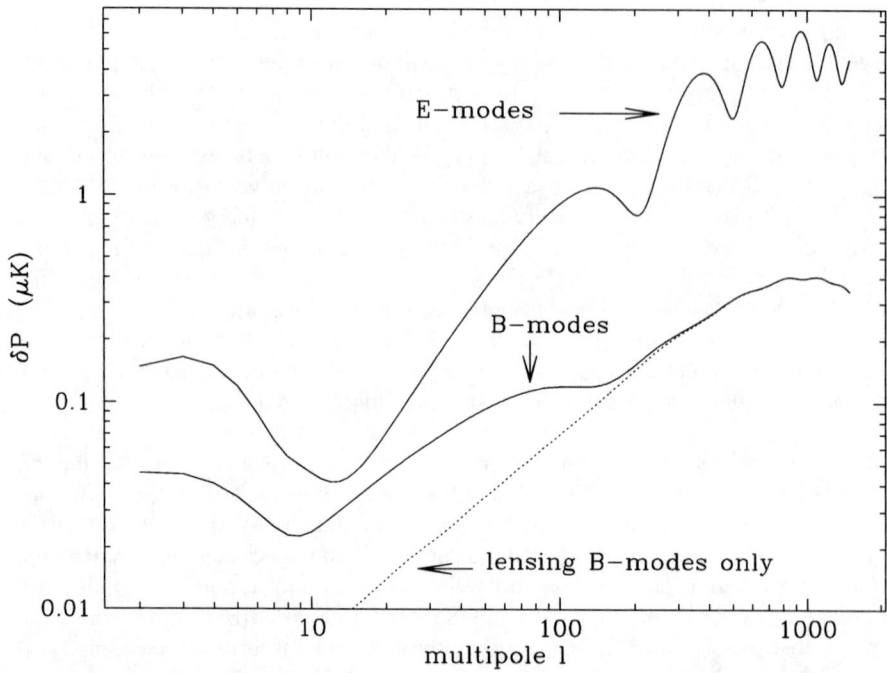

FIGURE 2. Predictions for the CMB polarization power spectra for E-modes and B-modes. Note the log scale for the vertical axis. For the lower curves, the dashed line represents the B-modes generated by lensing of E-modes only, while the solid line shows the total expected B-mode signal. We consider a large contribution from gravity waves, as barely allowed by CMB temperature anisotropy data. Specifically, the tensor to scalar ratio is 0.1. The bumps at $\ell \approx 3.5$ are typical for reionization at a redshift of 7. Figure inspired by [1] and generated with CMBFAST, courtesy of Zaldarriaga & Seljak.

The Expectation of Particular Polarization Patterns: E-modes

Let us for simplicity consider a small part of the spherical shell of last-scattering electrons, so that we may treat it as the xy plane, and then limit consideration to Fourier wavevectors in that plane. Consider a wavevector in the $\hat{x}$ direction. Fluid flow (in the $\pm\hat{x}$ directions) and associated Doppler shifts for this Fourier mode will give rise to polarizations in either the $\hat{x}$ or the $\hat{y}$ direction, but never in the direction $\pm 45°$ from $\hat{x}$. This observation lead two groups of authors ([3] and [4]) to suggest distinguishing two types of Fourier modes of polarization. The kind associated with the acoustic oscillations picture are called the *E*-modes (or sometimes *G*-modes), in which the polarization direction is either parallel or perpendicular to the wavevector of the Fourier mode. If the polarization is at $\pm 45°$ to the wavevector, the mode is a *B*-mode (sometimes called a *C*-

mode).[3] Both [3] and [4] provide explicit directions for decomposing raw experimental data into E- and B- modes. Predictions for the amplitude of the E-mode anisotropy are given in Figure 2.

The Tantalization of B-mode Polarization Anisotropies

The by-now "ordinary" picture of acoustic oscillations does not generate B-modes, so at first take, the B-modes are valuable as a control measurement, which should not show cosmologically significant amplitudes at anything like the level of the E-modes. Very local systematic effects on experiments are not all so constrained. (Hu, Hedman, & Zaldarriaga [7] present detailed predictions on the modes produced by a large set of experimental systematic effects, for example.) Two effects which could result in detection of B-mode polarization have stirred up great interest in the experimental community at least.

B-modes from Gravity Waves. Gravity waves leftover from inflation serve as a primordial source of anisotropies at the time of last scattering, distinct from the picture of small-amplitude superhorizon Fourier modes growing (and decreasing) under the effects of gravity and radiation pressure.[4] The largest angular scales in the CMB temperature anisotropy power spectrum ($\ell < 100$) probe Fourier modes which had wavelength larger than the horizon at the time of last scattering. The CMB power spectrum at the smallest ℓ thus indicates the primordial fluctuation level (before acoustic oscillations), including effects of gravity waves. However, since the fluctuation level *without* gravity waves is not known accurately, the large-angle CMB temperature anisotropy data only limit the size of any such gravity wave background. Proposed direct detection methods do not probe cosmological scales (see [8] for more discussion). Gravity waves rippling through the universe at the time of last scattering cause local quadrupolar anisotropies around scattering electrons by alternately stretching and squeezing the wavelengths of the plasma radiation in a plane normal to the propagation direction of a wave during its passage. This effect generates both E and B modes in the angular power spectrum [9]. Figure 2 indicates the size of the effect on the power spectrum for an inflation scenario generating the maximum quantity of gravity waves consistent with the extant CMB temperature anisotropy data. B-modes from gravity waves are small.

B modes from Lensing. The CMB photons escaping the surface of last scatter traverse an increasingly lumpy metric en route to detection at Earth, as the Universe fills with structure. Though no mechanisms for further polarizing (or depolarizing) the radiation are expected, the patterns of polarization are slightly distorted by gravitational lensing [10]. Imagine a physics demo in which electric fields lines are indicated by

---

[3] The nomenclature comes from analogy to vector fields, which may have a gradient but no curl, like electric fields (E or G modes) or have curl but no gradient, like magnetic fields (B or C).
[4] Such Fourier modes may be considered scalar perturbations of the metric; gravity waves represent tensor perturbations.

sprinkling grass seeds onto a viewgraph next to some electrodes; the long axes of the seeds lie along the field lines. A slight wrinkling of the viewgraph gently nudges the grass seeds around, so that the seeds no longer perfectly follow the field lines. The resultant distribution is unlikely to be completely curlfree. In the same way, E-modes are generically partially converted to B-modes by gravitational lensing. Hu & Takamoto 2002 provide excellent figures illustrating the effect in a greatly exaggerated situation. Loosely, the correlation of the induced B-modes with their parent E-modes is indicative of the strength of the deflection field. The lensing also correlates Fourier modes of the primordial (E-mode) polarization, an effect which may be measured by looking at higher order moments of the E-mode distribution than just its power spectrum (see for example [12] and [13]).

The typical deflection of a CMB photon (which retains its polarization state during deflection, of course) by the intervening matter is a staggering 3′, but the coherence size is even larger, a few degrees [11]. Thus, the perturbation to the primordial E-mode patterns, which are predicted to have maximal anisotropy near 3′, is small. Figure 2 illustrates typical predictions for the lensing signal. The B-mode lensing signal peaks at higher ℓ than the gravity wave effect, has a characteristic shape in most cosmological models, and may in principle be independently evaluated from the higher order E-mode correlations. Its existence may be viewed as a bonus: the CMB polarization encodes information about the matter distribution in the nonlinear phase of structure evolution too.

EXPERIMENTS

In the thirty years prior to 1995, five experiments were carried out to measure or limit the CMB polarization. More than a dozen are underway today, with more proposed nearly every month. In December, 2002, the DASI collaboration published the first detection of polarization anisotropy[14]. The results come from some two hundred days of excellent weather at the South Pole, using thirteen receivers interferometrically: the CMB polarization anisotropy is small (as predicted)! The DASI team detected E-mode polarization at intermediate angular scales, but not B-modes, in agreement with the prediction that cosmological B-modes (as opposed to local effects) are much smaller than E-modes.

Prospects

By the time these proceedings are published, the first MAP satellite results will be public, and are likely to include new information about the CMB polarization[5], with all-sky coverage, a huge frequency range (18-100 GHz), and the enviable control of systematics only achieved in the space environment. MAP uses correlation techniques

---

[5] Unfortunately, the author doesn't know what, not being a member of the collaboration.

to achieve polarization sensitivity and reduce systematic effects, not unlike the interferometry employed by DASI [16]. DASI continues to collect data at 30 GHz, with its proven system (and tested analysis procedures). The sister interferometer to DASI, the CBI, has collected higher resolution data at 30 GHz from a good site in Chile. The CAPMAP collaboration is deploying four correlation polarimeters (like the ones used in PIQUE[15]) on a 7 m telescope, providing a beam size of $3'$ to allow probing of the E-modes where they are expected to peak; this is the highest resolution experiment of the present crop. The four polarimeters, operating at 90 GHz, will collect data over the winter of 2003, and be joined by ten more receivers (six at 90 GHz and four at 40 GHz) in late 2003. The BOOMerANG collaboration has recompiled its experiment's focal plane with polarization-sensitive bolometers at 150 GHz and launched a circumpolar balloon flight in early 2003. The MAXIMA collaboration also plans a North American balloon flight in 2003 using bolometers. Bolometer-based experiments offer the possibility of higher sensitivity than the lower-frequency correlation methods, but effecting polarization sensitivity with bolometers in the microwave is a new business. These two experiments will test the waters. In summary, we may expect plenty more data on the CMB polarization in the next year. A number of other experiments should come online in the subsequent few years, including several large bolometer arrays. Finally, the European Space Organization's Planck Surveyor mission will fly in the second half of the decade, with excellent sensitivity to polarization.

CONCLUSIONS

Theoretical predictions for the polarization of the CMB have galvanized scores of experimentalists into attempting to measure it. Prospects for learning new things about the earliest-yet-detected structures in the universe from the polarization abound, and that is assuming the present theoretical picture is complete. In the perhaps more interesting possible world in which not every CMB puzzle piece fits together with other cosmological data, the polarization data may be even more crucial.

I categorize the present goals for CMB polarization into six as follows, and conclude with a parochial survey of how we might achieve them.

1. Detect the E-modes implied by extant measurements of the CMB temperature anisotropy.
2. Detect the predicted "T-E" cross-correlation between the temperature anisotropy and the E-modes.
3. Further characterize the E-modes en route to final goals, with narrower bands in ℓ-space, multiple frequencies, and at higher values of ℓ.
4. Detect reionization through low-ℓ polarization and T-E data. (Traversing through reionized plasma can impart non-primordial polarization to the CMB photons, but the effects show up only at very large angles; see Figure 2.)
5. Detect lensing and characterize the deflection field: this requires exquisite E and B maps at $4'$.
6. Detect gravity waves in the low-ℓ B-modes.

The first of these goals has already been achieved by DASI[14]. The next two will be addressed in the nearest future by MAP, but also by many ground-based and balloon-based experiments in the next few years. The last three are quite likely to require a satellite, at least for the final story.

ACKNOWLEDGMENTS

Thanks to the CAPMAP collaboration, Lyman Page, and Wayne Hu for many helpful conversations.

REFERENCES

1. Hu, W., and Dodelson, S., *ARA&A*, **40**, 171-216 (2002).
2. Hu, W., and White, M., *NewA*, **2**, #4, 323-344 (1997).
3. Zaldarriaga, M., and Seljak, U., *Phys. Rev. D.*, **55**, 1830 (1997).
4. Kamionkowski, M., Kosowsky, A., and Stebbins, A., *Phys. Rev. D.*, **55**, 7368 (1997).
5. Wang, X., Tegmark, M., and Zaldarriaga, M., *Phys. Rev. D.*, **65**, 123001 (2002).
6. Hu, W., Seljak, U., White, M., and Zaldarriaga, M., *Phys. Rev. D.*, **57**, 3290-3301 (1998).
7. Hu, W., Hedman, M., and Zaldarriaga, M., (2003), astro-ph/0210096.
8. Caldwell, R. R., Kamionkowski, M., and Wadley, L., *Phys. Rev. D.*, **59**, 027101 (1999).
9. Seljak, U., and Zaldarriaga, M., *Phys. Rev. Let.*, **78**, 2054-2057 (1997).
10. Zaldarriaga, M., and Seljak, U., *Phys. Rev. D.*, **58**, 023003 (1998).
11. Hu, W., and Okamoto, T., *Ap. J.*, **574**, 566-570 (2002).
12. Guzik, J., Seljak, U., and Zaldarriaga, M., *Phys. Rev. D*, **62**, 043517 (2000).
13. Hu, W., *Phys. Rev. D*, **66**, 083515 (2002).
14. Kovac, J., Leitch, E., Pryke, C., Carlstrom, J., Halverson, N., and Holzapfel, W., *Nature*, **420**, 772-787 (2002).
15. M. M. Hedman *et al.*, *Ap. J.*, **548**, L111-L114 (2001).
16. Leitch, E., Kovac, J., Pryke, C., Carlstrom, J., Halverson, N., and Holzapfel, W., *Nature*, **420**, 763-771 (2002).

The Absence of the Integrated Sachs-Wolfe Effect: Constraints on a Cosmological Constant

S.P. Boughn[*] and R.G. Crittenden[†]

[*]*Department of Astronomy, Haverford College, Haverford, PA 19041*
[†]*Institute of Cosmology and Gravitation, Portsmouth PO1 2EG*

Abstract. The hard X-ray background (XRB) provides a tracer of matter out to a redshift of $z \simeq 4$. If the universe is open or has a significant cosmological constant, then the cosmic microwave background (CMB) and XRB should be correlated due to the integrated Sachs-Wolfe (ISW) effect. A comparison of the COBE CMB map with HEAO1 A2 2-10 keV map of the XRB shows no such correlation. If the X-ray bias factor is on the order of 2-3 as implied by the clustering of the XRB, the implied 95% CL upper limit on a cosmological constant is $\Omega_\Lambda \lesssim 0.60$.

In a remarkably short time, the standard paradigm of cosmology seems to have shifted from a Friedmann universe (open or flat) to one with a cosmological constant or some other form of "dark energy". Recent observations of supernovae light curves (Perlmutter et al. 1999) suggest that the expansion of the universe is accelerating rather than decelerating. Combined with the determination of the spectrum of fluctuations in the cosmic microwave background (CMB) and a number of other observations (Bahcall et al. 1999), this suggests that the universe is spatially flat and dominated by a cosmological constant with an equivalent density parameter, $\Omega_\Lambda \equiv \Lambda/3H_0^2 \simeq 0.7$ (where H_0 is the Hubble constant). While future supernovae observations and the anticipated results from NASA's *MAP* satellite mission will provide critical tests of this model, it is important that the ΛCDM paradigm be subjected to other, independent tests.

One such test involves the formation of large-scale structures that are still in the linear regime ($\delta\rho/\rho \ll 1$) and are, therefore, relatively well understood. As CMB photons fall into collapsing structures they gain energy and as they climb out, they lose energy. In a flat, matter dominated universe the two effects cancel and there is no net redshift or blueshift if the structures are in the linear regime. However, if the universe possesses a significant cosmological constant, then the collapse of these structures will be retarded and the emerging CMB photons will possess a net blueshift (Crittenden & Turok 1996). This effect is generally known as the "late-time, integrated Sachs-Wolfe effect" (Sachs & Wolfe 1967). The CMB anisotropies created in this way are naturally correlated with the gravitational potential. Thus, we expect to see correlations between the CMB and tracers of local ($z \lesssim 2$) mass concentrations. For a $\Omega_\Lambda \sim 0.7$ universe, these correlations are primarily on scales of 1 to 10 degrees.

In previous work (Boughn, Crittenden & Turok 1998), we searched for correlations of the *COBE* CMB maps with the *HEAO1 A2* satellite observations of the 2-10 *keV* X-ray background (XRB). We found no correlations; however, translating this into a limit on Ω_Λ was frustrated by a lack of the knowledge of the X-ray bias factor, i.e., the factor by which the clustering of X-ray sources exceeds the clustering of the underlying matter distribution. The large-scale clustering of the hard X-ray background has since been (marginally) detected (as will be discussed in the next section) and a concomitant constraint on Ω_Λ can be determined.

The hard X-ray background (XRB) was discovered before the cosmic microwave background (CMB), but only now is its origin being fully understood. While there have been several attempts to measure large-scale, correlated fluctuations in the hard XRB, these have only yielded upper limits or, at best, marginal detections (e.g. Scharf et al. 2000). The best observations relevant to large-scale structure are still those from the HEAO1 A2 all-sky experiment that measured the surface brightness of the X-ray background in the $0.1-60\ keV$ band (Boldt 1987).

A standard way to detect the clustering of sources (or of the emission of these sources) is to compute the auto-correlation function (ACF), defined by

$$ACF(\gamma) = \int (I(\theta,\phi) - \bar{I})(I(\theta',\phi') - \bar{I}) d\Omega/\bar{I}^2 \qquad (1)$$

where I is the X-ray intensity, $\bar{I}$ is the mean X-ray intensity, γ is the angle between θ,ϕ and θ',ϕ', and the integral is over all solid angles. The map was corrected for a variety of systematic large-scale structures and known point sources were removed, as was the plane of the Galaxy (Boughn 1999). The ACF of this "cleaned" map is shown in Figure 1 (Boughn, Crittenden, & Koehrsen 2002). The dashed curve in Figure 1 is that expected from an uncorrelated signal (i.e., no clustering) convolved with the point spread function (PSF) of the telescope beam.

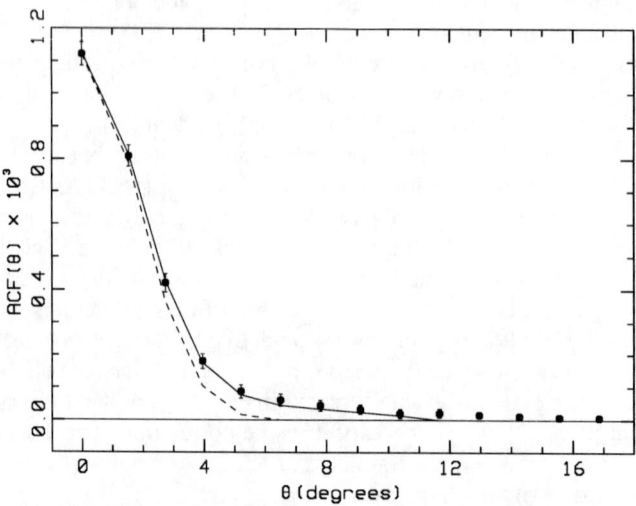

FIGURE 1. The auto-correlation function of the HEAO1 A2 map with bright sources and the Galactic plane removed and corrected for large-scale, high Galactic latitude structure. The dashed curve is that expected from beam smearing due to the PSF of the map while the solid curve includes a contribution due to clustering in the XRB.

It is clear that for $\theta \gtrsim 3°$, the correlations in the X-ray background are greater than can be accounted for by beam smearing. The solid curve in Figure 1 is a two-parameter fit of beam smearing plus intrinsic clustering proportional to $1/\theta$ convolved with the beam. Such clustering provides acceptable fits to the ACF of both radio and X-ray surveys on somewhat smaller angular scales (Cress & Kamionkowski 1998, Giacconi et al. 2001). The bottom line is that there is fairly strong evidence for intrinsic clustering on angular scales of $\theta \sim 4°$ to $8°$ at the level of $ACF \sim 3.6 \times 10^{-4}\theta^{-1}$.

For a given level of intrinsic clustering of the X-ray background it is relatively straightforward to compute the implied X-ray bias for a given cosmological model. We normalized the fluctuations to the COBE power spectrum as determined by Bond, Jaffe & Knox (1998). Typical biases appear to be $b_X = 2.3 \pm 0.3(0.7/h)^{0.9}$ where $h = H_0/100\ km\ s^{-1}Mpc^{-1}$ (Boughn, Crittenden, & Koehrsen 2002).

Given the level of the X-ray ACF in Figure 1 and the implied X-ray bias factor, it is possible to interpret the lack of a cross-correlation between the CMB and the hard X-ray background as a constraint on the value of the cosmological constant, Ω_Λ. The calculation of the HEAO-COBE cross correlation was discussed in Boughn, Crittenden, & Turok (1998) and has not changed. The results are shown in Figure 2, along with predictions for three different values of Ω_Λ. While the X-ray bias depends strongly on the Hubble parameter, the predicted cross correlation is only weakly dependent on it, changing only 10% for reasonable values of H_0. The cross correlation depends primarily on Ω_Λ; no correlation is expected if there is no cosmological constant and the ISW effect increases as Ω_Λ grows. The error bars in Figure 2 were calculated from Monte Carlo simulations and arise primarily due to cosmic variance in the observed correlation. They are significantly correlated. The data are most consistent with there being no intrinsic cross correlation ($\Omega_\Lambda = 0.0$). A limit set by calculating the likelihood of a model relative to the "no correlation" model is $\Omega_\Lambda \leq 0.60$ at the 95%. C.L.

The ISW is relatively insensitive to the exact shape of the redshift distribution of the sources of the XRB; however, our constraint on Ω_Λ is quite sensitive to the value of the

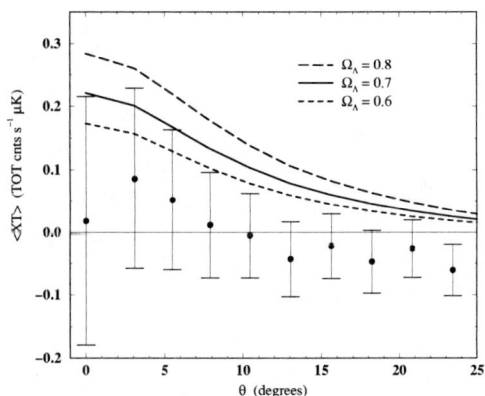

FIGURE 2. The X-ray/CMB cross correlation. Error bars are highly correlated. Also shown are the predictions for three $\Lambda - CDM$ models with varying Ω_Λ ($H_0 = 70\ km\ s^{-1}Mpc^{-1}$).

bias parameter. If the sources of the XRB should turn out to be unbiased, i.e., $b_X = 1$, then the constraint on Ω_Λ would be weakened considerably to $\Omega_\Lambda \lesssim 0.9$. We hasten to add that such a low bias would require that the instrinsic ACF be reduced by more than a factor of four, which seems unlikely.

To summarize, with the observed X-ray clustering, current $\Lambda - CDM$ models predict a detectable correlation of the cosmic microwave and X-ray backgrounds arising via the integrated Sachs-Wolfe effect. That we have not observed this effect suggests $\Omega_\Lambda < 0.60$. This is beginning to conflict with models preferred by a combination of CMB, LSS and SNIA data (e.g., de Bernardis et al. 2000 & Bahcall et al. 1999).

We would like to acknowledge K. Jahoda, N. Turok, E. Groth, S. Raible, and G. Koehrsen for various contributions. RC acknowledges support from a PPARC Advanced Fellowship. Supported in part by NASA grant NAG5-9285.

REFERENCES

1. Bahcall, N. Ostriker, J.P., Perlmutter, S. & Steinhardt, P. 1999, Science, 284, 1481
2. Boldt, E. 1987, Phys. Rep., 146, 215
3. Bond, J. R., Jaffe, A. & Knox, L. 1998, Phys Rev D, 57, 2117B
4. Boughn, S. 1999, ApJ, 526, 14
5. Boughn, S., Crittenden, R. & Koehrsen, G. 2002, ApJ, 580, 672
6. Boughn, S., Crittenden, R. & Turok, N. 1998, New Astron., 3, 275 (BCT)
7. Cress, C. M. & Kamionkowski, M. 1998, MNRAS, 297, 486
8. Crittenden, R. & Turok, N. 1996, PRL 76, 575
9. de Bernardis et al. 2000, Nature, 404, 955
10. Giacconii, R. et al. 2001, ApJ, 551, 624
11. Perlmutter, S. et al. 1999, ApJ, 517, 565
12. Sachs, R. & Wolfe, A. 1967, ApJ, 147, 1
13. Scharf, C. A., Jahoda, K., Treyer, M., Lahav, O., Boldt, E. & Piran, T. 2000, ApJ, 544, 49
14. Treyer, M. A., Scharf, C. A., Lahav, O., Jahoda, K., Boldt, E. & Piran, T. 1998, ApJ, 509, 531

THE FIRST STARS, REIONIZATION, AND INDUCED STRUCTURE

The First Sources of Light

Volker Bromm* and Abraham Loeb*†

*Astronomy Department, Harvard University, 60 Garden St., Cambridge, MA 02138
†Institute for Advanced Study, Princeton, NJ 08540; Guggenheim Fellow

Abstract. We review recent theoretical results on the formation of the first stars and quasars in the universe, and emphasize related open questions. In particular, we list important differences between the star formation process at high redshifts and in the present-day universe. We address the importance of heavy elements in bringing about the transition from an early star formation mode dominated by massive stars, to the familiar mode dominated by low mass stars, at later times. We show how gamma-ray bursts can be utilized to probe the first epoch of star formation. Finally, we discuss how the first supermassive black holes could have formed through the direct collapse of primordial gas clouds.

INTRODUCTION

The first sources of light ionized [1, 2] and metal-enriched [3] the intergalactic medium (IGM) and consequently had important effects on subsequent galaxy formation [4] and on the large-scale polarization anisotropies of the cosmic microwave background [5]. *When did the cosmic dark ages end?* In the context of popular cold dark matter (CDM) models of hierarchical structure formation, the first stars are predicted to have formed in dark matter halos of mass $\sim 10^6 M_\odot$ that collapsed at redshifts $z \simeq 20 - 30$ [4, 6]. The first quasars, on the other hand, are likely to have formed in more massive host systems, at redshifts $z \geq 10$ [7], and certainly before $z \sim 6.4$, the redshift of the most distant quasar known [8].

Results from recent numerical simulations of the collapse and fragmentation of primordial clouds suggest that the first stars were predominantly very massive, with typical masses $M_* \geq 100 M_\odot$ [9, 10, 11, 12]. Despite the progress already made, many important questions remain unanswered; the purpose of this brief review is to discuss these open questions and to put them in perspective. An example for an open question is: *How does the primordial initial mass function (IMF) look like?* Having constrained the characteristic mass scale, still leaves undetermined the overall range of stellar masses and the power-law slope which is likely to be a function of mass. In addition, it is presently unknown whether binaries or, more generally, clusters of zero-metallicity stars, can form. Evidently, the observational signature as well as the fate of the first stars depend sensitively on whether primordial star formation is predominantly clustered or isolated. This in turn is affected by the nature of the feedback that the first stars exert on their surroundings. The first stars are expected to produce copious amounts of UV photons and to possibly explode as energetic hypernovae. *How effective will their negative feedback be in suppressing star formation in neighboring high-density clumps?*

Predicting the properties of the first stars is important for the design of upcoming instruments, such as the *James Webb Space Telescope* (JWST) [1], or the next generation of large (> 10m) ground-based telescopes. The hope is that over the upcoming decade, it will become possible to confront current theoretical predictions about the properties of the first sources of light with direct observational data. The increasing volume of new data on high redshift galaxies and quasars from existing ground-based telescopes, signals the emergence of this new frontier in cosmology.

STAR FORMATION THEN AND NOW

Currently, we do not have direct observational constraints on how the first stars, the so-called Population III stars, formed at the end of the cosmic dark ages. It is, therefore, instructive to briefly summarize what we have learned about star formation in the present-day universe, where theoretical reasoning is guided by a wealth of observational data (see [13] for a recent review).

Population I stars form out of cold, dense molecular gas that is structured in a complex, highly inhomogeneous way. The molecular clouds are supported against gravity by turbulent velocity fields and pervaded on large scales by magnetic fields. Stars tend to form in clusters, ranging from a few hundred up to $\sim 10^6$ stars. It appears likely that the clustered nature of star formation leads to complicated dynamics and tidal interactions that transport angular momentum, thus allowing the collapsing gas to overcome the classical centrifugal barrier [14]. The IMF of Pop I stars is observed to have the approximate Salpeter form (e.g., [15])

$$\frac{dN}{d\log M} \propto M^x, \qquad (1)$$

where

$$x \simeq \begin{cases} -1.35 & \text{for } M \geq 0.5 M_\odot \\ 0.0 & \text{for } 0.007 \leq M \leq 0.5 M_\odot \end{cases}. \qquad (2)$$

The lower cutoff in mass corresponds roughly to the opacity limit for fragmentation. This limit reflects the minimum fragment mass, set when the rate at which gravitational energy is released during the collapse exceeds the rate at which the gas can cool (e.g., [16]). The most important feature of the observed IMF is that $\sim 1 M_\odot$ is the characteristic mass scale of Pop I star formation, in the sense that most of the mass goes into stars with masses close to this value. In Figure 1, we show the result from a recent hydrodynamical simulation of the collapse and fragmentation of a molecular cloud core [17, 18]. This simulation illustrates the highly dynamic and chaotic nature of the star formation process[2].

The metal-rich chemistry, magnetohydrodynamics, and radiative transfer involved in present-day star formation is complex, and we still lack a comprehensive theoretical framework that predicts the IMF from first principles. Star formation in the high redshift universe, on the other hand, poses a theoretically more tractable problem due to a

---

[1] See http:// ngst.gsfc.nasa.gov.
[2] See http:// www.ukaff.ac.uk/starcluster for an animation.

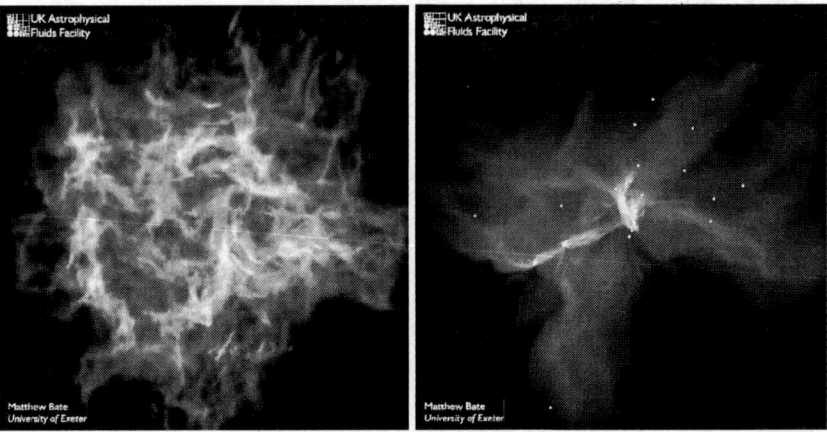

FIGURE 1. A hydrodynamic simulation of the collapse and fragmentation of a turbulent molecular cloud in the present-day universe (from [18]). The cloud has a mass of $50 M_\odot$. The panels show the column density through the cloud, and span a scale of 0.4 pc across. *Left:* The initial phase of the collapse. The turbulence organizes the gas into a network of filaments, and decays thereafter through shocks. *Right:* A snapshot taken near the end of the simulation, after 1.4 initial free-fall times of 2×10^5 yr. Fragmentation has resulted in ~ 50 stars and brown dwarfs. The star formation efficiency is $\sim 10\%$ on the scale of the overall cloud, but can be much larger in the dense sub-condensations. This result is in good agreement with what is observed in local star-forming regions.

number of simplifying features, such as: (i) the initial absence of heavy metals and therefore of dust; and (ii) the absence of dynamically-significant magnetic fields, in the pristine gas left over from the big bang. The cooling of the primordial gas does then only depend on hydrogen in its atomic and molecular form. Whereas in the present-day interstellar medium, the initial state of the star forming cloud is poorly constrained, the corresponding initial conditions for primordial star formation are simple, given by the popular ΛCDM model of cosmological structure formation. We now turn to a discussion of this theoretically attractive and important problem.

PRIMORDIAL STAR FORMATION

How did the first stars form? A complete answer to this question would entail a theoretical prediction for the Population III IMF, which is rather challenging. Let us start by addressing the simpler problem of estimating the characteristic mass scale of the first stars. As mentioned before, this mass scale is observed to be $\sim 1 M_\odot$ in the present-day universe. To investigate the collapse and fragmentation of primordial gas, we have carried out numerical simulations, using the smoothed particle hydrodynamics (SPH) method. We have included the chemistry and cooling physics relevant for the evolution of metal-free gas (see [10] for details). Improving on earlier work [9, 10] by initializing our simulation according to the ΛCDM model, we focus here on an isolated overdense region that corresponds to a $3\sigma-$peak: a halo containing a total mass of $10^6 M_\odot$, and

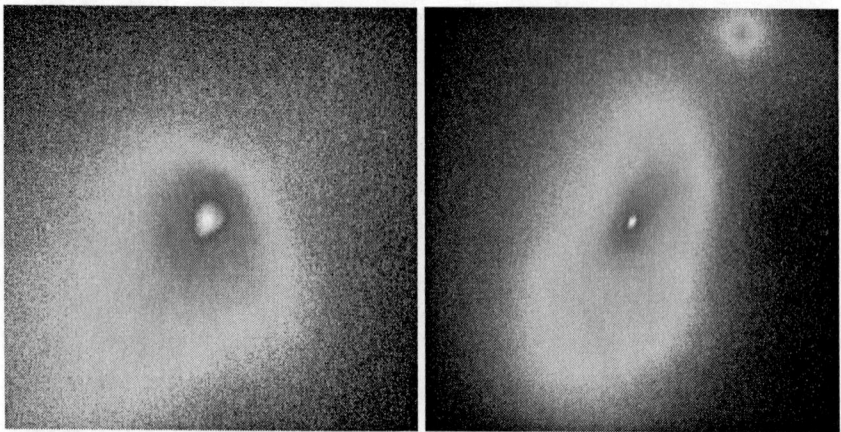

FIGURE 2. Collapse and fragmentation of a primordial cloud. Shown is the projected gas density at a redshift $z \simeq 21.5$, briefly after gravitational runaway collapse has commenced in the center of the cloud. *Left:* The coarse-grained morphology in a box with linear physical size of 23.5 pc. At this time in the unrefined simulation, a high-density clump (sink particle) has formed with an initial mass of $\sim 10^3 M_\odot$. *Right:* The fine-grain morphology in a box with linear physical size of 0.5 pc. The central density peak, vigorously gaining mass by accretion, is accompanied by a secondary clump.

collapsing at a redshift $z_{\text{vir}} \simeq 20$.

In Figure 2 (*left panel*), we show the gas density within the central ~ 25 pc, briefly after the first high-density clump has formed as a result of gravitational runaway collapse. Once the gas has exceeded a threshold density of 10^7 cm^{-3}, a sink particle is inserted into the simulation to replace it. This choice for the density threshold ensures that the local Jeans mass is resolved throughout the simulation. The clump (i.e., sink particle) has an initial mass of $M_{\text{Cl}} \sim 10^3 M_\odot$, and grows subsequently by ongoing accretion of surrounding gas. High-density clumps with such masses result from the chemistry and cooling rate of molecular hydrogen, H_2, which imprint characteristic values of temperature, $T \sim 200$ K, and density, $n \sim 10^4$ cm^{-3}, into the metal-free gas (see [10]). Evaluating the Jeans mass for these characteristic values results in $M_J \sim 10^3 M_\odot$, which is close to the initial clump masses found in the simulations.

The high-density clumps are clearly not stars yet. To probe the subsequent fate of a clump, we have re-simulated the evolution of the central clump with sufficient resolution to follow the collapse to higher densities (see [19] for a description of the refinement technique). In Figure 2 (*right panel*), we show the gas density on a scale of 0.5 pc, which is two orders of magnitude smaller than before. Several features are evident in this plot. First, the central clump does not undergo further sub-fragmentation, and is likely to form a single Population III star. Second, a companion clump is visible at a distance of ~ 0.25 pc. If negative feedback from the first-forming star is ignored, this companion clump would undergo runaway collapse on its own approximately ~ 3 Myr later. This timescale is comparable to the lifetime of a very massive star (VMS)[20]. If the second clump was able to survive the intense radiative heating from its neighbor, it could become a star before the first one explodes as a supernova (SN). Whether more

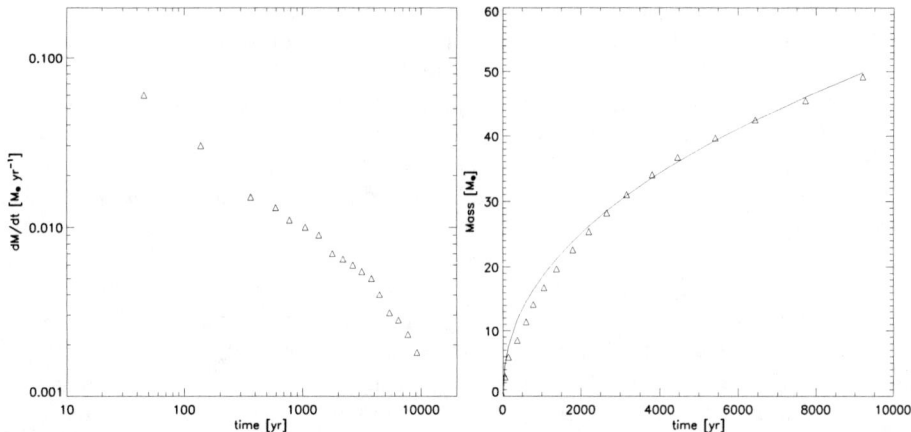

FIGURE 3. Accretion onto a primordial protostar. The morphology of this accretion flow is shown in Fig. 2. *Left:* Accretion rate (in $M_\odot$ yr^{-1}) vs. time (in yr) since molecular core formation. *Right:* Mass of the central core (in $M_\odot$) vs. time. *Solid line:* Accretion history approximated as: $M_* \propto t^{0.45}$. Using this analytical approximation, we extrapolate that the protostellar mass has grown to $\sim 150 M_\odot$ after $\sim 10^5$ yr, and to $\sim 700 M_\odot$ after $\sim 3 \times 10^6$ yr, the total lifetime of a very massive star.

than one star can form in a low-mass halo thus crucially depends on the degree of synchronization of clump formation. Finally, the non-axisymmetric disturbance induced by the companion clump, as well as the angular momentum stored in the orbital motion of the binary system, allow the system to overcome the angular momentum barrier for the collapse of the central clump (see [14]).

The recent discovery of the star HE0107-5240 with a mass of $0.8 M_\odot$ and an iron abundance of [Fe/H] $= -5.3$ [21] shows that at least a few low mass stars could have formed out of extremely low-metallicity gas. Our simulations show that although the majority of clumps are very massive, a few of them, like the secondary clump in Fig. 2, are significantly less massive. Alternatively, low-mass fragments could form in the dense, shock-compressed shells that surround the first hypernovae [22].

How massive were the first stars? Star formation typically proceeds from the 'inside-out', through the accretion of gas onto a central hydrostatic core. Whereas the initial mass of the hydrostatic core is very similar for primordial and present-day star formation [23], the accretion process – ultimately responsible for setting the final stellar mass, is expected to be rather different. On dimensional grounds, the accretion rate is simply related to the sound speed cubed over Newton's constant (or equivalently given by the ratio of the Jeans mass and the free-fall time): $\dot{M}_{\rm acc} \sim c_s^3/G \propto T^{3/2}$. A simple comparison of the temperatures in present-day star forming regions ($T \sim 10$ K) with those in primordial ones ($T \sim 200-300$ K) already indicates a difference in the accretion rate of more than two orders of magnitude.

Our refined simulation enables us to study the three-dimensional accretion flow around the protostar (see also [24, 25, 26]). We now allow the gas to reach densities of 10^{12} cm^{-3} before being incorporated into a central sink particle. At these high densities, three-body reactions [27] have converted the gas into a fully molecular form. In Figure 3, we show how the molecular core grows in mass over the first $\sim 10^4$ yr after its formation. The accretion rate (*left panel*) is initially very high, $\dot{M}_{\rm acc} \sim 0.1 M_\odot$ yr^{-1}, and subsequently declines according to a power law, with a possible break at ~ 5000 yr. The mass of the molecular core (*right panel*), taken as an estimator of the proto-stellar mass, grows approximately as: $M_* \sim \int \dot{M}_{\rm acc} dt \propto t^{0.45}$. A rough upper limit for the final mass of the star is then: $M_*(t = 3 \times 10^6 {\rm yr}) \sim 700 M_\odot$. In deriving this upper bound, we have conservatively assumed that accretion cannot go on for longer than the total lifetime of a VMS.

Can a Population III star ever reach this asymptotic mass limit? The answer to this question is not yet known with any certainty, and it depends on whether the accretion from a dust-free envelope is eventually terminated by feedback from the star (e.g., [24, 25, 26, 28]). The standard mechanism by which accretion may be terminated in metal-rich gas, namely radiation pressure on dust grains [29], is evidently not effective for gas with a primordial composition. Recently, it has been speculated that accretion could instead be turned off through the formation of an H II region [28], or through the radiation pressure exerted by trapped Lyα photons [26]. The termination of the accretion process defines the current unsolved frontier in studies of Population III star formation. Current simulations indicate that the first stars were predominantly very massive, and consequently rather different from present-day stellar populations. The crucial question then arises: *How and when did the transition take place from the early formation of massive stars to that of low-mass stars at later times?* We address this problem next.

THE SECOND GENERATION OF STARS

The very first stars, marking the cosmic Renaissance of structure formation, formed under conditions that were much simpler than the highly complex environment in present-day molecular clouds. Subsequently, however, the situation rapidly became more complicated again due to the feedback from the first stars on the IGM. Supernova explosions dispersed the nucleosynthetic products from the first generation of stars into the surrounding gas (e.g., [30, 31, 32]), including also dust grains produced in the explosion itself [33, 34]. Atomic and molecular cooling became much more efficient after the addition of these metals. Moreover, the presence of ionizing cosmic rays, as well as of UV and X-ray background photons, modified the thermal and chemical behavior of the gas in important ways (e.g., [35, 36]).

Early metal enrichment was likely the dominant effect that brought about the transition from Population III to Population II star formation. Recent numerical simulations of collapsing primordial objects with overall masses of $\sim 10^6 M_\odot$, have shown that the gas has to be enriched with heavy elements to a minimum level of $Z_{\rm crit} \simeq 10^{-3.5} Z_\odot$, in order to have any effect on the dynamics and fragmentation properties of the system [37, 38]. Normal, low-mass (Population II) stars are hypothesized to only form out of gas with metallicity $Z \geq Z_{\rm crit}$. Thus, the characteristic mass scale for star formation is expected to be a function of metallicity, with a discontinuity at $Z_{\rm crit}$ where the mass scale changes by $\sim$ two orders of magnitude. The redshift where this transition occurs has important implications for the early growth of cosmic structure, and the resulting observational signature (e.g., [1, 3, 22, 39]).

Important caveats, however, remain. The determination of the critical metallicity mentioned above [38] implicitly assumes that the gas at temperatures below ~ 8000 K is maintained in ionization equilibrium by cosmic rays, with an ionization rate that is scaled from the Galactic value by the factor $Z/Z_\odot$. The cosmic-ray flux in the early universe might well have not obeyed this simple relation, and it is not clear whether cosmic rays could have successfully 'activated' the metals.

GAMMA-RAY BURSTS AS PROBES OF THE FIRST STARS

Gamma-ray bursts (GRBs) are the brightest electromagnetic explosions in the universe, and they should be detectable out to redshifts $z > 10$ [40, 41]. Although the nature of the central engine that powers the relativistic jets is still debated, recent evidence indicates that GRBs trace the formation of massive stars [42, 43, 44, 45, 46]. Since the first stars are predicted to be predominantly very massive, their death might possibly give rise to GRBs at very high redshifts. A detection of the highest-redshift GRBs would probe the earliest epochs of star formation, one massive star at a time. The upcoming *Swift* satellite[3], planned for launch in late 2003, is expected to detect about a hundred GRBs per year. The redshifts of high-z GRBs can be easily measured through infrared photometry, based on the Gunn-Peterson trough in their spectra due to Lyα absorption by neutral intergalactic hydrogen along the line of sight. *Which fraction of the detected bursts will originate at redshifts $z \geq 5$?*

To assess the utility of GRBs as probes of the first stars, we have calculated the expected redshift distribution of GRBs [47]. Under the assumption that the GRB rate is simply proportional to the star formation rate, we find that about a quarter of all GRBs detected by *Swift*, will originate from a redshift $z \geq 5$ (see Fig. 4). This estimate is rather uncertain because of the poorly determined GRB luminosity function. We caution that the rate of high-redshift GRBs may be significantly suppressed if the early massive stars fail to launch a relativistic outflow. This is conceivable, as metal-free stars may experience negligible mass loss before exploding as a supernova. They would then retain their massive hydrogen envelope, and any relativistic jet might be quenched before escaping from the star [48].

If high-redshift GRBs exist, the launch of *Swift* later this year will open up an exciting new window into the cosmic dark ages. In difference from quasars or galaxies that fade with increasing redshift, GRB afterglows maintain a roughly constant observed flux at different redshifts for a fixed observed time lag after the γ-ray trigger [41]. The increase in the luminosity distance at higher redshifts is compensated by the fact that a fixed observed time lag corresponds to an intrinsic time shorter by a factor of $(1+z)$ in the source rest-frame, during which the GRB afterglow emission is brighter. This quality makes GRB afterglows the best probes of the metallicity and ionization state of the

---

[3] See http://swift.gsfc.nasa.gov.

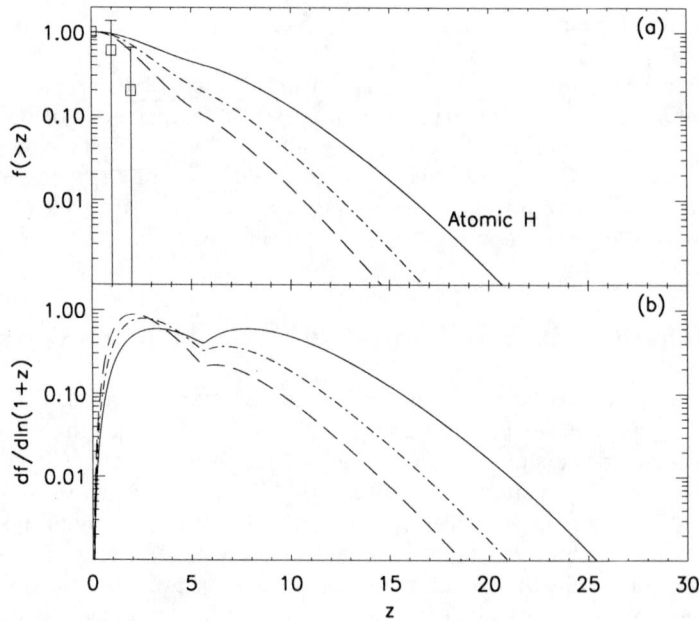

FIGURE 4. Redshift distribution of all GRBs in comparison with measurements in flux-limited surveys (from [47]). (a) Fraction of bursts that originate at a redshift higher than z vs. z. The data points reflect ~ 20 observed redshifts to date. (b) Fraction of bursts per logarithmic interval of $(1+z)$ vs. z. *Solid lines:* All GRBs for star formation in halos massive enough to allow cooling via lines of atomic hydrogen. The calculation assumes that the GRB rate is proportional (with a constant factor) to the star formation rate at all redshifts. *Dot-dashed lines:* Expected distribution for *Swift*. *Long-dashed lines:* Expected distribution for BATSE. The curves for the two flux-limited surveys are rather uncertain because of the poorly-determined GRB luminosity function.

intervening IGM during the epoch of reionization. In difference from quasars, the UV emission from GRBs has a negligible effect on the surrounding IGM (since $\sim 10^{51}$ ergs can only ionize $\sim 4 \times 10^4 M_\odot$ of hydrogen). Moreover, the host galaxies of GRBs induce a much weaker perturbation to the Hubble flow in the surrounding IGM, compared to the massive hosts of the brightest quasars [49]. Hence, GRB afterglows offer the ideal probe (much better than quasars or bright galaxies) of the damping wing of the Gunn-Peterson trough [50] that signals the neutral fraction of the IGM as a function of redshift during the epoch of reionization.

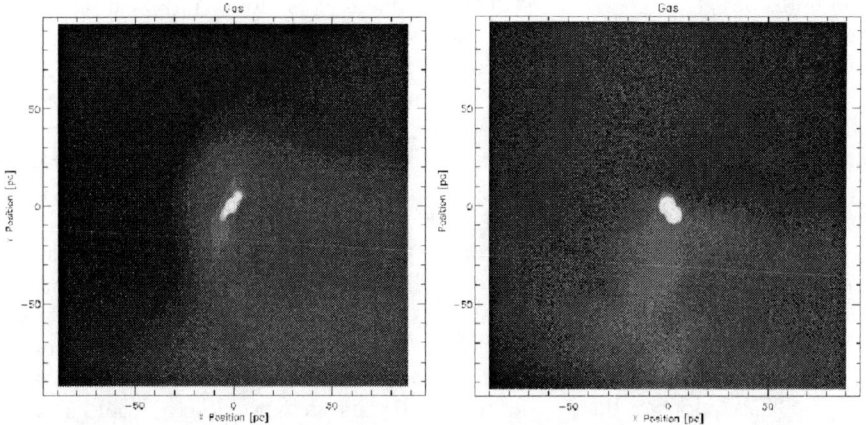

FIGURE 5. Central gas densities in simulations of two dwarf galaxies just above the atomic cooling threshold at $z \sim 10$ with different initial spins and no H_2 molecules (from [19]). Shown is the projection in the $x-y$ plane, and the box size is 200 pc on a side. *Left:* Case with zero initial spin. One compact object has formed in the center with a mass of $2.7 \times 10^6 M_\odot$ and a radius of ≤ 1 pc. *Right:* Case with an initial spin of $\lambda = 0.05$. Here, two compact objects have formed with masses of $2.2 \times 10^6 M_\odot$ and $3.1 \times 10^6 M_\odot$, respectively, and radii ≤ 1 pc.

THE FIRST QUASARS

Quasars are believed to be powered by the heat generated during the accretion of gas onto supermassive black holes (SMBHs; see e.g., [51]). The existence of SMBHs with inferred masses of $\geq 10^9 M_\odot$, less than a billion years after the big bang, as implied by the recent discovery of quasars at redshifs $z \geq 6$ [8, 52, 53, 54], provides important constraints on the SMBH formation scenario [7]. The brightest quasars at $z \geq 6$ are most likely hosted by rare galaxies, more massive than $\sim 10^{12} M_\odot$ [49], that end up today as the most massive elliptical galaxies [55]. *Can the seeds of SMBHs form through the direct collapse of primordial gas clouds at high redshifts?* Previous work [56] has shown that without a pre-existing central point mass, this is rendered difficult by the negative feedback resulting from star formation in the collapsing cloud. The input of kinetic energy due to supernova explosions prevents the gas from assembling in the center of the dark matter potential well, thus precluding the direct formation of a SMBH. If, however, star formation was suppressed in a cloud that could still undergo overall collapse, such an adverse feedback would not occur.

We have carried out SPH simulations [19] of isolated 2σ−peaks with total masses of $10^8 M_\odot$ that collapse at $z_{\rm vir} \sim 10$. The virial temperature of these dwarf galaxies exceeds $\sim 10^4$K, so as to allow collapse of their gas through cooling by atomic hydrogen transitions [57]. Since structure formation proceeds in a bottom-up fashion, such a system would encompass lower-mass halos that would have collapsed earlier on. These subsystems have virial temperatures below 10^4 K, and consequently rely on the presence of H_2 for their cooling. Molecular hydrogen, however, is fragile and readily destroyed by photons in the Lyman-Werner bands with energies $(11.2-13.6$ eV$)$ just below the

Lyman limit [58]. These photons are able to penetrate a predominantly neutral IGM.

We first consider the limiting case in which H_2 destruction is complete. Depending on the initial spin, which is a measure of the degree of rotational support in the cloud, we find that either one (for zero initial spin) or two compact objects form with masses in excess of $10^6 M_\odot$ and radii < 1 pc (see Fig. 5). In the case of nonzero spin a binary system of clumps has formed with a separation of ~ 10 pc. Such a system of two compact objects is expected to efficiently radiate gravitational waves that could be detectable with the planned *Laser Interferometer Space Antenna*[4] (LISA)[59].

What is the further fate of the central object? Once the gas has collapsed to densities above $\sim 10^{17}$ cm^{-3} and radii $< 10^{16}$ cm, Thomson scattering traps the photons, and the cooling time consequently becomes much larger than both the free-fall and viscous timescales (see [19] for details). The gas is therefore likely to settle into a radiation-pressure supported configuration resembling a rotating supermassive star. Recent fully-relativistic calculations of the evolution of such stars predict that they would inevitably collapse to a massive black hole [60]. Under a wide range of initial conditions, a substantial fraction ($\sim 90\%$) of the mass of the supermassive star is expected to end up in the black hole.

Is such a complete destruction of H_2 possible? When we include an external background of soft UV radiation in our simulation, we find that a flux level comparable to what is expected close to the end of the reionization epoch is sufficient to suppress H_2 molecule formation. This is the case even when the effect of self-shielding is taken into account. The effective suppression of H_2 formation crucially depends on the presence of a stellar-like radiation background. It is therefore likely that stars existed before the first quasars could have formed.

ACKNOWLEDGMENTS

VB thanks the Institute for Advanced Study for its hospitality during part of the work on this review, and expresses his gratitude to Paolo Coppi and Richard Larson for the many discussions on the first stars. AL acknowledges support from the Institute for Advanced Study at Princeton, the John Simon Guggenheim Memorial Fellowship, and NSF grants AST-0071019, AST-0204514.

REFERENCES

1. Wyithe, J. S. B., & Loeb, A. 2003a, ApJ, in press (astro-ph/0209056)
2. Cen, R. 2003, ApJ, submitted (astro-ph/0210473)
3. Furlanetto, S. R., & Loeb, A. 2003, ApJ, in press (astro-ph/0211496)
4. Barkana, R., & Loeb, A. 2001, Physics Reports, 349, 125
5. Kaplinghat, M., Chu, M., Haiman, Z., Holder, G., Knox, L., & Skordis, C. 2002, ApJ, submitted (astro-ph/0207591)
6. Yoshida, N., Abel, T., Hernquist, L., & Sugiyama, N. 2003, ApJ, submitted

---

[4] See http://lisa.jpl.nasa.gov/

7. Haiman, Z., & Loeb, A. 2001, ApJ, 552, 459
8. Fan, X., et al. 2003, AJ, in press (astro-ph/0301135)
9. Bromm, V., Coppi, P. S., & Larson, R. B. 1999, ApJ, 527, L5
10. Bromm, V., Coppi, P. S., & Larson, R. B. 2002, ApJ, 564, 23
11. Nakamura, F., & Umemura, M. 2001, ApJ, 548, 19
12. Abel, T., Bryan, G. L., & Norman, M. L. 2002, Science, 295, 93
13. Pudritz, R. E. 2002, Science, 295, 68
14. Larson, R. B. 2002, MNRAS, 332, 155
15. Kroupa, P. 2002, Science, 295, 82
16. Rees, M. J. 1976, MNRAS, 176, 483
17. Bate, M. R., Bonnell, I. A., & Bromm, V. 2002, MNARS, 332, L65
18. Bate, M. R., Bonnell, I. A., & Bromm, V. 2003, MNRAS, in press (astro-ph/0212380)
19. Bromm, V., & Loeb, A. 2003, ApJ, submitted (astro-ph/0212400)
20. Bromm, V., Kudritzki, R. P., & Loeb, A. 2001, ApJ, 552, 464
21. Christlieb, N., et al. 2002, Nature, 419, 904
22. Mackey, J., Bromm, V., & Hernquist, L. 2003, ApJ, in press (astro-ph/0208447)
23. Omukai, K., & Nishi, R. 1998, ApJ, 508, 141
24. Omukai, K., & Palla, F. 2001, ApJ, 561, L55
25. Ripamonti, E., Haardt, F., Ferrara, A., & Colpi, M. 2002, MNRAS, 334, 401
26. Tan, J. C., & McKee, C. F. 2003, these proceedings
27. Palla, F., Salpeter, E. E., & Stahler, S. W. 1983, ApJ, 271, 632
28. Omukai, K., & Inutsuka, S. 2002, MNRAS, 332, 59
29. Wolfire, M. G., & Cassinelli, J. P. 1987, ApJ, 319, 850
30. Madau, P., Ferrara, A., & Rees, M. J. 2001, ApJ, 555, 92
31. Mori, M., Ferrara, A., & Madau, P. 2002, ApJ, 571, 40
32. Thacker, R.J., Scannapieco, E., & Davis, M. 2002, ApJ, 581, 836
33. Loeb, A.,& Haiman, Z. 1997, ApJ, 490, 571
34. Todini, P., & Ferrara, A. 2001, MNRAS, 325, 726
35. Machacek, M. E., Bryan, G. L., & Abel, T. 2001, ApJ, 548, 509
36. Machacek, M. E., Bryan, G. L., & Abel, T. 2003, MNRAS, 338, 27
37. Omukai, K. 2000, ApJ, 534, 809
38. Bromm, V., Ferrara, A., Coppi, P. S., & Larson, R. B. 2001, MNRAS, 328, 969
39. Schneider, R., Ferrara, A., Natarajan, P., & Omukai, K. 2002, ApJ, 571, 30
40. Lamb, D. Q., & Reichart, D. E. 2000, ApJ, 536, 1
41. Ciardi, B., & Loeb, A. 2000, ApJ, 540, 687
42. Bloom, J. S., Kulkarni, S. R., & Djorgovski, S. G. 2002, AJ, 123, 1111
43. Kulkarni, S. R., et al. 2000, Proc. SPIE, 4005, 9
44. Totani, T. 1997, ApJ, 486, L71
45. Wijers, R. A. M. J., Bloom, J. S., Bagla, J. S., & Natarajan, P. 1998, MNRAS, 294, L13
46. Blain, A. W., & Natarajan, P. 2000, MNRAS, 312, L35
47. Bromm, V., & Loeb, A. 2002, ApJ, 575, 111
48. Heger, A., Fryer, C. L., Woosley, S. E., Langer, N., & Hartmann, D. H. 2003, ApJ, submitted (astro-ph/0212469)
49. Barkana, R., & Loeb, A. 2003, Nature, in press (astro-ph/0209515)
50. Miralda-Escudé, J. 1998, ApJ, 501, 15
51. Rees, M. J. 1984, ARA&A, 22, 471
52. Becker, R. H., et al. 2001, AJ, 122, 2850
53. Djorgovski, S.G., Castro, S., Stern, D., & Mahabal, A. A. 2001, ApJ, 560, L5
54. Fan, X., Narayanan, V. K., Strauss, M. A., White, R. L., Becker, R. H., Pentericci, L., & Rix, H.-W. 2002, AJ, 123, 1247
55. Loeb, A., & Peebles, P. J. E. 2002, ApJ, submitted (astro-ph/0211465)
56. Loeb, A., & Rasio, F. A. 1994, ApJ, 432, 52
57. Oh, S.P., & Haiman, Z. 2002, ApJ, 569, 558
58. Haiman, Z., Rees, M. J., & Loeb, A. 1997, ApJ, 476, 458; erratum 484, 985
59. Wyithe, J. S. B., & Loeb, A. 2003b, ApJ, submitted (astro-ph/0211556)
60. Baumgarte, T. W., & Shapiro, S. L. 1999, ApJ, 526, 941

On the Detectability of the Cosmic Dark Ages: 21-cm Lines from Minihalos

Hugo Martel[*], Paul R. Shapiro[*], Ilian T. Iliev[†], Evan Scannapieco[†], and Andrea Ferrara[†]

[*]*Department of Astronomy, University of Texas at Austin*
[†]*Osservatorio Astrofisico di Arcetri, Italy*

Abstract. In the standard Cold Dark Matter (CDM) theory of structure formation, virialized minihalos (with $T_{\rm vir} \leq 10,000{\rm K}$) form in abundance at high redshift ($z > 6$), during the cosmic "dark ages." The hydrogen in these minihalos, the first nonlinear baryonic structures to form in the universe, is mostly neutral and sufficiently hot and dense to emit strongly at the 21-cm line. We calculate the emission from individual minihalos and the radiation background contributed by their combined effect. Minihalos create a "21-cm forest" of emission lines. We predict that the angular fluctuations in this 21-cm background should be detectable with the planned LOFAR and SKA radio arrays, thus providing a direct probe of structure formation during the "dark ages." Such a detection will serve to confirm the basic CDM paradigm while constraining the background cosmology parameters, the shape of the power-spectrum of primordial density fluctuations, the onset and duration of the reionization epoch, and the conditions which led to the first stars and quasars. We present results here for the currently-favored, flat ΛCDM model, for different tilts of the primordial power spectrum. These minihalos will also cause a "21-cm forest" of absorption lines, as well, in the spectrum of radio continuum sources at high redshift, if the latter came into existence before the end of reionization.

INTRODUCTION

No direct observation of the universe during the period between the recombination epoch at redshift $z \simeq 10^3$ and the reionization epoch at $z > 6$ as yet been reported. While a number of suggestions for the future detection of the reionization epoch, itself, have been made, this period prior to the formation of the first stars and quasars – the cosmic "dark ages" – has been more elusive. Standard Big-Bang cosmology in the CDM model predicts that nonlinear baryonic structure first emerges during this period, with virialized halos of dark and baryonic matter which span a range of masses from less than $10^4 M_\odot$ to about $10^8 M_\odot$ which are filled with neutral hydrogen atoms. The atomic density $n_{\rm H}$ and kinetic temperature T_K of the gas are high enough that collisions populate the hyperfine levels of the ground state of these atoms in a ratio close to that of their statistical weight (3:1), with a spin temperature T_S that greatly exceeds the excitation temperature $T_* = 0.0681{\rm K}$. Since $T_S > T_{\rm CMB}$, the temperature of the Cosmic Microwave Background (CMB), as well, for the majority of halos, these "minihalos" can be a detectable source of redshifted 21-cm emission. The direct detection of minihalos at such high redshift would be an unprecedented measure of the density fluctuations in the baryons and of the total matter power spectrum at small scales. The results presented here are described in more detail in [3, 4, 6].

21-CM EMISSION AND ABSORPTION

The 21-cm emission from a single halo depends upon its internal atomic density, temperature, and velocity structure. We model each CDM minihalo here as a nonsingular, truncated isothermal sphere ("TIS") of dark matter and baryons in virial and hydrostatic equilibrium, in good agreement with the results of gas and N-body simulations from realistic initial conditions [1, 2, 7]. The minihalos which contribute significantly to the 21-cm emission span a mass range from M_{min} to M_{max} which varies with redshift. M_{min} is close to the Jeans mass M_J of the uncollapsed IGM prior to reionization, while M_{max} is the mass for which $T_{vir} = 10^4$K according to the TIS model [1]. Typically, at $z = 9$, $M_{min} \sim$ few $\times 10^3 M_\odot$ and $M_{max} \sim$ few $\times 10^7 M_\odot$, depending upon the cosmological model.

Our results for individual minihalos are summarized in Figure 1a. Line profiles of different minihalos along the same line of sight should not typically overlap. The proper mean free path for photons to encounter minihalos in ΛCDM is 160 kpc at $z = 9$ [5], corresponding to a frequency separation $\Delta \nu_{sep} \sim 0.1$ MHz $\gg \Delta \nu_{eff} < 10$ kHz. These results predict a "21-cm forest" of minihalo emission lines. At $z = 9$, for example, there are about 160 lines per unit redshift along a typical line of sight[5]. Detecting the stronger lines would require sub-arcsecond spatial resolution, ~ 1 kHz frequency resolution, and $\sim$nJy sensitivity. SKA is expected to have sufficient resolution for such

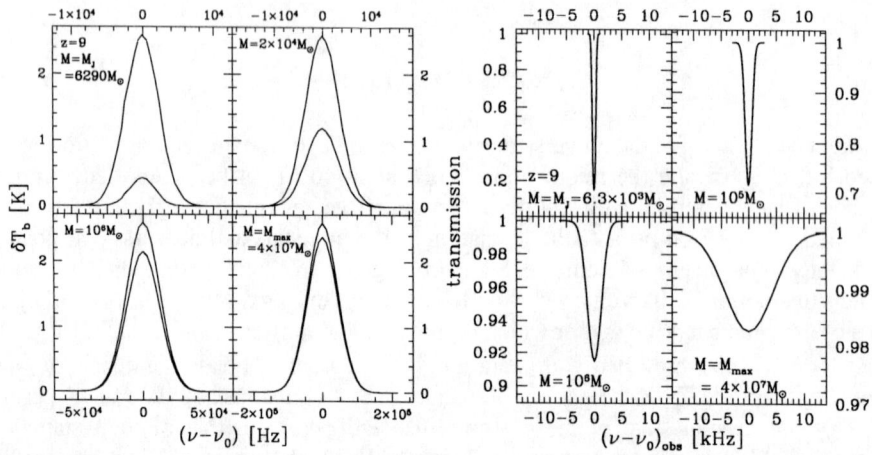

FIGURE 1. (a) (left) 21-cm emission line profiles for individual minihalos of different mass at $z = 9$. Differential antenna temperature $\overline{\delta T_b}$[K] versus emitted frequency ν_{em}, from detailed radiative transfer calculation (lower curves) and optically thin approximation (upper curves). Optical depth is important, particularly for smaller halos. (b) (right) 21-cm absorption line profiles for individual minihalos of different mass at $z = 9$. Transmission factor versus received frequency ν_{rec} at $z = 0$ for line of sight through minihalo center (i.e. zero impact parameter).

observations, but probably not sufficient sensitivity.

The minihalos which produce the "21-cm forest" of emission lines described above have appreciable optical depth to 21-cm absorption. The same minihalos will therefore produce a "21-cm forest" of *absorption* lines, too, in the spectra of radio continuum sources if the latter are discovered at high redshift ($z > 6$), prior to the end of the reionization epoch. We plot illustrative 21-cm absorption line profiles for absorbing minihalos of different masses at $z_{abs} = 9$ in Figure 1b. These absorption features should be observable provided sufficiently bright (~ 1 mJy) background source exists.

21-CM RADIATION BACKGROUND

The beam-averaged differential temperature $\overline{\delta T_b}$ within a given beam of angular size $\Delta \theta_{beam}$ is calculated by integrating our results for individual minihalos over the Press-Schechter mass function of minihalos sampled by the beam. We consider the currently-favored flat ΛCDM model ($\Omega_0 = 0.3$, $\lambda_0 = 0.7$, COBE-normalized, $\Omega_b h^2 = 0.02$, $h = 0.7$), with primordial power spectrum index $n_p = 0.9$, 1, and 1.1.

In principle, the variation of $\overline{\delta T_b}$ with observed frequency should permit a discrimination between the 21-cm emission from minihalos and the CMB and other backgrounds, due to their very different frequency dependences. However, the average differential

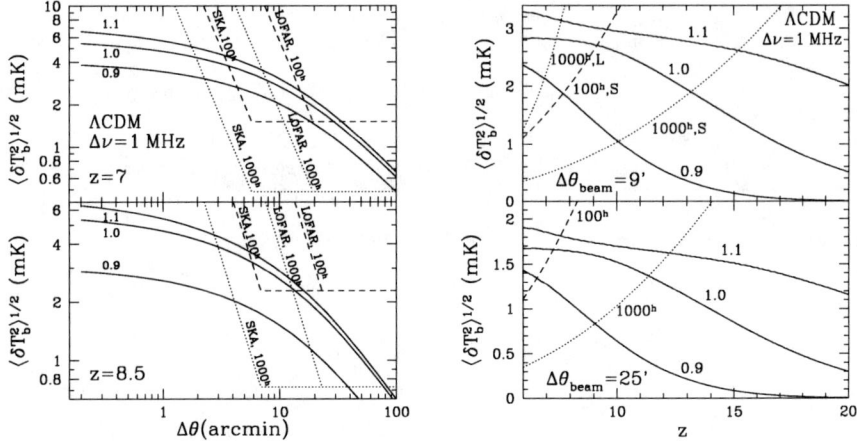

FIGURE 2. (a) (left) Predicted 3-σ differential antenna temperature fluctuations at $z = 7$ ($\nu_{rec} = 177.5$ MHz, top panel) and $z = 8.5$ ($\nu_{rec} = 150$ MHz, bottom panel) for bandwidth $\Delta \nu_{obs} = 1$ MHz vs. angular scale $\Delta \theta_{beam}$ for ΛCDM models with tilt $n_p = 0.9$, 1.0, and 1.1, as labeled (solid curves). Also indicated is the predicted sensitivity of LOFAR and SKA for a confidence level of 5 times the noise level after integration times of 100 hr (dashed lines) and 1000 hr (dotted lines), with compact subaperture (horizontal lines) and extended configuration needed to achieve higher resolution (diagonal lines). (b) (right) Predicted 3-σ differential antenna temperature fluctuations at $\Delta \theta_{beam} = 9'$ (top panel) and 25' (bottom panel). Symbols have same meaning as in the left panels. Letters "L" and "S" correspond to LOFAR and SKA, respectively.

brightness temperature of the minihalo background is very low and its evolution is fairly smooth, so such measurement may be difficult in practice with currently planned instruments like LOFAR and SKA. The angular fluctuations in this emission, on the other hand, should be much easier to detect. The amplitude of q-σ angular fluctuations (i.e. q times the rms value) in the differential antenna temperature is given in the linear regime by $\langle \delta T_b^2 \rangle^{1/2}/\overline{\delta T_b} = qb(z)\sigma_p$, where σ_p is the rms density fluctuation at redshift z in a randomly placed cylinder which corresponds to the observational volume defined by the detector angular beam size, $\Delta\theta_{\text{beam}}$, and frequency bandwidth, Δv_{obs}, and $b(z)$ is the bias factor which accounts for the clustering of rare density peaks relative to the mass.

Illustrative results are plotted for 3-σ fluctuations as a function of $\Delta\theta_{\text{beam}}$ for $z = 7$ and 8.5, in Figure 2a, along with the expected sensitivity limits for the planned LOFAR (300 m filled aperture) and SKA (1 km filled aperture) arrays. We plot in Figure 2b the predicted spectral variation of these fluctuations vs. redshift z for illustrative beam sizes of $\Delta\theta_{\text{beam}} = 9'$ and $25'$. These 3-σ fluctuations should be observable with both LOFAR and SKA with integration times of between 100 and 1000 hours. For a $25'$ beam, for example, 3-σ fluctuations can be detected for untilted ΛCDM by both with a 100 hours integration for $z \sim 6-7.5$ and a 1000 hours integration for $z < 11.5$, while for a $9'$ beam, SKA can detect them after 100 hours for $z < 9$ and after 1000 hours for $z < 13$.

REFERENCES

1. Iliev, I. T., and Shapiro, P. R., *M.N.R.A.S.*, **188**, 791 (2001).
2. Iliev, I. T., and Shapiro, P. R., in *The Mass of Galaxies at Low and High Redshift* (ESO Astrophysics Symposia), eds. R. Bender and A. Renzini, Heidelberg: Springer-Verlag, 2002, in press (astro-ph/0112427).
3. Iliev, I. T., Scannapieco, E., Martel, H., and Shapiro, P. R., *M.N.R.A.S.*, submitted (2002b) (astro-ph/0209216).
4. Iliev, I. T., Shapiro, P. R., Ferrara, A., and Martel, H., *Ap.J.*, **572**, L123 (2002a).
5. Shapiro, P. R., in *Proceedings of the XX*[th] *Texas Symposium on Relativistic Astrophysics and Cosmology*, eds. J. C. Wheeler and H. Martel (AIP Conference Series, Vol 586), 2001, pp. 219–232.
6. Shapiro, P. R., and Iliev, I. T. 2003, in preparation
7. Shapiro, P. R., Iliev, I. T, and Raga, A. C., *M.N.R.A.S*, **307**, 203 (1999).

Photoevaporation of Minihalos during Reionization

Paul R. Shapiro*, Ilian T. Iliev†, Alejandro C. Raga** and Hugo Martel*

Department of Astronomy, University of Texas at Austin
†*Osservatorio Astrofisico di Arcetri, Italy*
**Instituto de Ciencias Nucleares, UNAM, Mexico*

Abstract. We present the first gas dynamical simulations of the photoevaporation of cosmological minihalos overtaken by the ionization fronts which swept through the IGM during reionization in a ΛCDM universe, including the effects of radiative transfer. We demonstrate the phenomenon of I-front trapping inside minihalos, in which the weak, R-type fronts which traveled supersonically across the IGM decelerated when they encountered the dense, neutral gas inside minihalos, becoming D-type I-fronts, preceded by shock waves. For a minihalo with virial temperature $T_{vir} \lesssim 10^4$K, the I-front gradually burned its way through the minihalo which trapped it, removing all of its baryonic gas by causing a supersonic, evaporative wind to blow backwards into the IGM, away from the exposed layers of minihalo gas just behind the advancing I-front.

Such hitherto neglected feedback effects were widespread during reionization. N-body simulations and analytical estimates of halo formation suggest that sub-kpc minihalos such as these, with $T_{vir} \lesssim 10^4$K, were so common as to dominate the absorption of ionizing photons. This means that previous estimates of the number of ionizing photons per H atom required to complete reionization which neglected this effect may be too low. Regardless of their effect on the progress of reionization, however, the minihalos were so abundant that random lines of sight thru the high-z universe should encounter many of them, which suggests that it may be possible to observe the processes described here in the absorption spectra of distant sources.

I-FRONTS AND MINIHALOS AT HIGH REDSHIFT

The first sources of ionizing radiation to condense out of the dark, neutral, opaque IGM reheated and reionized it between $z \sim 30$ and $z \sim 6$. Weak, R-type ionization fronts surrounding each source swept outward thru the IGM, overtaking smaller-mass virialized halos, called *minihalos*, and photoevaporating them. High-redshift sources of ionizing photons may have found their sky covered by these minihalos. The photoevaporation of minihalos may therefore have dominated consumption of ionizing photons during reionization[1]. In this paper, we focus on the currently favored ΛCDM model ($\Omega_0 = 0.3$, $\lambda_0 = 0.7$, $h = 0.7$, $\Omega_b h^2 = 0.02$). In this model, the universe at $z > 6$ was already filled with minihalos capable of trapping a piece of the global, intergalactic I-fronts, photoevaporating their gaseous baryons back into the IGM. Prior to their encounter with these I-fronts, minihalos with $T_{vir} < 10^4$K were neutral and optically thick to hydrogen ionizing radiation. N-body simulations in a cubic volume 100 kpc on a side (see Fig. 1a,b) reveal that at $z = 9$ minihalos with $T_{vir} < 10^4$K (i.e. $M < 10^{7.6} M_\odot$) were separated on average by only $d \sim 7$ kpc while their geometric cross section together covered $\sim 16\%$ of the area along every 100 kpc of an average line of sight [5]. If the sources of reionization were large mass halos with $T > T_{\text{vir}}$, then these were well-enough separated that typical reionization photons were likely to have been absorbed by intervening minihalos.

To demonstrate this in a statistically meaningful way with more dynamic range than N-body results, we combine the Press-Schechter (PS) prescription for deriving the average number density of halos with the truncated isothermal sphere (TIS) halo model[2, 4], to determine which halos are subject to photoevaporation and how common they are, as function of their redshifts. We then compute the fraction of the sky $F_{\text{cover,source}}$, as seen by a source halo of a given mass, which is covered by opaque minihalos located within the mean volume per source halo. If halos with $M > 10^8 M_\odot$ are the reionization sources, their minihalo covering fraction is close to unity and increases by a factor of a few if we

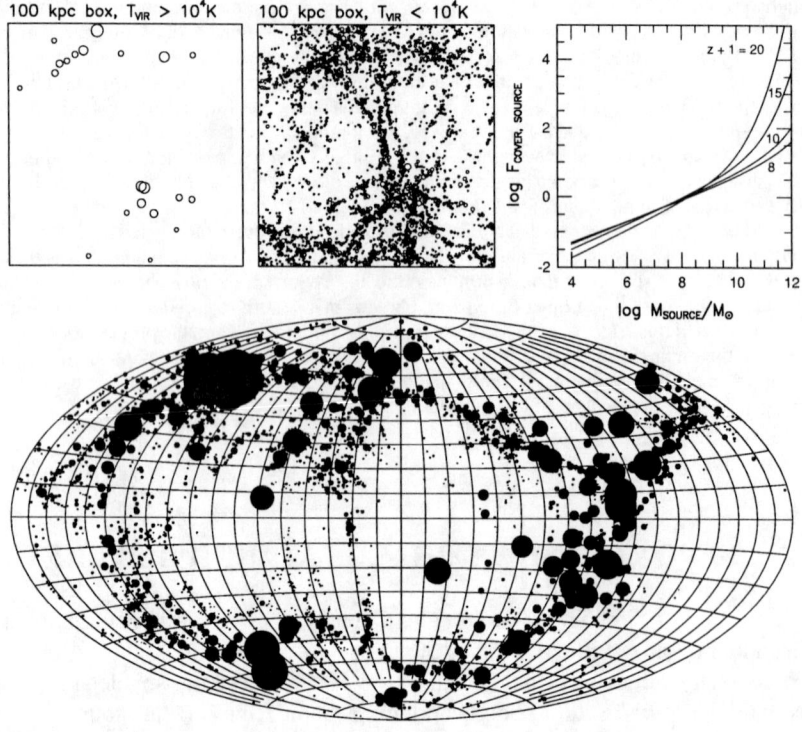

FIGURE 1. Dark matter halos of mass $M > 10^{7.6} M_\odot$ (i.e. $T_{\text{vir}} > 10^4 K$) (a) (top left) and $10^{7.6} M_\odot > M > 10^{5.6} M_\odot$ (i.e. $T_{\text{vir}} < 10^4 K$) (b) (top center) found in P^{3}M simulation of ΛCDM universe in 100 kpc (proper) box, at $z = 9$. (c) (top right) Fraction of sky covered by minihalos located within the mean volume per source halo versus halo mass, at different redshifts, computed using the TIS model and the PS approximation, corrected for linear bias. (d) (bottom) Sky covering by minihalos as seen from a $1.1 \times 10^8 M_\odot$ source, from P^{3}M simulation with 50 kpc box at $z = 9$. All minihalos within 25 kpc are plotted. The covering fraction is 23.6%.

take account of the statistical bias by which minihalos tend to cluster around the source halos (see Fig. 1c,d).

THE PHOTOEVAPORATION OF MINIHALOS

We have performed radiation-hydrodynamical simulations of the photoevaporation of a cosmological minihalo overrun by a weak, R-type I-front in the surrounding IGM, created by an external source of radiation[3]. Minihalos are modeled as TIS of CDM + baryons [2, 4]. We consider 3 different source spectra: (1) QSO-like: $F_\nu \propto \nu^{-1.8}$ ($\nu > \nu_H$); (2) Stellar blackbody: $T_{\text{eff}} = 50,000\,\text{K}$; (3) "No Metals" Stellar $T_{\text{eff}} = 100,000\,\text{K}$. The I-front encounters the minihalo at redshifts $z_{\text{initial}} = (6, 7, 9, 11)$. The flux levels are $F_0 = N_{\text{ph},56}(\nu > \nu_H)/r_{\text{Mpc}}^2 = (0.1, 0.5, 1, 2, 5, 10, 1000)$ (where $N_{\text{ph}} = N_{\text{ph},56} \times 10^{56}$ ionizing photons cm^{-2}s^{-1}, and $r = r_{\text{Mpc}} \times 1\,\text{Mpc}$). The halo masses are $M_{\text{halo}} = (10^4, 10^5, 10^6, 10^7, 2 \times 10^7, 4 \times 10^7) M_\odot$ (where $4 \times 10^7 M_\odot$ corresponds to $T_{\text{vir}} = 10^4 K$ at $z = 9$). The minihalo shields itself against ionizing photons, traps the R-type I-front which enters the halo, causing it to decelerate inside the halo to close to the sound speed of the ionized gas and transforms itself into a D-type front, preceded by a shock. The side facing the source expels a supersonic wind backward toward the source, which shocks the IGM outside the minihalo. The wind grows more isotropic with time as the remaining neutral halo material is photoevaporated. Since the gas itself was initially bound to a

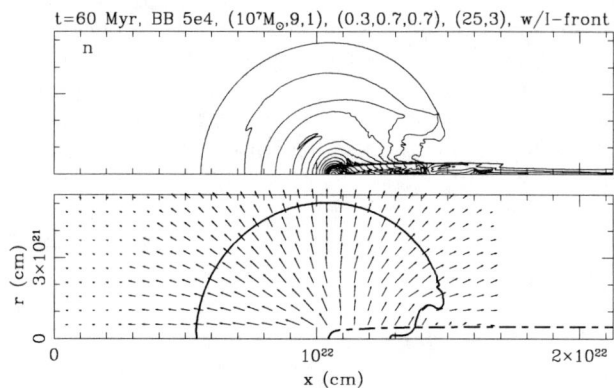

FIGURE 2. One time-slice, 60 Myr after I-front caused by source (located far to the left, along the x-axis) overtakes a $10^7 M_\odot$ minihalo [centered at $(r,x) = (0, 1.06 \times 10^{22}\,\text{cm})$] at $z = 9$ in the ΛCDM universe, for source with $F_0 = 1$, a stellar BB spectrum $T_* = 50,000\,\text{K}$. (a) (top) Isocontours of atomic density, logarithmically spaced; (b) (bottom) Velocity arrows are plotted with length proportional to gas velocity. An arrow of length equal to the spacing between arrows has velocity $25\,\text{km}\,\text{s}^{-1}$; minimum velocities plotted are $3\,\text{km}\,\text{s}^{-1}$. Solid line shows current extent of gas initially inside minihalos at $z = 9$. Dashed line is I-front (50% H-ionization contour).

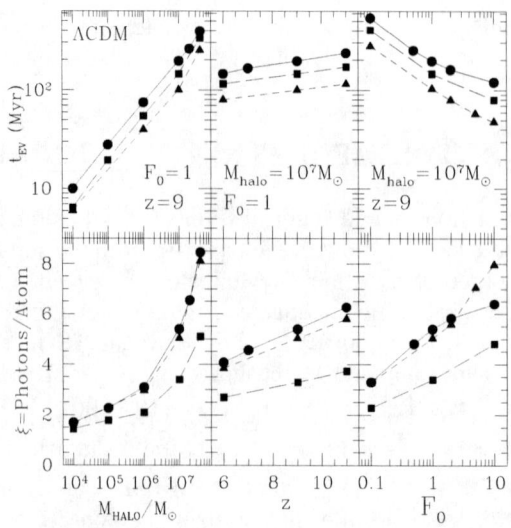

FIGURE 3. Photoevaporation times t_{ev} for individual minihalos (top panels) and total number of ionizing photons absorbed per minihalo H atom during this photoevaporation (bottom panels). Different panels show the variation in these quantities with the input parameters which label the horizontal axes: M_{halo} = minihalo mass, z = redshift when I-front first encounters minihalo, and F_0 = dimensionless ionizing photon flux, with spectrum which is either a stellar blackbody, with $T_* = 50,000$K, (circles) or $T_* = 100,000$K (squares), or else QSO-like (triangles).

dark halo with $\sigma_V < 10\,\mathrm{km\,s^{-1}}$, photoevaporation proceeds unimpeded by gravity. Figure 2 shows the structure of the photoevaporation flow 60 Myr after the global I-front first overtakes a $10^7 M_\odot$ minihalo.

The evaporation time per halo, t_{ev}, and the number, ξ, of ionizing photons absorbed per minihalo H atom during this evaporation time have important implications for the reionization of the universe. We summarize in Figure 3 the dependence of these quantities on the value of the M_{halo}, $z_{initial}$, F_0, and the source spectrum, based on our simulation results.

REFERENCES

1. Haiman, Z., Abel, T., and Madau, P., *ApJ.*, **551**, 599 (2001).
2. Iliev, I. T., and Shapiro, P. R., *M.N.R.A.S.*, **325**, 468 (2001).
3. Shapiro, P. R., in *Proceedings of the XXth Texas Symposium on Relativistic Astrophysics and Cosmology*, eds. J. C. Wheeler and H. Martel (AIP Conference Series), 2001, pp. 219–232.
4. Shapiro, P. R., Iliev, I. T, and Raga, A. C., *M.N.R.A.S*, **307**, 203 (1999).
5. Shapiro, P. R., Martel, H., and Iliev, I. T., 2002, in preparation.

Mass Limits to Primordial Star Formation from Protostellar Feedback

Jonathan C. Tan* and Christopher F. McKee[†]

*Princeton University Observatory, Peyton Hall, Princeton, NJ 08544, USA.
[†]Departments of Physics & Astronomy, University of California, Berkeley, CA 94720, USA.

Abstract. How massive were the first stars? This question is of fundamental importance for galaxy formation and cosmic reionization. Here we consider how protostellar feedback can limit the mass of a forming star. For this we must understand the rate at which primordial protostars accrete, how they and their feedback output evolve, and how this feedback interacts with the infalling matter. We describe the accretion rate with an "isentropic accretion" model: $\dot{m}_*$ is initially very large ($0.03 M_\odot$ yr^{-1} when $m_* = 1 M_\odot$) and declines as $m_*^{-3/7}$. Protostellar evolution is treated with a model that tracks the total energy of the star. A key difference compared to previous studies is allowance for rotation of the infalling envelope. This leads to photospheric conditions at the star and dramatic differences in the feedback. Two feedback mechanisms are considered: HII region breakout and radiation pressure from Lyman-α and FUV photons. Radiation pressure appears to be the dominant mechanism for suppressing infall, becoming dynamically important around 20 $M_\odot$.

THE COLLAPSE OF PRIMORDIAL GAS CLOUDS

Recent numerical studies have followed the gravitational collapse of perturbations from cosmological to almost stellar dimensions[1, 2]. Baryon-dominated clouds, cooled to about 200-300 K by trace amounts of molecular hydrogen, form at the centers of dark matter halos. For $n_H > 10^4$ cm^{-3} the cooling rate becomes independent of density, and so the dissipation of gravitational energy in the densest regions gradually raises the temperature. In the simulation of Abel et al.[1], the gas cloud is centrally-concentrated ($\rho \propto r^{-k_\rho}$; $k_\rho \simeq 2.2$) and is contracting quasi-hydrostatically with infall speeds about one third of the sound speed. In addition to thermal support, the cloud is filled with a turbulent cascade of weak shocks (T. Abel, private comm.). The structure can be described by an approximately hydrostatic polytrope with $\gamma_p = 1 + 1/n = k_P/k_\rho = 2(1 - 1/k_\rho) = 1.1$, where $P \propto r^{-k_P}$. The contraction is akin to the maximally sub-sonic Hunter[3] settling solution, for which the accretion rate is a factor $\phi_* \simeq 2.6$ greater than the classic Shu[5] solution. This accretion rate can be expressed in terms of the entropy parameter of the polytrope, $K = P/\rho^{\gamma_p}$ and the collapsed mass, $M \simeq m_*$,

$$\dot{m}_* = \frac{8\phi_*}{\sqrt{3}} \left[\frac{(3-k_\rho)k_P^3 K^3}{2(2\pi)^{5-3\gamma_p} G^{3\gamma_p - 1}} \right]^{\frac{1}{2(4-3\gamma_p)}} M^j \rightarrow 0.026 K'^{15/7} \left(\frac{M}{M_\odot} \right)^{-3/7} M_\odot \text{ yr}^{-1}, \quad (1)$$

where $j \equiv 3(1 - \gamma_p)/(4 - 3\gamma_p)$. The numerical evaluation assumes $\phi_* = 2.6$, since γ_p is not too different from one, and $K = 1.88 \times 10^{12} (T/300 \text{K}) n_{H,4}^{-0.1}$ cgs $\equiv 1.88 \times 10^{12} K'$ cgs.

We set the temperature normalization a factor 4/3 higher than is seen in simulations[1], to allow for partial pressure support from sonic and isotropic turbulent motions; i.e. T is an effective temperature. The small ratio of turbulent to thermal support is in marked contrast to contemporary massive star formation[6]. In primordial clouds it is the microphysics of H_2 cooling that determines both the evolution (via γ_p) and normalization (via K' and ϕ_*) of the accretion rate. Collapse assuming constant K - "isentropic accretion" - agrees with 1-D numerical studies[7, 8] (Fig. 1a).

The mass-averaged rotation speed is a fraction, $f_{\text{Kep}} \simeq 0.5$, of Keplerian, approximately independent of radius[1]. Assuming angular momentum is conserved inside the sonic point, r_{sp}, leads to a disk size $r_d = f_{\text{Kep}}^2 r_{\text{sp}} = 3.4 (f_{\text{Kep}}/0.5)^2 (M/M_\odot)^{9/7} K'^{-10/7}$ AU. Matter falls onto this disk at all radii $r \leq r_d$, as well as directly to the star. We follow Ulrich[9] in describing the density distribution of the rotating, freely-falling envelope.

PROTOSTELLAR EVOLUTION

At densities $n_H \sim 10^{16}$ cm^{-3} an optically thick protostellar core forms[7], bounded by an accretion shock. The size of the protostar then changes as it accretes matter and radiates energy. For spherical geometry, the high accretion rates typical of primordial star formation lead to optically thick conditions above the accretion shock[10, 11]. Accretion energy is advected, which swells the star. However, for collapse with angular momentum, a disk forms and we expect photospheric conditions over much of the stellar surface. We employ a polytropic stellar structure and extend the energy equation model of Nakano et al.[12], that includes gravitational, ionization-dissociation, and D-burning energies, to allow for optical depth in front of the accretion shock. We also allow for some fraction of disk accretion, depending on f_{Kep}. We model the "luminosity wave" expansion feature[10, 11] as a relaxation to a more compact ($n = 3$) state after the star is older than its Kelvin time: to conserve energy the outer radius of the star grows by a factor, which we estimate from Omukai & Palla[11]. We set $n = 1.5$ or 3, depending on convective stability. During optically thick accretion, before the luminosity wave, we derive an effective value for n by comparing to the results of Stahler et al.[10]. Figure 1b shows this case and isentropic models with $f_{\text{Kep}} = 0$ and 0.5. The two spherical models undergo similar evolution, while the rotating model forms a smaller protostar once the photosphere retreats to the stellar surface. The protostar is then supported by D core and shell burning, as in present-day star formation. The large accretion rate causes the star to join the main sequence only at relatively high masses. The photospheric temperature is much hotter than in the spherical models, leading to significant fluxes of ionizing and FUV radiation. These form the inputs for the feedback model.

FEEDBACK VERSUS ACCRETION

The lifetimes of primordial massive stars converge to about 2 Myr[13] so eq.(1) implies an upper limit to stellar masses of $\sim 2000 M_\odot$. However, other feedback processes are likely to intervene before this. Once the flux of ionizing photons from the protostar

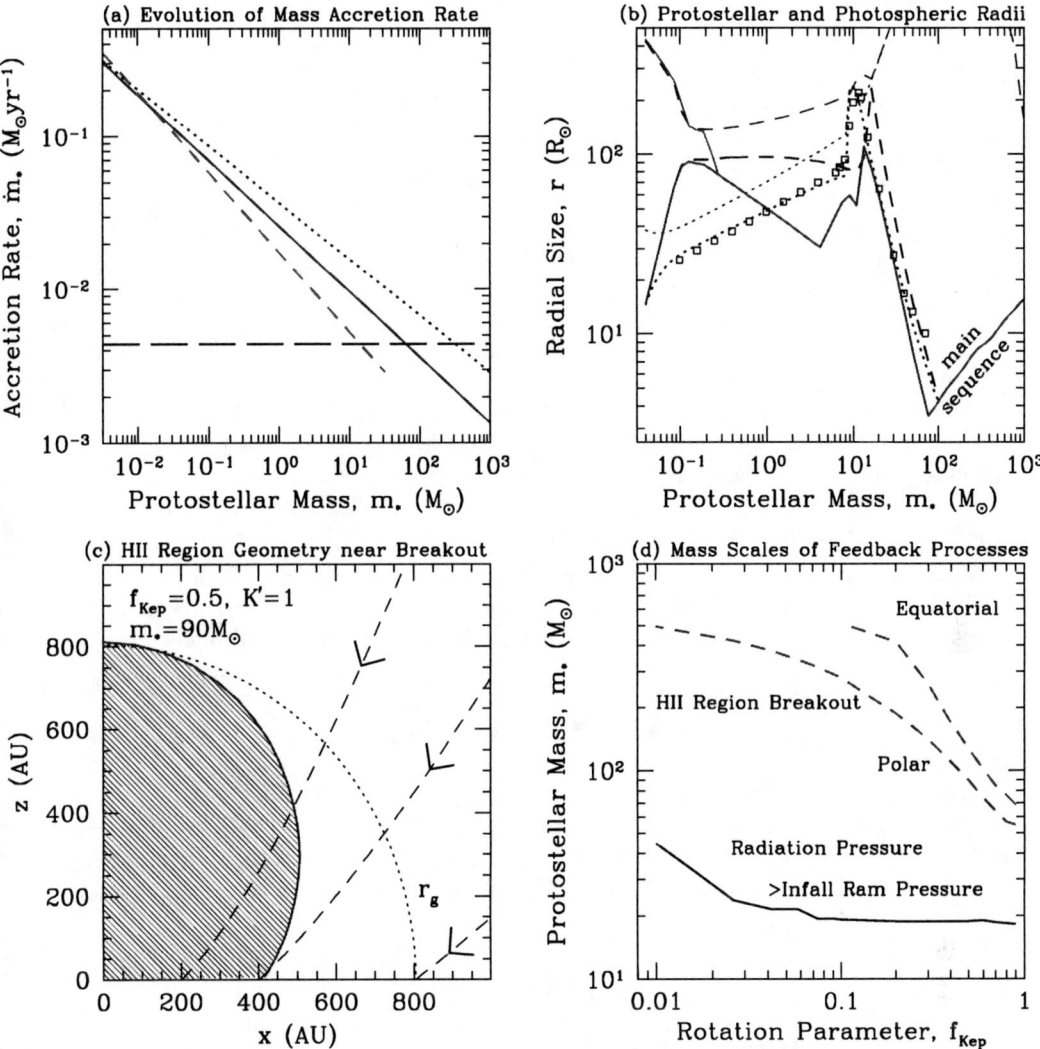

FIGURE 1. (a) Protostellar accretion rate as a function of the collapsed mass ($\simeq m_*$ in these models). *Solid* line: fiducial isentropic accretion model ($K' = 1$) from eq. (1); dotted[7] and dashed lines extrapolated from 1-D numerical studies; long-dashed line is $\dot{m}_* = 4.4 \times 10^{-3} M_\odot$ yr^{-1} used in the protostellar evolution models of Stahler et al.[10] and Omukai & Palla[11]. (b) Evolution of protostellar radius (lower, thick lines), which is the location of the accretion shock, and photospheric radius (upper, thin lines). The spherical constant accretion rate test case (dotted) is compared to other calculations[10, 11] of r_* (squares) for the evolution before H burning. Also shown are the spherical (dashed lines) and rotating ($f_{Kep} = 0.5$, solid lines) isentropic accretion models. The initial condition is taken from 1-D hydrodynamical collapse simulations[8]. Note that the photospheric and protostellar radii are the same in the rotating model for $m_* > 0.3 M_\odot$. (c) Geometry of the HII region (shaded) at polar breakout when $r_{HII} = r_g$, the gravitational radius for the ionized gas sound speed. The protostar is at (0,0) and the disk is in the $z = 0$ plane. Dashed streamlines show infall. (d) Mass scales of feedback processes versus the rotation parameter, f_{Kep}. HII region breakout at the pole and just above the equator occur at masses traced by the lower and upper dashed lines, respectively. Ly-α and FUV radiation pressure becomes greater than twice the radial infall ram pressure (evaluated at angle $\pi/3$ from the pole) for masses above the solid line.

is greater than that of neutral H to its surface, an HII region forms. Approximating sectors as independent, we calculate the extent of the ionized region. Accretion may be suppressed if the HII region expands to distances greater than r_g, where the escape speed equals the ionized gas sound speed, $\simeq 10\,\mathrm{km\,s^{-1}}$. For the fiducial model this occurs at the poles when $m_* = 90 M_\odot$ (Fig. 1c) and at the equator when $m_* = 140 M_\odot$. In this calculation we assumed a free-fall density distribution. In reality the ionizing radiation force[14] decelerates and deflects the flow. For collapse with angular momentum, most streamlines do not come too close to the star, so the effect discussed by Omukai & Inutsuka[14] of HII region quenching due to enhanced densities is greatly weakened. In fact, we anticipate that deflection is more important in reducing the concentration of inflowing gas near the star, so that the HII region becomes larger.

A second feedback effect is radiation pressure from Ly-α photons created in the HII region and FUV photons emitted by the star. These photons are trapped by the Lyman series damping wings of the neutral gas infalling towards the HII region. The energy density builds up until the escape rate, set by diffusion in frequency as well as in space[4], equals the input rate. The resulting pressure acts against the infall ram pressure. We use the results of Neufeld[15] to aid our numerical calculations. Radiation pressure becomes greater than twice the ram pressure at $m_* \simeq 20 M_\odot$ for typical f_{Kep}. The enhancement of radiation pressure above the optically thin limit is by a factor $\sim 10^3$. Infall is first reversed at the poles, which would allow photons to leak out and reduce the pressure acting in other directions. We shall examine this scenario in a future study.

ACKNOWLEDGMENTS

We thank T. Abel, V. Bromm, B. Draine, J. Goodman, D. McLaughlin, and J. Ostriker for helpful discussions. JCT is supported by a Spitzer-Cotsen fellowship from Princeton University and NASA grant NAG5-10811. The research of CFM is supported by NSF grant AST-0098365 and by a NASA grant funding the Center for Star Formation Studies.

REFERENCES

1. Abel, T., Bryan, G. L, and Norman, M. L. 2002, *Science*, 295, 93
2. Bromm, V., Coppi, P. S., & Larson, R. B. 1999, *Astrophys. J.*, 527, L5
3. Hunter, C. 1977, *Astrophys. J.*, 218, 834
4. Adams, T. F. 1972, *Astrophys. J.*, 174, 439
5. Shu, F. H. 1977, *Astrophys. J.*, 214, 488
6. McKee, C. F., & Tan, J. C. 2002, *Nature*, 416, 59
7. Omukai, K., & Nishi, R. 1998, *Astrophys. J.*, 508, 141
8. Ripamonti, E., Haardt, F., Ferrara, A., & Colpi, M. 2002, *Mon. Not. R.A.S.*, 334, 401
9. Ulrich, R. K. 1976, *Astrophys. J.*, 210, 377
10. Stahler, S. W., Palla, F., & Salpeter, E. E. 1986, *Astrophys. J.*, 302, 590
11. Omukai, K., & Palla, F. 2001, *Astrophys. J.*, 561, L55
12. Nakano, T., Hasegawa, T., Morino, J.-I., & Yamashita, T. 2000, *Astrophys. J.*, 534, 976
13. Schaerer, D. 2002, *Astron. & Astrophys.*, 382, 28
14. Omukai, K., & Inutsuka, S. 2002, *Mon. Not. R.A.S.*, 332, 59
15. Neufeld, D. A. 1990, *Astrophys. J.*, 350, 216

The Number of Supernovae from Primordial Stars in the Universe

John H. Wise* and Tom Abel*

*Department of Astronomy & Astrophysics, Pennsylvania State University, University Park, PA 16802, USA

Abstract. We explore the consequences of radiative feedback on early generations of star formation. The main cooling mechanism in primordial star formation is from rotational line emission of molecular hydrogen. However, copious amounts of UV photons produced by these first massive stars effectively dissociate H_2. A consequence of this negative radiative feedback is to raise the required halo mass that can form a primordial star within it. Using results from Eulerian AMR cosmological hydrodynamics simulations, we construct a semi-analytic method to trace the minimum mass of a dark matter halo required to form stars as a function of redshift. Provided this minimum mass, Press-Schechter formalism allows us to calculate the density of these halos and predict Pop III SNe rates. We present preliminary results for the differential and total SNe rates per sky area. If a fixed fraction of these supernovae would also lead to long duration gamma ray bursts, our results can be scaled appropriately. We also compute the evolution of the metallicity of the IGM from supernovae of primordial stars alone. If the first stars die in pair instability supernovae the metals they expelled could pre-enrich the IGM to an average as high as one ten-thousandth of solar metallicity.

MOTIVATION

Primordial stars are the first luminous objects in the universe and are at least partly responsible for ionizing the Universe. With infrared space observatories, such as SIRTF, PRIME, and JWST, we can expect to catch a glimpse of these objects. Although supernovae (SNe) are short compared to the lifetime of a star, they may be the best chance of observing light from primordial stars because of their intrinsic large luminosity. Also recent numerical simulations illustrate that fragmentation does not occur within the first cosmological objects that host cooling gas ($M_{tot} \sim 10^6 M_\odot$). Instead they form single isolated very massive stars ($M \sim 100 M_\odot$) [1]. In these first luminous pre–galactic objects, H_2 cooling is the main mechanism in which gas condenses when $T_{vir} < 10^4 K$ [2, 3]. In the currently favored flat vacuum dominated cold dark matter models of structure formation the first stars form at redshifts $10 \lesssim z \lesssim 50$. Combining simulations that included the negative radiative feedback of H_2 photo–dissociation from UV photons in the Lyman Werner bands [4] with the analytic Press Schechter formalism [5] allows to predict the number of primordial stars as a function of redshift. Using recent metal-free stellar and SNe models [6, 7], we also calculate the radiation backgrounds and average metallicities created by primordial stars alone. A similar recent study [8] did not employ the results from high resolution numerical simulations including radiative feedbacks [4].

SUPERNOVAE RATES WITH RADIATIVE FEEDBACK

Molecular hydrogen is responsible for cooling in primordial star formation, but it can be dissociated very easily by photons between 11.26–13.6 eV, the Lyman-Werner (LW) bands. An ultraviolet background (UVB) in this wavelength regime created by the first stars will influence subsequent primordial star formation by increasing the minimum halo mass necessary to form a cold, dense gas core [9, 4]. From semi-empirical relations [4], we consider the minimum mass of a star-forming halo to be

$$M = \exp\left[\frac{f_c}{0.139}\right]\left[1.25 \times 10^5 + 8.7 \times 10^5 \left(\frac{F_{LW}}{10^{-21}}\right)^{0.47}\right] M_\odot, \quad (1)$$

where $f_c = 0.05$ is the fraction of cold, dense gas in the halo and F_{LW} is the flux in the LW band in units of erg cm^{-2} s^{-1} Hz^{-1}. With no prior star formation, molecular cores should form in $1.8 \times 10^5 M_\odot$ halos. If a single star forms per halo, we can exploit the combination of PS formalism and luminosities from massive, metal-free stellar models to determine the volume-averaged emissivity at a particular redshift [6]. With this emissivity in hand, the UVB evolution is determined via the cosmological radiative transfer equation [10],

$$\left(\frac{\partial}{\partial t} - \nu H(z)\frac{\partial}{\partial \nu}\right) J = -3H(z)J - c\kappa J + \frac{c}{4\pi}\varepsilon, \quad (2)$$

where J is the specific intensity in units erg cm^{-2} s^{-1} Hz^{-1} sr^{-1} and is a function of frequency and redshift, H(z) is the Hubble parameter, κ is the continuum absorption coefficient per unit length along the line of sight, and ε is the proper volume-averaged emissivity. In our case, we can simplify this equation by ignoring absorption since $\kappa \sim 0$ in the LW band. Also, we can drop the derivative with frequency, which accounts for the redshifting of J since the IGM is optically thick above 13.6eV; therefore, no redshifted radiation will contribute to the LW band.

By evolving the UVB and minimum mass of a star forming halo, we can calculate the number density of these halos via PS formalism. We start at $z = 75$ and consider the time derivative of this number density to be the supernovae rate since these low mass halos only harbor one primordial star in this epoch. Therefore, multiple instances of star formation can be ignored. The binding energy of these halos are much less than the kinetic energy released by a primordial pair instability SNe, which indicates that all gas will be expelled from its the host halo thus prohibiting a second epoch of primordial star formation within the same dark matter halo for at least one Hubble time. We repeat this calculation for a wide range of primordial stellar masses that range from 5–500 $M_\odot$. Figure 1 depicts the evolution of the density of star forming halos, differential SNe rate, minimum mass of a star forming halo, and UVB. The cumulative SNe rates for different primordial stellar masses between 50 and 500$M_\odot$ vary surprisingly little from 2 to 1.2 per year per square arcminute.

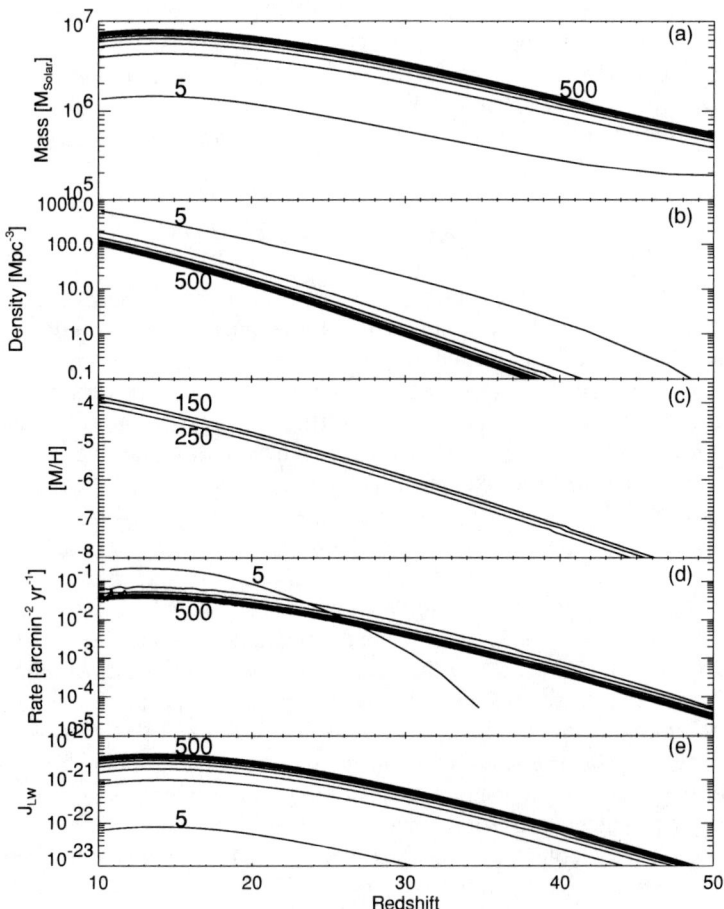

FIGURE 1. Lines represent evolution of (a) Minimum mass of star-forming halos, (b) density of star-forming halos, (c) metallicity (in solar units) (d) SNe Rates, and (e) UVB Background (in units of erg cm^{-2} s^{-1} Hz^{-1} sr^{-1}) in the LW band when we consider primordial stellar masses of 5, 50, 100, 150, 200, 250, 300, 400, 500 $M_\odot$. The labels are the stellar masses on the boundary contour.

METALLICITY EVOLUTION

For metal-free stars in the mass range of 140–260 $M_\odot$, pair instability SNe occur, which completely disrupts the star and a black hole does not form [7]. All metals are ejected into the IGM. In our volume-averaged calculation, we can apply these methods to the evolution of metallicity in the Universe. Figure 1c illustrates the increase of metallicity for star masses of 150, 200, and 250 $M_\odot$. Significant spread in metallicity is not apparent for different choices of star mass. However, it should be noted that these values are *upper limits* since we do not expect every star to reside in this mass range. Surprisingly when primordial star formation likely ceases around $z \sim 10$, the volume-averaged metallicity $[M/H]_\odot \simeq -4$ coincides with the metallicity of most metal deficient stars that are sparsely scattered in the Galactic halo and the VMS (Very Massive Star) component of metallicity fits of metal-poor ($-4 < [M/H]_\odot < -1$) stars [11]. Lastly, metallicities found in $z \sim 5$ Lyα clouds are slightly larger at $[M/H]_\odot \simeq -3.5$ than our calculated values [12]. However, the clustering of the first objects will have to be taken into account before we can more reliably predict metallicities of the Lyman alpha forest.

ADDITIONAL CONSIDERATIONS

Other factors could alter the feasiblity of observing primordial SNe. For instance, a fraction of these events could be accompanied by gamma-ray bursts (GRBs). In most GRB models, the radiation is beamed to small opening angles due to relavistic effects, which would render a fraction of GRBs to be unobservable in our perspective. Secondly, we did not consider the ionization fraction of the Universe for only the halos in the neutral fraction would produce primordial stars. Furthermore, only the pair instability SNe are visible since the other stellar masses result in direct black hole formation in which no radiation escapes. Thus, only the rates in the mass range 140–260 $M_\odot$ are to be considered for observations. Although the initial mass function (IMF) of primordial stars is unknown, simulations have hinted that a fraction of metal-free stars exist in the pair instability SNe mass range. When observational rates are determined, we can compare these results to theoretical values to calculate the fraction of primordial stars that results in visible SNe, which can constrain the elusive primordial IMF. Gravitational lensing might sginificantly increase these observable magnitudes of primordial SNe [13]. Our results are preliminary since we have not yet taken into account the reprocessing of the soft UVB from the Lyman series of intergalactic hydrogen [9], nor the effects of photo–ionization or the possible non–local feedback of the first supernovae.

In summary, first calculations for the rates of high redshift supernovae from primordial stars are encouragingly large with roughly one per arcminute per year. Given sufficiently sensitive instruments these should easily be found. A more detailed study of their observable magnitudes and rates is in preparation.

REFERENCES

1. Abel, T., Bryan, G. L., & Norman, M. L. 2002, *Science*, **295**, 93
2. Abel, T. 1995, Ph.D. thesis, Univ. Regensburg
3. Tegmark, M., Silk, J., Rees, M. J., Blanchard, A., Abel, T., & Palla, F. 1997, *Astrophys. J.*, **474**, 1
4. Machacek, M. E., Bryan, G. L., & Abel, T. 2001, *Astrophys. J.*, **548**, 509.
5. Press, W. H. & Schechter, P. 1974, *Astrophys. J.*, **187**, 425
6. Schaerer, D. 2002, *Astron. & Astrophys.*, **382**, 28
7. Heger, A. & Woosley, S. E. 2002, *Astrophys. J.*, **567**, 532.
8. Mackey, J., Bromm, V., & Hernquist, L. 2002, *Astrophys J.*, submitted (astro-ph/0208447)
9. Haiman, Z., Abel, T., & Rees, M. J. 2000, ApJ, **534**, 11
10. Peebles, P. J. E. 1993, *Principles of Physical Cosmology* (Princeton: Princeton University Press)
11. Qian, Y.-Z. & Wasserburg, G. J. 2002, *Astrophys J.*, **567**, 515.
12. Songaila, A. 2001, *Astrophys. J. Letters*, **561**, 153.
13. Marri, S. & Ferrara, A. 1998, *Astrophys. J.*, **509**, 43

THE NATURE OF DARK MATTER

CDM Substructure in Gravitational Lenses: Tests and Results

C.S. Kochanek* and N. Dalal[1†]

*Center for Astrophysics, MS-51, Cambridge MA 02138
†School of Natural Sciences, Institute for Advanced Study, Princeton NJ 08540

Abstract. We use a simple statistical test to show that the anomalous flux ratios observed in gravitational lenses are created by gravitational perturbations from substructure rather than propagation effects in the interstellar medium or incomplete models for the gravitational potential of the lens galaxy. We review current estimates that the substructure represents $0.006 < f_{sat} < 0.07$ (90% confidence) of the lens galaxy mass, and outline future observational programs which can improve the results.

1. INTRODUCTION

It is a generic feature of CDM (cold dark matter) halo simulations that a significant fraction of the halo mass remains in the form of satellites (e.g. Kauffmann et al. 1993, Moore et al. 1999, Klypin et al. 1999). The exact mass fraction remains somewhat unclear, but the global mass fraction is of order 5–10%, and the projected fraction inside cylinders of radius $R \sim 5h^{-1}$kpc is of order 1%. These mass fractions are significantly higher than are observed in satellites of the Galaxy, suggesting a conflict between CDM models and observations. Three general classes of solutions have been proposed. First, the satellites can be made invisible by suppressing star formation (Kauffmann et al. 1993; Bullock et al. 2000). Second, they can be destroyed by normal dynamical processes or abnormal ones such as self-interacting dark matter (Spergel & Steinhardt 2000; Yoshida et al. 2000; Colín et al. 2002; D'Onghia & Burkert 2002). Third, their formation might be avoided by significantly reducing the amplitude of the power spectrum on the relevant scales (Kamionkowski & Liddle 2000; Colín et al. 2000; Bode et al. 2001; Avila-Reese et al. 2001). The problem, of course, is that it is difficult to distinguish between undetectable and absent satellites.

It was realized early in the debate (see Moore et al. 1999) that gravitational lensing provided a means of resolving the issue because it could detect the gravitational perturbations created by substructure. It was already known that satellites provided a means of solving the "anomalous flux ratio" problem seen in some gravitational lenses (Mao & Schneider 1998). An example of such a problem is shown in Fig. 1, where the close image pair would be expected to have very similar fluxes for any lens model where the gravitational potential can be well-represented by a low-order Taylor series expan-

---

[1] Hubble Fellow

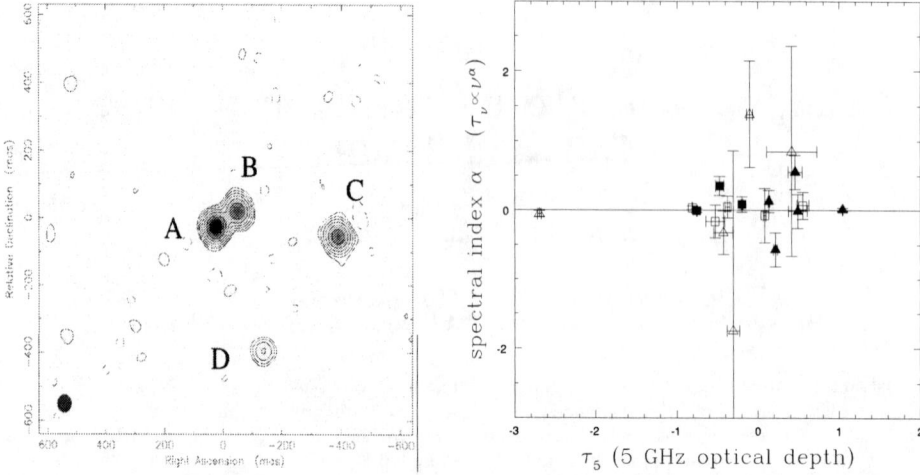

FIGURE 1. (LEFT) Example of an anomalous flux ratio. For a smooth potential we would expect the A and B images in B1555+375 to have the same flux (Marlow et al. 1999).
(RIGHT) ISM properties needed to explain anomalous flux ratios. For an optical depth function $\tau = \tau_5(\nu/5\mathrm{GHz})^\alpha$ we show the estimated spectral index α as a function of the optical depth τ_5 at 5 GHz for the radio lenses used in DK02 with published flux ratios at both 5 GHz and either 8 or 15 GHz. The points are coded by the image type: minima–squares, saddle points–triangles, brightest–filled, and faintest–open.

sion near the images. Metcalf & Madau (2001) and Chiba (2002) pointed out that such anomalies should be common given the predicted CDM substructure fractions. Mao & Schneider (1998), Keeton (2001), Bradac et al. (2002), and Chiba (2002) explored how substructure could explain the anomalous flux ratios in several lens systems. Metcalf & Zhao (2001), Keeton, Gaudi & Petters (2002) and Evans & Witt (2002) explored whether the model for the primary lens galaxy could be modified to explain the anomalous flux ratios, with mixed results which we will discuss in detail below. Our contribution in Dalal & Kochanek (2002, DK02 hereafter) was to analyze the data to make an experimental determination of the substructure fraction, finding it to be in the range $0.006 < f_{sat} < 0.07$ (90% confidence) based on a sample of radio lenses. This is in good agreement with the expectations for CDM, and well above standard estimates for the mass fractions in normal satellites.

Our review of the problem will cover three basic topics. First, we will discuss the problem of distinguishing CDM substructure from other possible origins for the anomalous flux ratios. In particular, we introduce a simple statistical test for substructure in the gravity as compared to either propagation effects in the interstellar medium of the lens galaxy or poorly modeled contributions to the smooth gravitational potential of the lens galaxy. In §3 we review our estimate of the substructure mass fraction and its relation to simulations. Finally, in §4 we discuss the future of the method.

2. TESTING FOR SUBSTRUCTURE

Most of the existing studies of the anomalous flux ratio problem have focused on demonstrating that the problem can be explained by CDM substructure. The next problem is to demonstrate that they cannot be explained by other effects. We can divide the other possibilities into three categories. First, propagation effects in the interstellar medium of the lens galaxy could produce the observed anomalies. Second, the flux ratio anomalies could be created by problems in the models for the smooth potential of the primary lens. Third, we could be misinterpreting a microlensing effect created by the normal stellar populations of the lenses with the effects of more massive satellites. Since we will focus on radio lenses, where microlensing effects must be small because of the large source sizes (Koopmans & de Bruyn 2000), we will not discuss this effect in detail.

Here we explore the first two problems – distinguishing substructure from propagation effects or modeling errors. The first approach we could take is to argue individual cases. For example, almost all propagation effects should show a strong frequency dependence. One way to explore the required properties of the ISM is to assume an optical depth, $\tau = \tau_5 (\nu/5\text{GHz})^\alpha$, for the radio lenses normalized by the optical depth τ_5 at 5 GHz and with a spectral index α for the frequency dependence. Fig. 1 shows the results of fitting such an ISM model to the radio lenses in DK02 where flux measurements were available at both 5 GHz and either 8 or 15 GHz. Common ISM effects, such as refractive scattering or free-free absorption, would show a spectral index of $\alpha \sim -2$, while the optical depth function needed to explain the data has almost no frequency dependence ($\alpha \simeq 0$). In short, explaining the anomalous flux ratios with the ISM requires the radio equivalent of the "gray dust" sometimes suggested to change the cosmological conclusions from Type Ia supernovae (Aguirre 1998).

Similarly, Metcalf & Zhao (2001), Keeton, Gaudi & Petters (2002) and Evans & Witt (2002) explore whether changes to the smooth potential can explain the problem. The basic result from these studies is that they cannot. Although Evans & Witt (2002) give an anti-substructure tenor to their results, we would argue that they have actually produced further arguments in favor of substructure. First, for the two lenses from DK02 they analyze, they can only explain the anomalous flux ratio of one system despite using lens models with essentially arbitrary angular structure. In fact, when we use similar models to analyze the full sample from DK02, we find that of the 6 systems (out of 7) which arguably have anomalous flux ratios, the more complicated models can successfully fit 2–3, at the price of having amplitudes for the higher order perturbations that are significantly larger than are generally observed for either the stars or in dark matter simulations. Second, the other two lenses Evans & Witt (2002) analyze are known from time variability studies to be microlensed (i.e. containing substructure but on a smaller mass scale), so the success of the Evans & Witt (2002) models at explaining the flux ratios in these systems shows that sufficiently complex macro models can mask the presence of substructure even when it is known to be present. This leaves us with a basic ambiguity of course, since standard models (ellipsoidal lens models combined with external tidal shear fields) cannot explain the anomalous flux ratios, while models with very complicated potentials can explain some, but not all, anomalous flux ratios, but can also do so in systems where they should not.

Fortunately, we need not live with these ambiguities, because low optical depth substructure has a unique property that allows us to statistically distinguish substructure from either the interstellar medium or problems in the smooth potential. The images of a lens can be assigned a parity depending on whether they are saddle points or minima of the virtual time delay surface (e.g. Schneider et al. 1992), and the four images alternate their parities as we go around the Einstein ring (saddle-minimum-saddle-minimum). Given a sample of lenses, we can divide the images into four separate classes (brightest saddle, faintest saddle, brightest minimum, faintest minimum) based on their parities and fluxes. As we now discuss, the effects of substructure depend on the image type, while the effects of the ISM and errors in the macro model generally do not.

Both microlensing by the stars (Schechter & Wambsganss 2002) and lensing by extended substructures (Keeton 2002) distinguishes between images based on their parities when the optical depth is low. The sense of the effect is to preferentially demagnify saddle points (negative total parity) compared to minima (positive total parity). The most magnified images are also affected more than the least magnified images because the high magnification makes them sensitive to smaller perturbations in the potential (Mao & Schneider 1998). We can study this effect by examining the distributions of the residuals, $\log(f_{obs}/f_{mod})$, between the model fluxes f_{mod} and observed fluxes f_{obs} for both the data and for different theories as to the origin of the anomalous flux ratios for the 4 different image types found in a quad lens (brightest saddle, faintest saddle, brightest minimum, faintest minimum). The top panel of Fig. 2 shows the distributions we find after fitting the 8 available radio quads using our standard model for the smooth potential (one or more singular isothermal ellipsoids in an external tidal shear field), and the middle panel shows the distribution predicted in a Monte Carlo simulation of a lens sample with a 5% substructure mass fraction. As expected from the previous theoretical studies, the simulation shows an offset of the distribution for the brightest saddle points from the distribution for the other images, in the sense of preferentially demagnifying the saddle point.

The ISM, for example, makes a very different prediction. It is the clumpy, high density components of the ISM which will modify flux ratios, and this has two important consequences. First, propagation effects should preferentially modify the fluxes of the *least* magnified images, because they have the smallest intrinsic angular sizes. The more magnified images have larger angular sizes and will more effectively smooth out any effects of a clumpy ISM. Second, a clumpy ISM cannot distinguish between images of differing parities because the ISM properties are locally determined while the image parity is not – the magnification tensor depends on the projected surface density, the projected shear component of the gravity and the source and lens redshifts. Thus, in a statistical sample, the ISM might systematically perturb the fainter images more than the brighter images, but it will not distinguish between saddle points and minima.

The macro model also will have difficulty systematically perturbing images of a particular parity. Qualitatively this can be understood by the symmetry of merging image pairs from the point of view of the central potential – any slope in the curvature needed to produce a change in the magnification of the saddle point can just as easily appear with the opposite sign so as to produce the opposite change. While we lack a mathematical proof to this effect, it certainly holds in Monte Carlo simulations of lenses produced by potentials with complicated, higher order angular structures (as in Evans & Witt 2002) that are then modeled using standard ellipsoidal potentials. In Fig. 2 we

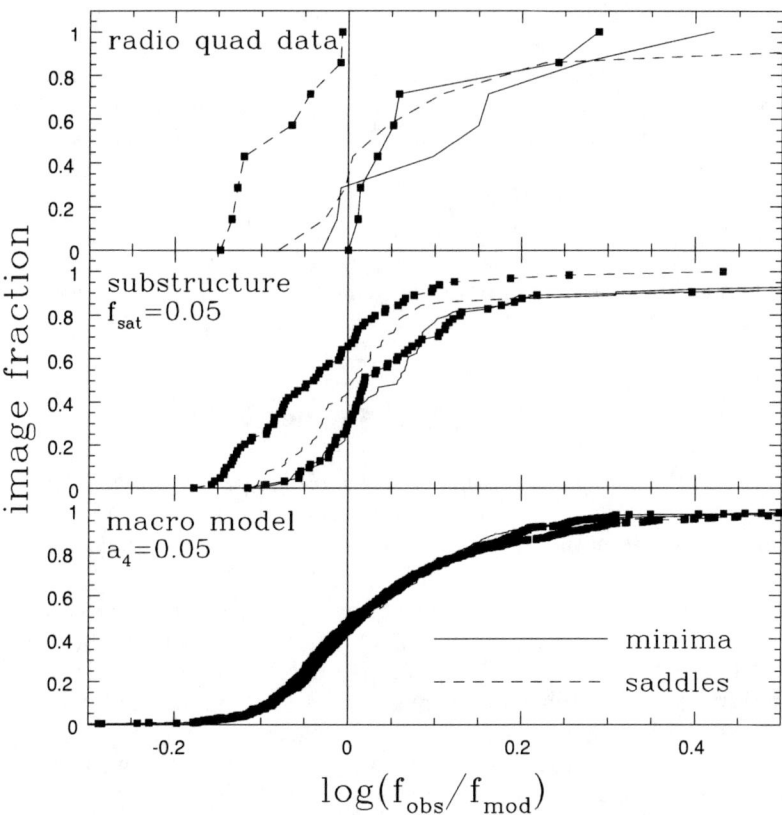

FIGURE 2. Residual distributions after fitting standard models. Each panel shows the integrated distributions of the flux residuals $\log(f_{obs}/f_{mod})$ given the observed f_{obs} and best fitting model f_{mod} image fluxes. The distributions are shown separately for the brightest (points) and faintest (no points) minima (solid lines) and saddle points (dashed lines). The top panel shows the distribution for the real data on 8 four-image radio lenses. The middle panel shows a Monte Carlo simulation of the distributions for lenses with a $f_{sat} = 0.05$ mass fraction in substructure modeled as tidally truncated singular isothermal spheres. The lower panel shows a Monte Carlo simulation of the distributions for lenses with a large amplitude, randomly oriented $\cos 4\theta$ term in the gravitational potential of the lens that is not included in the lens model used to interpret the data. The mode amplitude of $a_4 = 0.05$ is the standard Fourier component used to analyze the photometry of elliptical galaxies. Note how the data (top) shows the same shift to fainter fluxes for the brightest saddle points as is expected from low optical depth substructure (middle), while an error in the macro model (bottom) does not distinguish between the image types.

show an example of the distribution of residuals expected for a population of lenses with a large, unmodeled $\cos 4\theta$ perturbation to the potential. Unlike the distributions expected for substructure shown in Fig. 2, we see that there is no distinction in the model residuals for the different images.

This then leads to a simple test for substructure – do the distributions of residuals from standard ellipsoidal models distinguish between the image types or do they not? We can immediately see the answer in Fig. 2 – the distribution of residuals for the brightest saddle point differs from that of the other three images just as expected for substructure. Quantitatively, the Kolmogorov-Smirnov test probability that the model residuals for the brightest saddle points have the same distribution as for the other images is $< 0.1\%$. Other phrasings of the test, bootstrap resampling of the data, null tests with the image identifications randomly assigned all support the results that the statistical properties of the saddle points are fundamentally different from that of the other images. And this distinctiveness is the tell-tale sign that the anomalous flux ratios are due to substructure in the gravity rather than the ISM or problems in the smooth potential model for the lens galaxy.

3. IMPLICATIONS FOR CDM

If substructure is the explanation, then the next objective is to estimate the satellite mass fraction f_{sat}. Here we review the method and results of DK02. We modeled the substructure as tidally truncated singular isothermal spheres, with a critical radius scale $b = 0\rlap{.}''001$ and a tidal radius $a = (bb_0)^{1/2}$ where $b_0 \simeq 1\rlap{.}''0$ is the critical radius of the primary lens galaxy. When the dominant effect of the substructure is to perturb magnifications rather than image positions, we can only measure the mass fraction f_{sat} of the satellites with reasonable accuracy. The mass scale or mass function of the satellites is difficult to constrain unless there are significant astrometric perturbations. In outline, our approach was to take each lens and its standard model and then add random substructure realizations in order to determine the probability of finding an improved fit as a function of the parameters describing the substructure (principally the mass fraction, f_{sat}). Because the "macro" model for the smooth potential masks some of the effects of substructure, it is necessary to reoptimize the parameters of the "macro" model for every trial. We applied the method to a sample of 7 four-image lenses.

We can illustrate our method with Monte Carlo simulations. Fig. 3 shows the results for Monte Carlo simulations of our sample either with or without substructure. If we add no substructure, then we typically obtain an upper limit of $f_{sat} \lesssim 0.004$ given the properties of our lens sample. This sets a lower limit for our detection threshold somewhat above the substructure fraction which would be associated with the visible satellites in the Galaxy. If we put $f_{sat} = 0.05$ of the mass into substructure, then we recover the input fraction reasonably accurately. Of the eight Monte Carlo trials shown in Fig. 3, four agree with the input value to within the 68% (1σ) confidence region, and six agree to within the 90% confidence region. If we combine all 8 simulations into a synthetic sample of 56 lenses, we estimate that $f_{sat} = 0.034$ with a 90% confidence range of $0.023 \leq f_{sat} \leq 0.048$ that marginally excludes the true value. Similarly, if we attempt to recover the deflection scale of the substructure, the error bars are worse but the results do converge to the input value when we model a large enough sample of lenses.

Fig. 4 shows the results for the real data. The biggest systematic uncertainty in the data is the level of systematic uncertainty in the flux ratio measurements, so we show the results for 5%, 10% and 20% uncertainties in flux measurements. We adopt 10%

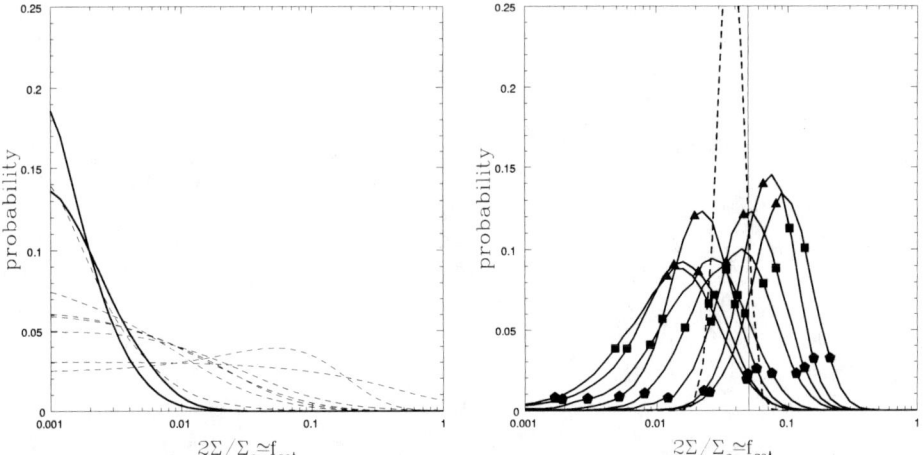

FIGURE 3. Monte Carlo simulations of the DK02 method. The left (right) panel shows the likelihood distributions for the substructure mass fraction produced by our method when $f_{sat} = 0$ ($f_{sat} = 0.05$). In the left panel, the dashed lines show the probabilities for the individual lenses and the solid lines show the joint likelihoods for two simulations of a sample of 7 lenses. For these simulations where the true $f_{sat} = 0$, we find an upper bound of $f_{sat} \lesssim 0.004$. In the right panel we show the joint likelihoods for eight simulations of a sample of 7 lenses (solid lines) in which the true substructure fraction is $f_{sat} = 5\%$. The recovered satellite fractions are statistically consistent with the input fraction, albeit with broad uncertainties due to the small sample size. The heavy dashed line simulates a sample of 56 lenses (the product of all the solid lines), and the recovered value is slightly lower than the input value given the distribution width. The points on the curves indicate the median (triangles), 68% confidence (squares) and 95% confidence (pentagons) regions.

as our standard value – the measurements probably are not as accurate as 5%, they probably are more accurate than 10%, and they certainly are more accurate than 20%. With the 10% flux uncertainties we find a substructure fraction of $0.006 \lesssim f_{sat} \lesssim 0.07$ (90% confidence) that is in good agreement with the expectations for CDM models. We obtain a very poor estimate of the characteristic deflection scale, finding $0\rlap{.}''0001 < b < 0\rlap{.}''007$, which for a substructure mass function $dn/dM \propto M^{-1.7}$ implies an upper end to the substructure mass function of $10^6 M_\odot$-$10^9 M_\odot$ that is in crude accord with our expectations. The degenerate direction in the error contours of Fig. 4 correspond to keeping the magnification perturbations nearly constant while varying the astrometric perturbations.

4. IN THE FUTURE

The future of the substructure question will be driven by further observations, both to clarify the origins of the anomalous flux ratios and to obtain improved estimates of the substructure mass fraction and mass function.

It is relatively straight forward to finish eliminating the ISM as a source of concern

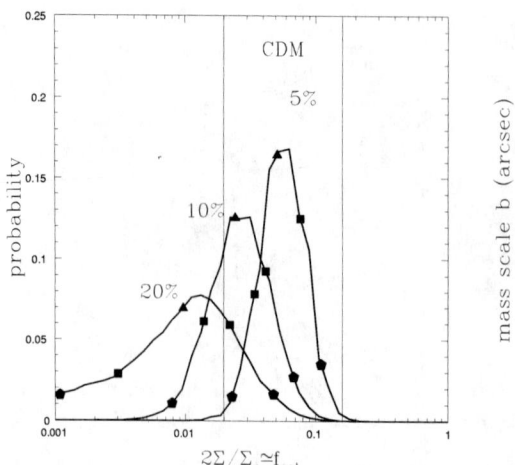 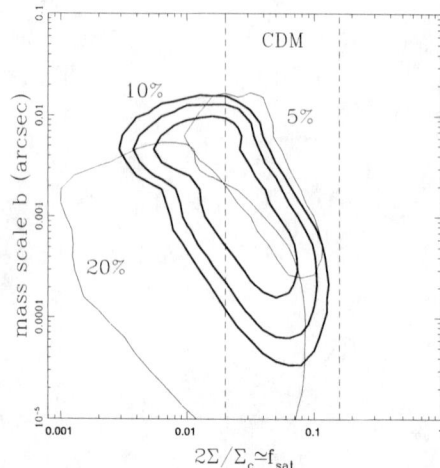

FIGURE 4. Results for the DK02 sample of 7 lenses. The left panel shows the results for estimating f_{sat} with $b = 0\rlap{.}''001$ fixed, and the right panel shows the results estimating the deflection scale b as well. Distributions are shown assuming flux measurement errors of 5%, 10% and 20%, where we adopted 10% as our standard estimate.

with new observations. In the radio this means measuring flux ratios at still higher frequencies (e.g. 43 GHz flux ratio measurements at the VLA) to constrain the frequency dependence of any propagation effect still more tightly. In the optical this means measuring flux ratios over long wavelength baselines to measure any dust extinction (e.g. Falco et al. 1999). Mid-infrared (5–10μm) flux ratios, where the wavelength is far to short to be bothered by electrons and far too long to be bothered by dust, are difficult to measure but completely insensitive to the ISM. Observations to find additional lensed structures in the systems with anomalous flux ratios are the best direct route to determining whether more complicated lens potentials are needed. In particular, very clean constraints on the strengths of any more complicated angular structure than is included in the standard ellipsoidal models can be obtained by analyzing the shapes of the Einstein ring images of the host galaxies (see Kochanek et al. 2001). Such data can be obtained for any lens through deep HST/NICMOS imaging of lens systems.

Improving estimates of the substructure parameters or the statistical case for (or against) substructure requires larger samples of lenses to include in the analysis. The primary problem in expanding the sample is the need to separate the effects of stars and satellites in the optically-selected lenses. The flux ratios of the optical quasars are affected by both substructure and stellar microlensing because the optical continuum emitting regions of accretion disks are so compact (see Schneider et al. 1992) for a general review of microlensing). By measuring the flux ratios of these lenses in either the mid-IR, where the emitting region is a large dust "torus," or in the emission lines, where the emitting region is the relatively large broad/narrow emission line region, we can separate the effects of the stars from the effects of substructure (e.g. Moustakas & Metcalf 2002). While mid-IR imaging is difficult except for the brightest quasar lenses

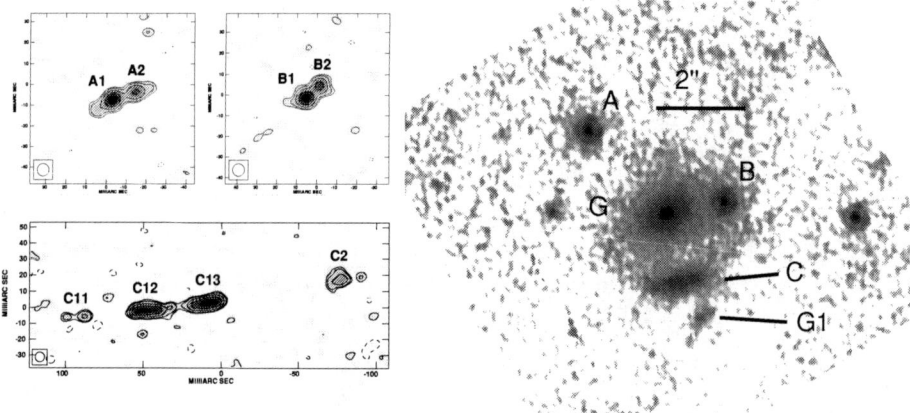

FIGURE 5. The Astrometric anomaly in MG2016+112. The left panels show the VLBI images from Koopmans et al. (2002) of the A, B and C images. The right panel shows the CASTLES HST/NICMOS H-band image of the system. For the same symmetry reasons that close image pairs should have the same fluxes, the C_{11}–C_{12} and C_{13}–C_2 image separations should be the same. The fact that they differ is the astrometric equivalent of a flux anomaly. In this case it is created by the small galaxy G1 sitting to the South of the C images.

(see Agol et al. 2000), the advent of high spatial resolution integral field spectrographs on many 8m-class telescopes will make it relatively easy to measure emission line flux ratios.

Our analysis methods also need to be improved. In particular, we need to properly treat the highest mass satellites, both to constrain the satellite mass function and to better estimate the mass fraction. In the DK02 analyses, we included the highest mass satellites as part of the macro model because they have such an enormous effect on the models that it is impossible to produce a reasonable model without including them. One example is the small satellite in MG0414+0534 (Object "X"; Schechter & Moore 1993), which at H-band has only 10% the luminosity of the main lens and in models has only 12% the critical radius of the main lens, but even when fitting only the positions of the quasar images produces a $\Delta\chi^2 \simeq 100$ improvement in the fit once it is included (Ros et al. 2000).

We can also search for the astrometric equivalents of anomalous flux ratios so as to provide better constraints on the mass scale of the substructure (e.g Wambsganss & Paczynski 1992, Metcalf 2002). One example is the lens MG2016+112 (see Fig. 5) where in VLBI maps the C image is seen to be a pair of merging images each of which is composed of two VLBI components (Koopmans et al. 2001). For the same reasons that we would expect a merging image pair to have similar fluxes, we would expect them to have similar separations, so the very asymmetric separations of the VLBI components C_{11}–C_{12} as compared to C_{13}–C_2 is the astrometric equivalent of an anomalous flux ratio. In this case, as in MG0414+0534, the culprit is a visible satellite G1 sitting to the south of the C image complex. It has only 1-2% the H-band luminosity and $\sim 8\%$ the deflection scale of the main lens, and is known to lie at the same redshift as the lens galaxy. The VLBI data has sufficient resolution to detect a satellite with a deflection scale nearly

10 times smaller. The holy grail of searching for CDM substructure in gravitational lenses would be to find an astrometric anomaly similar to that in MG2016+112, so that there is compelling evidence for the existence of a satellite, but for which no luminous counterpart can be detected.

ACKNOWLEDGMENTS

N.D. gratefully acknowledges the support of NASA through Hubble Fellowship grant #HST-HF-01148.01-A awarded by STScI, which is operated by AURA for NASA, under contract NAS 5-26555. CSK is supported by the Smithsonian Institution and NASA grant NAG5-9265.

REFERENCES

1. Agol, E., Jones, B., & Blaes, O., 2000, ApJ, 545, 657
2. Aguirre, A. N. 1998, ApJ, 512, L19
3. Avila-Reese, V., Colín, P., Valenzuela, O., D'Onghia, E., & Firmani, C. 2001, ApJ, 559, 516
4. Bode, P., Ostriker, J. P., & Turok, N. 2001, ApJ, 556, 93
5. Bradac, M., Schneider, P., Steinmetz, M., Lombardi, M., & King, L.J., 2002, A&A, 388, 373
6. Bullock, J. S., Kravtsov, A. V., & Weinberg, D. H. 2000, ApJ, 539, 517
7. Chiba, M. 2002, ApJ, 565, 17
8. Colín, P., Avila-Reese, V., & Valenzuela, O. 2000, ApJ, 542, 622
9. Colín, P., Avila-Reese, V., Valenzuela, O., & Firmani, C. 2002, astro-ph/0205322
10. Dalal, N. & Kochanek, C. S. 2002, ApJ, 572, 25 [DK02]
11. D'Onghia, E. & Burkert, A. 2002, astro-ph/0206125
12. Evans, N. W. & Witt, H. J. 2002, astro-ph/0212013
13. Falco, E. E. et al. 1999, ApJ, 523, 617
14. Kamionkowski, M. & Liddle, A. R. 2000, Physical Review Letters, 84, 4525
15. Kauffmann, G., White, S.D.M., & Guiderdoni, B., 1993, MNRAS, 264, 201
16. Keeton, C. R. 2001a, astro-ph/0111595
17. Keeton, C. R. 2002, astro-ph/0209040
18. Keeton, C.R., Gaudi, B.S., & Petters, A.O., 2002, ApJ submitted, astro-ph/0210318
19. Klypin, A., Kravtsov, A.V., Valenzuela, O., & Prada, F., 1999, ApJ, 522, 82.
20. Kochanek, C. S., Keeton, C. R., & McLeod, B. A. 2001, ApJ, 547, 50
21. Koopmans, L. V. E. and de Bruyn, A. G. 2000, A&A, 358, 793
22. Koopmans, L. V. E., Garrett, M. A., Blandford, R. D., et al., Porcas, R. W. 2002, MNRAS, 334, 39
23. Mao, S., & Schneider, P., 1998, MNRAS, 295, 587
24. Marlow, D.R., Myers, S.T., Rusin, D., et al., 1999, AJ, 118, 654
25. Metcalf, R. B. & Madau, P. 2001, ApJ, 563, 9
26. Metcalf, R. B. and Zhao, H. 2002, ApJ, 567, L5
27. Metcalf, R. B., 2002, ApJ, 580, 696
28. Moore, B. Ghigna, S., Governato, F., et al., 1999, ApJ, 524, L19
29. Moustakas, L. A. & Metcalf, R. B. 2002, astro-ph/0203012
30. Ros., E., Guirado, J.C., Marcaide, J.M., et al., 2000, A&A, 362, 845
31. Schechter, P. L. & Moore, C. B. 1993, AJ, 105, 1
32. Schechter, P. L. & Wambsganss, J. 2002, astro-ph/0202425
33. Schneider, P., Ehlers, J., & Falco, E.E., 1992, Gravitational Lenses, (Springer Verlag: Berlin)
34. Spergel, D. N. & Steinhardt, P. J. 2000, Physical Review Letters, 84, 3760
35. Wambsganss, J., & Paczynski, B., 1992, ApJ, 397, L1
36. Yoshida, N., Springel, V., White, S. D. M., & Tormen, G. 2000, ApJ, 544, L87

Probing the Distribution of Mass via Gravitational Lensing

Priyamvada Natarajan

Department of Astronomy, Yale University, 260 Whitney Avenue, New Haven, CT 06511, U. S. A.

Abstract. Gravitational lensing has been used to successfully map the distribution of mass on a wide range of scales. I review the techniques and recent results primarily from galaxy-galaxy lensing studies in massive intermediate redshift clusters and explore the implications for the nature of dark matter.

INTRODUCTION

Gravitational lensing provides a powerful tool to statistically measure the mass and probe the details of the mass distribution for field galaxies (Tyson et al. 1984; Brainerd et al. 1996). These studies confirm the existence of massive dark matter halos around typical field galaxies, extending to beyond 100 kpc[1](Brainerd et al. 1996; Fischer et al. 2000; Smith et al. 2001b; McKay et al. 2001). The same technique can be modified and implemented within clusters to constrain the masses of cluster galaxies (Natarajan & Kneib 1997, NK97; Geiger & Schneider 1998). Successful application of the same to the rich, lensing cluster AC 114 at $z = 0.31$, suggests that the average M/L ratio and spatial extents of the dark matter halos associated with early-type galaxies in such dense environments may differ significantly from those of comparable luminosity field galaxies (Natarajan et al. 1998).

The detailed mass distribution within massive clusters, and in particular the fraction of the total cluster mass that is associated with individual galaxies – has important implications for the frequency and nature of galaxy interactions. The characteristic halo sizes of cluster galaxies that survive tidal deformation and stripping in their dense environments offer tantalizing clues as to the nature of dark matter itself.

GALAXY-GALAXY LENSING IN CLUSTERS

The results of our long-term study of galaxy-galaxy lensing in massive cluster-lenses spanning $z = 0.17$ to 0.58, utilizing high-quality archival *Hubble Space Telescope* (*HST*) data are presented here. Local anisotropies in the shear maps are assumed to arise from

---

[1] We adopt h=H_o/100km s^{-1} Mpc^{-1}=0.5 and $q_o = 0.5$, $\Omega_o = 1$. Our results however, are not sensitive to values of the cosmological parameters.

dark matter substructure within these clusters. Associating such substructure with bright early-type cluster galaxies, we quantify the properties of typical L^* cluster members in a statistical fashion. The fraction of total mass associated with individual galaxies within the inner regions of these clusters ranges from 10–20% implying that the bulk of the dark matter in massive lensing clusters is smoothly distributed. Looking at the properties of the cluster galaxies, we find strong evidence (> 3-σ significance) that a fiducial early-type L^* galaxy in these clusters has a mass distribution that is tidally truncated compared to equivalent luminosity galaxies in the field. In fact, we exclude field galaxy scale dark halos for these cluster early-types at > 10-σ significance. We compare the tidal radii obtained from this lensing analysis with the central density of the cluster potentials and find a correlation which is in excellent agreement with theoretical expectations of tidal truncation: $\log[r_t*] \propto (-0.6 \pm 0.2)\log[\rho_0]$. Some details of the analysis and modeling are presented in subsequent sections.

Modeling cluster cores

The technique applied by NKSE98 quantifies the local weak distortions in the observed shear field of massive cluster-lenses, as perturbations arising from the massive halos of cluster galaxies (for details see NK97). By associating these perturbations with bright early-type cluster members, the relative mass fraction in their halos is constrained using a combined χ^2-maximum likelihood method. The strength of this approach is the simultaneous use of constraints from the observed strong and weak lensing features. To quantify the lensing distortion induced by the global potential, both the smooth component and individual galaxy-scale halos are modeled self-similarly using a surface density profile, $\Sigma(R)$, which is a linear superposition of two pseudo-isothermal elliptical components (see the PIEMD models derived by Kassiola & Kovner 1993),

$$\Sigma(R) = \frac{\Sigma_0 r_0}{1 - r_0/r_t} \left(\frac{1}{\sqrt{r_0^2 + R^2}} - \frac{1}{\sqrt{r_t^2 + R^2}} \right), \tag{1}$$

with a core radius r_0 and a truncation radius $r_t \gg r_0$. The free parameters of this profile are chosen for both the smooth component and the clumps so as to obtain the appropriate mass distributions on the relevant scales. The projected radius R is a function of the sky coordinates x and y and the ellipticity ε (see §2.2 of Natarajan & Kneib 1997). One of the attractive features of this model is that the total mass is finite ($\propto \Sigma_0 r_0 r_t$). With the additional assumption that light traces mass, galaxy halos in clusters are characterized by simple scaling laws of the 2 fiducial parameters with luminosity, r_t^* - the tidally truncated radius and σ^* - the central velocity dispersion. A maximum-likelihood method is used to obtain significance bounds on these parameters that characterize a typical L^* halo in the cluster. We have extended the formalism developed in NK97 to include the strong lensing data for the inner regions of the clusters, these are used to obtain the best model χ^2 fit followed by a likelihood method that incorporates the constraints from the shear field.

The HST cluster lenses

For our analysis we selected clusters at $z > 0.1$ for which deep, high-quality *HST* imaging is available and which contain spectroscopically-confirmed multiply-imaged high redshift galaxies. These lensed features are essential to construct a detailed mass distribution for the cluster cores (e.g. Kneib et al. 1996; Smith et al. 2001a; Smail et al. 2001), while the existence of spectroscopic redshifts allows us to calibrate these mass distributions onto an absolute scale. This selection yields five clusters with redshifts spanning $z = 0.17$–0.58 for our analysis: A 2218, A 2390, Cl 2244−02, Cl 0024+16, and Cl 0054−27. In addition to these five clusters, we also include our previous analysis of the $z = 0.31$ cluster AC 114 (NKSE98). These clusters do not constitute a well-defined sample, for instance, their X-ray luminosities span an order of magnitude and their central mass densities show a similarly large dispersion. It is this latter property which is of most interest for our analysis – and the large range spanned by the sample therefore provides a powerful test of the variation in characteristics of galaxy halos with the density of the local environment.

We obtain likelihood contours for the galaxy perturber models of each of the five clusters. In all cases we detect an unambiguous galaxy-galaxy lensing signal at the >3-σ level – confirming the existence of truncated dark halos associated with early-type galaxies in clusters. In Fig. 1, we show the best-fit lensing mass model for A2218 that includes in the optimization 25 galaxy-scale components. The likelihood analysis yields best-fit model parameters: σ^* the central velocity dispersion and truncation radius r_t^* for a typical L^* cluster member (see illustrative curves for A2218 in Fig. 2). [2] The mass-to-light ratios quoted here take passive evolution of the stellar content of elliptical/S0 galaxies into account modeled using the stellar population synthesis models of Bruzual & Charlot (1993).

RESULTS ON SUBSTRUCTURE IN CLUSTERS

We have statistically extracted characteristic parameters for typical L^* cluster galaxies that inhabit massive, dense lensing cluster-lenses ranging in redshift from 0.17–0.58. This has been achieved by combining strong and weak lensing *HST* observations in conjunction with an assumed parametric mass model. We find that the inferred mass distribution of a fiducial L^* is extremely compact, although the inferred r_t^*'s lie well outside the optical radii and correspond to roughly between 5–$10R_e$. Our analysis also shows that the halos of individual cluster galaxies contribute at most 10–20% of the total mass of the cluster within the central 1 Mpc, covered by the *HST WFPC2* imaging using the results of our likelihood analysis alongwith the best-fit parameters that characterize the smooth clump. Therefore, in the inner regions of these clusters the bulk of the dark matter is in fact smoothly distributed. Similar lensing studies of field galaxies, e.g. Wilson et al. (2001), typically find a non-zero signal for the radially averaged stacked

---

[2] The mass obtained for a typical bright cluster galaxy by Tyson et al. (1998) using only strong lensing constraints inside the Einstein radius of the cluster Cl 0024+16 is consistent with our results.

tangential shear out to 200 kpc. In contrast our study of the halos of galaxies in clusters detects a finite r_t^*, which we attribute to the tidal truncation induced by the motion of these cluster galaxies inside the potential well. From the contours in the likelihood plots, the presence of field galaxy scale dark halos can, in fact, be excluded at $> 10\text{-}\sigma$ significance (Natarajan, Kneib & Smail 2002).

The clusters we study here are all rich systems spanning a range in central density, which may explain why the best-fit values of r_t^* obtained vary by a factor of 2–3. To test this suggestion we plot in Fig. 3 the variation of the central density of the cluster dark matter with r_t^*/σ^* based on our lens models and evaluated at the cluster core radius. We see a good correlation and derive a best-fit slope of -0.6 ± 0.2. This compares well with the theoretical expected value from a tidal stripping model (Merritt 1983) of -0.5:

$$r_t^* \approx 40 \left(\frac{\sigma_*}{180\,\text{km\,s}^{-1}}\right)\left(\frac{\rho_0(r_c)}{3.95 \times 10^6 M_\odot\,\text{kpc}^{-3}}\right)^{-\frac{1}{2}} \text{kpc}. \qquad (2)$$

Dark halos of the scale detected here indicate a high probability of galaxy encounters over a Hubble time within a rich cluster. However, since the internal velocity dispersions of these cluster galaxies ($< 250\,\text{km\,s}^{-1}$) are much smaller than their orbital velocities, these interactions are unlikely to lead to mergers, suggesting that the encounters of the kind simulated in the 'galaxy harassment' picture (Moore et al. 1996) are frequent and likely. In fact, high resolution cosmological N-body simulations of cluster formation and evolution (Ghigna et al. 1998; Moore et al. 1996; Okamoto & Habe 1999), find that the dominant interactions are between the global cluster tidal field and individual galaxies after $z = 2$. The cluster tidal field significantly tidally strips galaxy halos in the inner 0.5 Mpc and the radial extent of the surviving halos is a strong function of their distance from the cluster center. Much of this modification is found to occur between $z = 0.5\text{--}0$.

The prospects for extending this technique to larger scales within clusters in order to study the efficiency of halo stripping as a function of radius (variation of r_t^* as a function of radius) and morphological type are very promising with data from the *Advanced Camera for Survey* on *HST*.

CONSTRAINTS ON THE NATURE OF DARK MATTER

The truncation radii of galaxy halos in clusters provides clues to the dark of dark matter, whether is collisionless or self-interacting (i.e. collisional). If the dark matter is collisionless, then the sizes of galaxy halos in clusters are primarily shaped by the dynamical effects of tidal truncation and collisional stripping (Ghigna et al. 1998). The dominant process of tidal truncation acts on the short orbital time-scale due to the tidal field of the cluster as a whole (Taylor & Babul 2001), while collisional stripping results from binary interactions among individual galaxies (Binney & Tremaine 1987). The global tidal field of the cluster truncates the dark matter halo of each galaxy at a radius inside of which the mean mass density of the galaxy is roughly equal to the mean interior density of the cluster,

$$<\rho_{gal}(r_t)> = <\rho_{clus}(r_i)>. \qquad (3)$$

However, if the dark matter is fluid-like then galaxy halos would be further stripped by ram pressure down to a radius which is significantly smaller than the tidal radius (roughly by the ratio between the velocity dispersion inside the galaxy halo and the velocity of the galaxy through the cluster; see Furlanetto & Loeb 2002 for further details),

$$<\rho_{clus}(r_i)> v_{g_i}^2 = \rho_{gal}(r_t)\sigma_{gal}^2 \qquad (4)$$

for FDM, where v_{g_i} is the velocity component relative to the cluster center-of-mass. This can be used to obtain an upper limit.

Here we use the truncation radii of a sample of cluster galaxies, inferred observationally from the galaxy-galaxy lensing in the cluster A 2218 (NKS02), to decide whether the dark matter is collisionless or fluid-like. We find that the inferred truncation radii in the cluster Abell 2218 are consistent with the tidal radii expected for collisionless dark matter, but rule-out fluid-like dark matter for which ram pressure stripping is effective (see Fig. 4 and Natarajan, Loeb, Kneib & Smail 2002). The transition between the collisionless and collisional regimes is set by the ratio between the mean-free-path of dark matter particles, λ, and the radius, r_t, of the galaxy halos under consideration. This ratio is given by,

$$\frac{\lambda}{r_t} \approx \frac{m_p}{\sigma_p \Sigma(r_t)}, \qquad (5)$$

where σ_p/m_p is the collisional cross-section per unit particle mass and $\Sigma(r_t)$ is the surface mass density of a galaxy halo at its truncation radius. The fluid regime is obtained for $\frac{\lambda}{r_t} \leq 1$. The characteristic surface mass densities of the galaxies in A 2218 can be directly inferred from the analysis of the lensing data. We find that for an L_* galaxy, $\Sigma_*(r_t) \approx 0.024$ g cm^{-2}. Since the fluid regime is ruled out, we exclude all values of $\sigma_p/m_p \geq 42$ cm^2 g^{-1}. Our constraints are complementary to those derived by Gnedin & Ostriker (2001) from considerations of thermal conduction in the mildly collisional regime. These authors exclude the regime $0.3 \leq \sigma_p/m_p \leq 10^4$ cm^2 g^{-1}, based on the consideration that cluster ellipticals will otherwise deviate from the fundamental plane beyond the observed scatter. Our new constraint allows us to rule out the high cross-section regime and hence we conclude that $\sigma_p/m_p \leq 0.3$ cm^2 g^{-1} as an absolute upper limit. Dark matter cross-sections higher than this upper limit were postulated by Spergel & Steinhardt (2000) in order to reconcile problems that cold dark matter models possess when compared to observational data (such as the abundance of galactic sub-halos and the slope of the inner mass density profile of galaxies; see further discussion in Dave et al. 2001; Yoshida et al. 2000; Stoehr et al. 2002; Miralda-Escude 2002).

FUTURE PROSPECTS FOR CLUSTER LENSING

The prospects for determining high-resolution mass maps of clusters by conducting systematic lensing analysis of X-ray selected high and low redshift clusters out to large radii are particularly promising with the ACS on HST. This will enable a better understanding of galaxy halo properties and in particular, their evolution in clusters. The currently underway wide-field surveys that target shear selected clusters, also enable

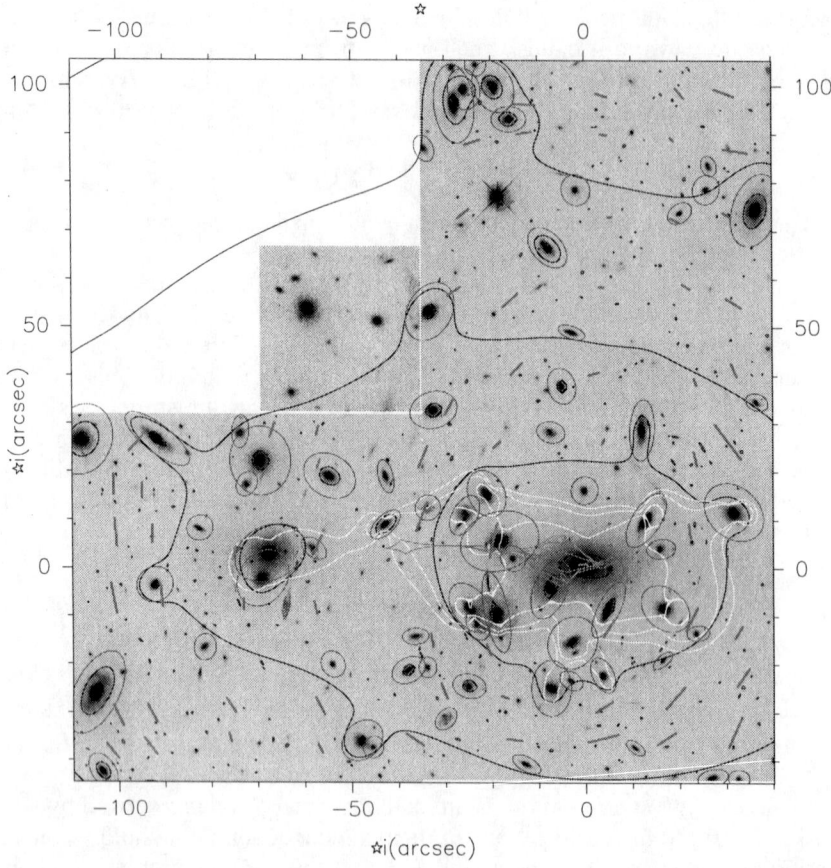

FIGURE 1. The best-fit mass model for the cluster A2218 at $z = 0.175$ overlaid on the HST-WFPC2 image. The equi-mass contours are over-plotted (solid lines), as are all the cluster galaxies that are used in the statistical galaxy-galaxy lensing analysis (ellipses).

probing the shapes of cluster density profiles and their 3D structure. With secure and tightly constrained mass models cluster cores can be deployed effectively as magnifying telescopes enabling zooming in on high redshift lensed galaxies to study their physical properties and dynamics which would otherwise be inaccessible.

ACKNOWLEDGMENTS

PN thanks her collaborators Jean-Paul Kneib, Ian Smail and Avi Loeb for permission to report results from joint work.

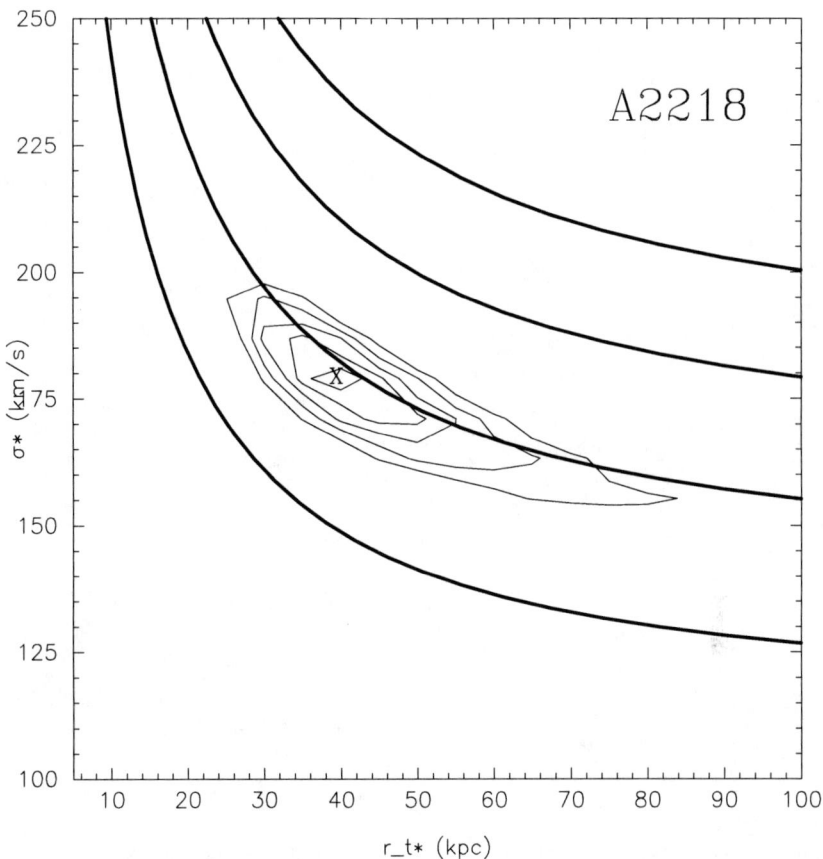

FIGURE 2. Results of maximum-likelihood analysis of lensing data for the cluster A 2218 (NKS02). The contour plot shows the best-fit values for the model parameters σ_* and r_{t*}, which are the central velocity dispersion and truncation radius for a typical L_* galaxy in the cluster. The likelihood contours are plotted in intervals of $1-\sigma$ starting from the inside out. The thick open curves are lines of constant enclosed mass.

REFERENCES

1. Binney, J., & Tremaine, S., 1987, Galactic Dynamics, (Princeton: Princeton U. Press), Ch. 7
2. Brainerd, T.G., Blandford, R.D., Smail, I., 1996, ApJ, 466, 623
3. Bruzual, G., Charlot, S., 1993, ApJ, 405, 538
4. Dave, R., Spergel, D., Steinhardt, P. J., & Wandelt, B. 2001, ApJ, 547, 574
5. Fischer et al., 2000, AJ, 120, 1198
6. Furlanetto, S., & Loeb, A. 2002, ApJ, 565, 854
7. Geiger, B., & Schneider, P., 1998, MNRAS, 295, 497
8. Ghigna, S., Moore, B., Governato, F., Lake, G., Quinn, T., Stadel, J., 1998, MNRAS, 300, 146
9. Gnedin, O., & Ostriker, J. P. 2001, ApJ, 561, 61
10. Kassiola, A., Kovner, I., 1993, ApJ, 417, 450
11. Kneib, J-P., Ellis, R.S., Smail, I., Couch, W.J., Sharples, R.M., 1996, ApJ, 471, 643
12. McKay, T., et al., 2002, ApJ, 571, L85

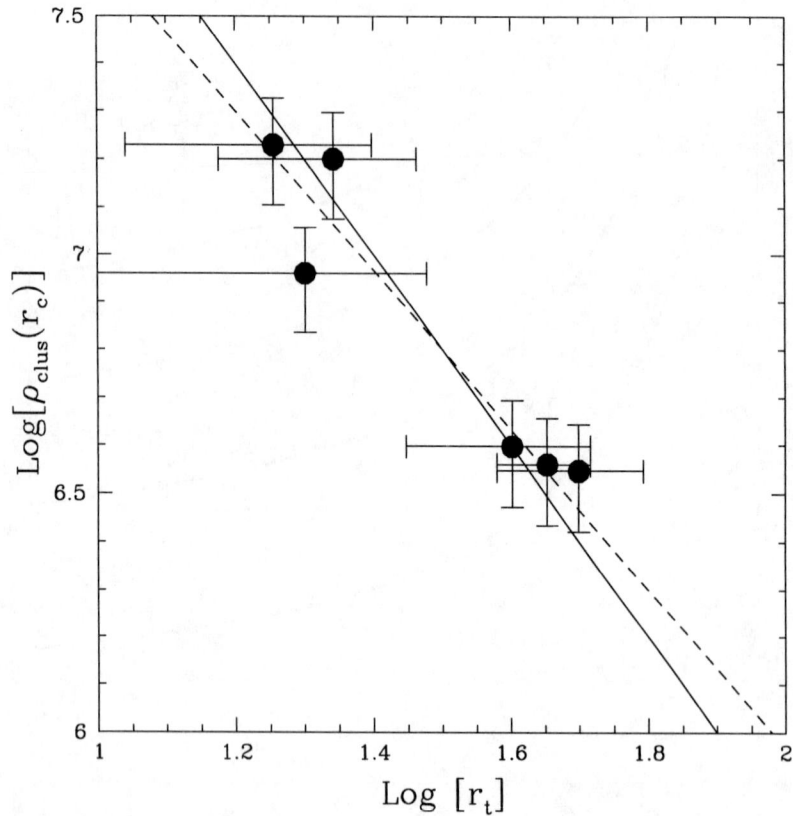

FIGURE 3. Scaling out the variation in the σ^*'s, the central density of the cluster evaluated at the core radius is plotted against the tidal radius. The errors plotted in r_t are the 3-σ values. Performing a least-squares fit to the data points, the value of the best-fit power-law index η, where $r_t^* \propto \rho^\eta$, is estimated to be -0.6 ± 0.2, (dashed line) in excellent agreement with theoretical expectations of $\eta = -0.5$ (solid line).

13. Merritt, D., 1983, ApJ, 264, 24
14. Miralda-Escude, J. 2002, ApJ, 564, 1019
15. Moore, B., Katz, N., Lake, G., Dressler, A., Oemler, A., 1996, Nature, 379, 613
16. Natarajan, P., Kneib, J.-P., 1997, MNRAS, 287, 833 [NK97]
17. Natarajan, P., Kneib, J.-P., Smail, I., Ellis, R.S., 1998, ApJ, 499, 600 [NKSE98]
18. Natarajan, P., Kneib, J.-P., & Smail, I., 2002, ApJ, 580, L11 [NKS02]
19. Natarajan, P., Loeb, A., Kneib, J.-P., & Smail, I., 2002, ApJ, 580, L17
20. Okamoto, T., Habe, A., 1999, ApJ, 516, 591
21. Smail, I., et al., 2001, ApJ, 323, 839
22. Smith, G.P., et al., 2001a, ApJ, 552, 493
23. Smith, D.R., Bernstein, G., Fischer, P., Jarvis, M., 2001b, ApJ, 551, 643
24. Spergel, D., & Steinhardt, P. J. 2000, Phys. Rev. Lett., 84, 17, 3760
25. Stoehr, F., et al. 2002, MNRAS, 335, L84
26. Taylor, J. E., & Babul, A. 2001, ApJ, 559, 716
27. Tyson, J.A., Valdes, F., Jarvis, J.F., Mills, A.P., 1984, ApJ, 281, L59

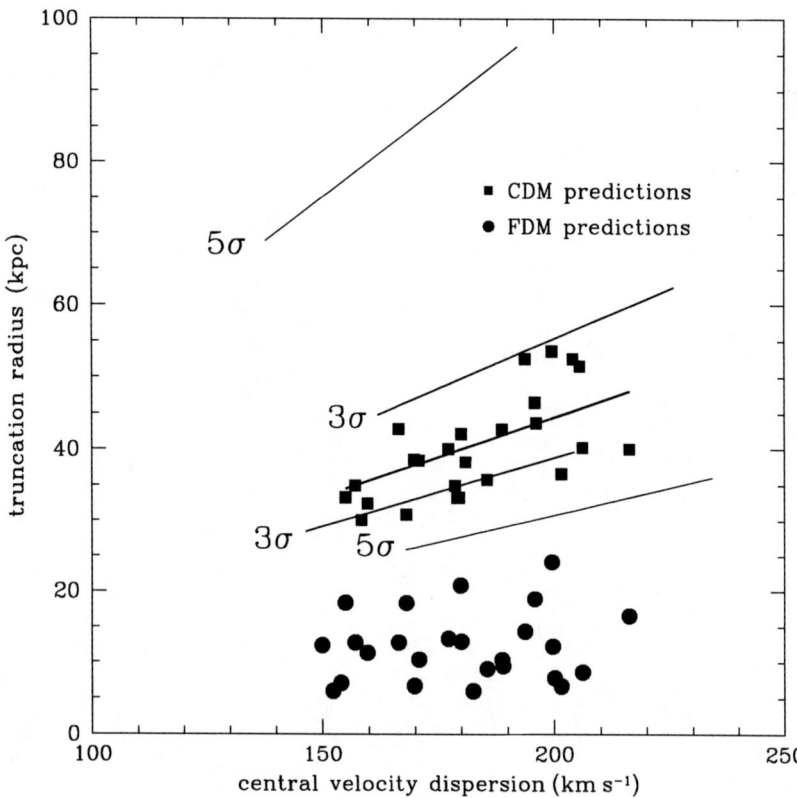

FIGURE 4. Distribution of truncation radii as inferred from the lensing analysis of 25 galaxies superimposed on the scaling relation (thick solid line) in the cluster A 2218. The 3-σ and 5-σ lines denote the corresponding confidence levels in the parameters r_{t*}, and σ^* obtained for a typical L_* galaxy in this cluster (derived from the confidence level contours in Figure 1). The expected distribution of tidal radii for fluid-like dark matter (FDM; solid circles) and collisionless cold dark matter (CDM; solid squares) are also shown. Note that the linear relation between truncation radius and velocity dispersion for the values inferred from lensing is a consequence of the assumed scaling laws with galaxy luminosity.

28. Yoshida, N., Springel, V., White, S. D. M., & Tormen, G. 2000, ApJ, 535, L103
29. Wilson, G., Kaiser, N., Luppino, G.A., Cowie, L.L., 2001, ApJ, 555, 572

Galaxies and Halos in the Sloan Digital Sky Survey

Timothy A. McKay* and the SDSS Collaboration[†]

*Department of Physics, University of Michigan, 500 East University, Ann Arbor, MI, 48109
[†]www.sdss.org

Abstract. Structure formation theory provides very effective predictions of the properties of dark matter halos, including their mass function, clustering, and internal structure. Observations of structure, however, rely on luminous galaxies as tracers. A detailed understanding of the way galaxies occupy dark matter halos is essential for connecting structure formation theory to observation. We describe some of the observables available for contraining the halo occupancy, illustrating each using data from the Sloan Digital Sky Survey. Many of these observables can now be measured with great statistical precision. Comparison of these observables to theory is now limited by systematic uncertainty in the relationship between observable quantities (like velocity dispersion vs. cluster richness) and theoretically favored quantities (like M_{200}). We argue for the use of carefully crafted simulations in making this connection, and illustrate their use in some example analyses.

PRELIMINARIES

We now have a reasonably well established framework for understanding the formation of structure in the universe. At the time of recombination we begin with a nearly uniform fluid composed of weakly interacting cold dark matter (perhaps 90%) and ordinary baryons (perhaps 10%). Very small density fluctuations are imprinted on this fluid, perhaps by quantum fluctuations generated in the Big Bang. These fluctuations are then amplified by the presence of gravitationally interacting matter. This amplification is dominated by dark matter.

As the Universe evolves dark matter halos begin to separate from the overall Hubble flow. These dark matter halos, in one way of accounting, *are* the structure. If we knew the mass, location, and internal structure of all the halos, we would have a complete description of the matter distribution. Identifying these dark halos, and probing their properties, is a major goal.

Within dark matter halos, a radiative differentiator acts. While the dark matter is thought to interact only gravitationally, the minority component of baryonic material can radiate away its energy: cooling and sinking to the center of the potential well until it is rotationally supported. Within this baryonic core stars form, evolve, and feed back material and entropy into the surrounding halo, forming galaxies. It is these galaxies, and not the halos, which we observe. In this way of thinking, they are labels, lovely markers hinting at the presence of much more extensive dark matter halos.

Connecting Galaxies with Halos

To connect our observable, the locations and motions of luminous galaxies, to theoretical predictions, mostly of the properties of dark matter halos (for example [1]), we need to understand the relationship between the luminous properties of galaxies and their dark matter environments.

To first order, this relationship can be described as bias. In one form the bias is defined as the ratio of the galaxy-galaxy and mass-mass correlation functions. On large scales, greater than a few Mpc, bias is observed to be simple [2], as expected. More complex behaviour is generically expected on smaller scales. A kind of conspiracy of effects converts the complex mass-mass correlation function to the simple unbroken power law observed in the galaxy-galaxy correlation function [3].

This is just the first order picture, and a more complete portrait, including the dependence of this bias on galaxy and halo properties, contains substantial information about galaxy formation, and perhaps about the nature of dark matter.

The Halo-Occupancy Distribution Function

Both analytic theory and N-body numerical experiments can predict the properties of dark matter halos, including their mass function, their clustering properties, and their internal structure. Comparing this theory to observation requires an understanding of how galaxies 'occupy' these halos.

One useful way to describe the relationship between halos and the galaxies which populate them relies on the 'halo occupancy distribution function' $P(N|M)$. This function describes the probability that a halo of mass M will host a total of N galaxies. This function, together with some information about spatial and velocity bias between the dark matter and galaxies within each halo, provides a useful way of thinking about bias [4]. It provides an essentially complete description of the relationship between the distribution of galaxies we observe, and the distribution of dark matter we are so eager to learn. Given the details of this HOD, it is possible to calculate essentially any observable of large- scale structure.

The approach outlined here is pretty sketchy. For example, if we're going to count galaxies, what will count? Is this a luminosity limited sample? In what band? How should we identify the dark matter halos, how should we define their masses? In the end, this initial HOD description will have to be expanded to accomodate the differing occupancy of different types of galaxies.

The tools available for constraining the HOD range from the N-body simulations of dark matter to observations of real galaxies. The simulations can inform our interpretations of observations at an important level, helping us to understand the effects of projection and to determine our selection functions. The observations obviously provide feedback to the simulations, providing direct HOD constraints, measuring various scaling relations on halo scales, and determining the dependence of these things on galaxy properties.

The process of constraining the HOD will have to include an iterative dance among all these elements, revising models and making new observations until all the pieces fit together.

IDENTIFYING HALOS AND MEASURING MASSES

To generate constraints on the HOD, we must first identify halos. We describe here measurements based on two approaches. In the first, we identify halos with individual galaxies, labeled with their properties including luminosity, type, and environment. This is surely a very first order method. While at low halo mass there is a roughly one-to-one correspondence between bright galaxies and halos, this is not true at higher mass, where individual halos host groups and clusters of galaxies.

A second way to identify halos is to use group and cluster finders, which label the halos with galaxy content. This method is closer to the proper spirit of the HOD, but has the drawback of being sensitive only to relatively massive halos.

The details of these halo identifications need to be understood. The efficiency and purity of their selection has to be known as a function of redshift, and its dependence on the selection of tracer galaxies must be known. It's also very important to understand how the centers for these halos are identified, and how their spatial extent is constrained. Analysis of simulated universes can provide important insight here.

Once halos are identified, we need to measure their masses. We have two basic probes of mass. In dynamical measurements of mass, the observables are the positions and velocities for a set of luminous test particles. These probe the dynamical effect of gravity on the test particles, which sample the velocity field around the halo. In lensing measurements of mass, the observable is a shear field and the geometry associated with it. Lensing probes the space-time geometry around the objects of interest. It provides a probe of the projected galaxy-mass correlation function around halos.

Note that neither measures mass very directly. Inferring masses from these measurements in either case requires careful consideration of a variety of effects. Of particular importance is our choice of probes. Since the relationship between galaxy properties and mass is strong, our choice of test particles, and of lens and source galaxies affects strongly what we observe. All of that rich behavior needs to be understood. In both cases, measurements of these halo mass probes can now be made with high signal-to-noise. Essentially all the work from now on lies in accurately understanding how these precisely determined observables relate to theoretically well determined quantities, like halo mass function and structure.

SDSS HALO STUDIES

The data which enable the studies we describe come from the Sloan Digital Sky Survey. The SDSS is a large collaboration, involving perhaps 200 scientists at a number of institutions. It is designed to make comprehensive astronomical observations. Over the coming few years the SDSS will complete an imaging survey of 10^4 square degrees of the sky. Images are obtained in 5 colors for about 10^8 galaxies. In addition to imaging, the SDSS will measure high quality spectra for about 10^6 galaxies and 10^5 quasars. This set of observations will support a very broad range of science goals, in much the same sense that data collected by large high energy physics experiments is useful for many purposes. We expect SDSS data to be an important tool for astrophysics for several decades. For the analyses described here, both the imaging and spectroscopic data play an important role.

Halos identified by individual galaxies

We begin with the lensing studies, measuring the correlation between locations of foreground galaxies and distortion in the shapes of distant sources. The SDSS data used for this study are drawn from the Early Data Release [5]. They include imaging and spectroscopic data for about 4% of the SDSS survey region. From these data, we select a sample of 34,693 foreground 'lens' objects. Every one of these objects has a spectroscopic redshift and highly accurate 5 color photometry. For this purpose they are drawn entirely from the SDSS 'main' galaxy sample. We also select a fainter background sample of 3,615,718 'source' objects.

Details of the lens and source sample selection, and the subsequent analysis, can be found in McKay et al. [6]. While the foreground redshift distribution is accurately measured, with a median redshift of 0.1, the background source galaxy redshift distribution is estimated from the magnitude distribution. Using these samples we measure the galaxy-mass correlation function $\xi_{GM}(r)$ around our lens galaxies. It is important to note that the signal we are measuring is extremely small. The peak distortion is only about 0.5%. Despite this tiny signal, the $\xi_{GM}(r)$ is detected at >13 σ in the g, r, and i bands. The observed $\xi_{GM}(r)$ is well fit by a power law of the form:

$$\xi_{GM}(r) = (2.5 \pm 0.7 h M_\odot pc^{-2}) \times (r/1Mpc)^{-0.8 \pm 0.2} \quad (1)$$

Now that we have detected $\xi_{GM}(r)$, we can begin to study how it varies with the luminous properties of the lens galaxies. As a first check, we divide all lens galaxies into four luminosity bins in each of the five SDSS colors. We then compare ξ_{GM} in each luminosity bin to probe mass-to-light scalings.

To characterize the variation of $\xi_{GM}(r)$ with lens luminosity, we fit the measured $\xi_{GM}(r)$ from 20-260 h^{-1} kpc with a singular isothermal sphere model. For this best fit model we integrate the associated mass out to 260 h^{-1} kpc and call this M_{260}. This outer radius is chosen because contributions to $\xi_{GM}(r)$ due to neighboring galaxies are estimated to be less than 10% at this radius. We then examine how this parameter varies with luminosity. Figure 1 shows for each color the actual M_{260} to light scaling, and then χ^2 contours for the best fit normalization and power law index in each color.

There is little relationship between M_{260} and luminosity in u. This is not surprising as the u luminosity of a galay is often dominated by recent, short-lived, bursts of star formation, and hence does not well reflect the galaxy's mass. But the relationship between M_{260} and luminosity in the other bands is strong, and in every case consistent with linear. In this case, the normalization can be described as a mass-to-light ratio. For the i band the best fit value for this is $M_{260}/L_i = 124 \pm 15 M_\odot/L_\odot$

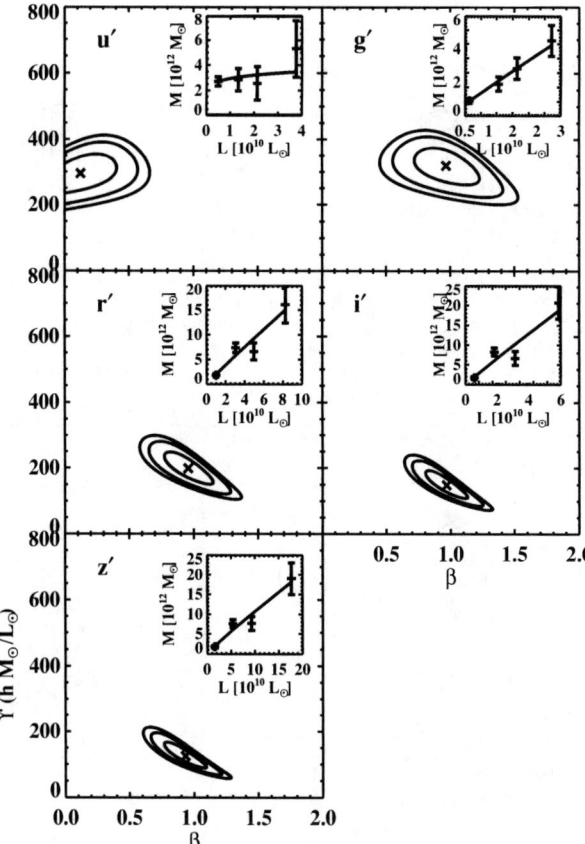

FIGURE 1. The five panels in this figure summarize the relation between M_{260} and luminosity in each of the five SDSS bands. For each band the small inset figure shows this directly. Points in these inset figures are the measured M_{260} and mean luminosity of galaxies in four luminosity bins. The line in these inset figures shows the best fit to a power law relation between M_{260} and luminosity of the form: $M_{260} = \Upsilon \times \left(L_{central}/10^{10}L_\odot\right)^\beta$. The larger figure shows 68%, 95%, and 99% confidence contours for the fit parameters Υ and β.

It is important to be cautious in interpreting any such measurement of a mass-to-light ratio. Mass is not the observable quantity. Masses are derived only under rather naive assumptions (a singular isothermal sphere mass model) which while consistent with the data, are not well constrained by it. What we really have is is a measurement of the scaling between luminosity and a fit parameter of a model. If we fit to different models, we may find different scalings. But it is clear at least that the mass of halos on these large scales varies with luminosity.

It is useful to test the conclusions of these lensing measurements using dynamical mass probes [7]. We begin by identifying a set of luminous test particles in orbit around galaxies. In order to identify simple systems, we look for a set of relatively isolated host

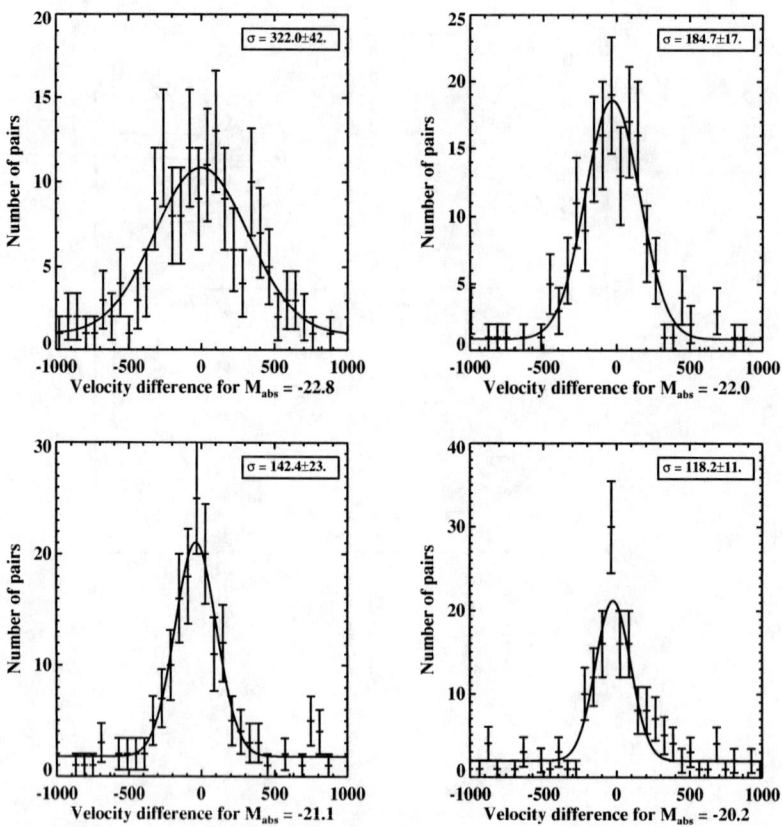

FIGURE 2. This figure shows velocity difference histograms for faint satellites of four groups of relatively isolated host galaxies. The hosts are grouped by absolute magnitude, ranging from rather $M_{abs} = -20.2$ at the lower right to $M_{abs} = -22.8$ at the upper left. The increase in satellite dispersion with host luminosity is clear.

galaxies surrounded by fainter, less luminous, satellites. Each host galaxy has only a few satellites, so we can only measure the average dynamical response of a class of satellites to their hosts.

By constructing a velocity difference histogram for a class of galaxies, we probe the velocity structure of the galaxy-galaxy correlation function $\xi_{GG}(r, \Delta v)$. This velocity structure represents the average dynamical effect of the hosts in the same way that $\xi_{GM}(r)$ represents their average projected surface mass density. We find that the velocity difference width increases significantly with host galaxy luminosity (see Figure 2). While deriving masses from these velocity distributions is quite model dependent, a simple virial mass estimate yields mass-to-light scalings consistent with those derived from lensing [7].

It is reassuring that these two completely different ways of probing mass reveal comparable relationships between mass and light. They are subject to totally different systematic errors, and to quite different problems in interpretation. While this comparison is only a first step, it shows that these combined methods hold great promise for quantifying the relationship between galaxies and their dark matter environments. The signals are there. We have only to understand the systematics.

Halos identified by groups and clusters

These studies can be expanded to halos identified by groups and clusters in a straightforward way. To do this requires a catalog of groups and clusters, something which is not a standard output of the SDSS analysis system.

For this purpose we have been using group and cluster catalogs derived using the maxBCG method [8, 9]. This method takes advantage of the very uniform colors (the E/S0 ridgeline) of galaxies found in many groups and clusters. Since this color shifts with redshift, groups and clusters appear as overdensities of objects in position color space in a way which is remarkably insensitive to projection effects. This tight correlation in color allows cluster members to be indentified and counted. This estimated number of cluster members, N_{gal}, provides an estimate of cluster richness. The color itself provides accurate estimates of redshift.

This method is observationally very robust. It's redshift estimates are very good, better than 0.02, and this N_{gal} richness measure is a very well defined count of easily identifiable objects. While it is not yet clear what fraction, or what subclass, of groups this method finds, we have confirmed, by comparison to x-ray selected and other optically selected cluster lists, that it is quite complete in its identification of clusters.

This maxBCG method, fits neatly into the halo occupancy distribution picture of structure formation. In this picture, N-body simulations are used to determine the halo mass function. Then galaxies are included via a halo occupancy distribution function P(N|M), which describes the probability of having N galaxies in a halo of mass M. The maxBCG catalog provides a very clean measure of N, nearly unaffected by projection. It's also complete to roughly z=0.4.

We can now use the same lensing and dynamical measures to probe the relationship between M_{avg} and N. We begin with measurements of the cluster-mass correlation function. Although clusters are rare and our samples are small, they are also massive. So the S/N in cluster shear measurements is very similar to the S/N in galaxy-galaxy lensing measurements [10]. This is generically true across the mass spectrum, so we can use lensing to probe halos of a wide range of masses.

Figure 3 shows the dependence of the cluster-mass correlation function on cluster richness. To determine this, we fix the r dependence by fitting $\xi_{cluster-mass}(r)$ to an SIS model, and examine changes in the normalization with cluster richness. This amounts to measuring an effective velocity dispersion for the clusters. As with the galaxies, the signal is clearly there to trace the variation in the cluster-mass correlation function with richness N_{gal}.

To supplement this with dynamical measurements we determine the cluster-galaxy

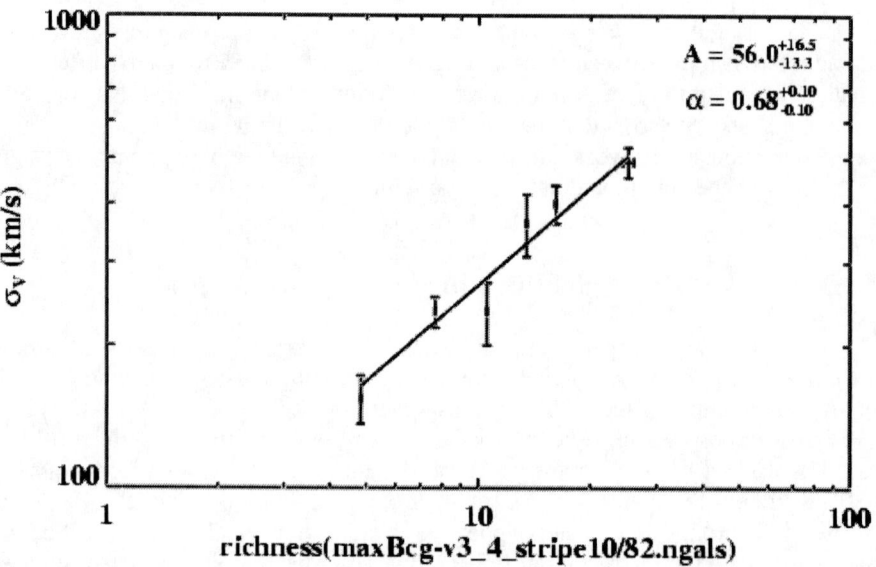

FIGURE 3. This figure shows the variation of an effective velocity dispersion derived from lensing estimates on cluster richness. The effective σ_v is derived by fitting $\xi_{cluster-mass}(r)$ to an SIS model out to a radius of 500 h^{-1} kpc. The parameters A and α represent best fits for the normalization and slope of a power law relation between σ_v and N_{gal}.

correlation function by finding maxBCG objects for which a spectrum was taken of the central galaxies. We then search around these BCGs for spectroscopic neighbors. Then, in a manner analogous to what we did for the galaxies, we make velocity difference histograms in narrow ranges of richness.

A clear variation of the width of this velocity difference histogram with richness is observed. Both of the cluster mass probes I described measure some kind of effective velocity dispersion. Both show a smooth transition from groups with ~200 km/s velocity dispersions to clusters with velocity dispersions of 900 km/s and more. The estimates from velocity dispersion and lensing are consistent at the 20% level. Both are detected at high S/N, so in principle they can yield precise constraints on richness-mass calibrations. In practice, details of the changing size of objects, their velocity structure, and possible velocity biases need to be understood to take full advantage of this.

Utilizing observable simulated universes

We clearly have a set of experimentally accessible observables which probe the relationship between halos and mass. Imagine that we have a simulation of the universe made with rich enough physics input to reasonably represent observable data. That is, imagine a simulated universe which contains not only dark matter, but luminous galaxies in something like their full variety.

Given such simulations, we can repeat the same 'observations' done in the real universe in an environment which contains all the physics we believe is relevant. We can then compare various predictions to reality at the level of the observables, rather than interposing models which involve untested or patently incorrect assumptions. The remainder of this proceeding describes some early examples of the kinds of comparisons between observations and simulations which we advocate.

The first example involves the GIF simulations [11]. These simulations are built on top of the VIRGO consortium N- body simulations. They add to the N-body outputs by identifying galaxies with the most massive subhalos. Semianalytic prescriptions are used to provide luminosities, colors, and stellar masses for all of these galaxies.

Most important for this study, each of these galaxies has velocity information derived from the full N-body simulation. This allows us to conduct the same dynamical analysis in the simulated data used in the real data. We can compare 'predictions' from the simulations to observations at the observable level (variation of velocity dispersion with luminosity) rather than at the level of model fits. Furthermore, we can use the simulations to tell us how the observables relate to the 'real' masses of the systems.

There are some important limitations, mostly that these simulations include only the most massive and luminous galaxies. So overlap with the observations is not as complete as we would like. Nevertheless, the variation in the width of the velocity difference histogram with host luminosity seen in the GIF simulation is consistent with that seen in SDSS data [7].

What is most useful about doing this in simulations is that we can directly probe the way in which an observable (like this M_{260}) relates to a theoretically interesting quantity, like M_{200}, the mass measured out to the point where the overdensity is 200 times the mean density. The GIF analysis suggests that these satellite dynamics are indeed probing masses on halo scales, at least up to a scale factor (about 0.7 in this case). There are many reasons to be cautious in asserting this, but especially because the GIF host luminosity range is quite narrow, only about a factor of three.

We are also developing simulated universes aimed at our measurements of halos identified by groups and clusters. For this work, Risa Wechsler has built simulations on the Hubble Volume simulations which are designed to very specifically match observed SDSS galaxies. That is, rather than use semianalytic prescriptions to 'grow' galaxies, she inserts the galaxies in ways which are constrained by data.

The basic algorithm has several steps: Choose an appropriate number of galaxies, with r-band luminosities drawn from the observed SDSS luminosity function. For each galaxy, choose a mass particle in the simulation in a way which matches the observed luminosity dependent clustering seen in the SDSS data. Add passive luminosity evolution, and assign galaxy colors by selecting a real SDSS galaxy with similar luminosity and local density, then translating the SED of this galaxy to the simulated galaxy redshift.

This method produces simulated clusters with properties (like the E/S0 ridgeline) which are remarkably similar to real clusters. By running the maxBCG algorithm on these simulated universes, we can in principle calibrate the relation between richness N_{gal} and mass. But to apply this, we must be certain that the N_{gal} counted in the simulation really reflects what we count in the SDSS data. To check this we are conducting a number of tests. First, we can try matching the halo number distribution and space density to translate N_{gal}^{sim} to N_{gal}^{data}. This is somewhat problematic though, as we'd like to use the cluster number density to constrain cosmology.

As a result it's better to use the cluster-galaxy and cluster-mass correlation functions as constraints. We note in passing that a number of these comparisons are complex. Our simulations do not currently include any special status for a brightest cluster galaxy, despite evidence that they have very unusual dynamics and locations. The simulations also lack higher order galaxy property correlations, velocity bias and so on. Much work remains, but progress is being made pretty rapidly.

CONCLUSIONS

If we are to determine the distribution of matter in the universe, we must understand in some detail the relationship between the dark matter halos which dominate the mass budget and the luminous galaxies which illuminate them. We must use the locations of galaxies to identify halos, after which we can study their properties by both lensing and dynamical means.

Relating the observables, $\xi_{GM}(r)$ and $\xi_{GG}(r,\Delta v)$, to halo mass requires modeling. Even with this small subset of SDSS data, uncertainty in this modeling already dominates our ability to interpret these results. As we make ever more precise, systematic uncertainty in this modeling will continue to limit our ability to interpret the results.

The most straightforward predictions of structure formation theory describe the halo mass function, clustering, and structure. Since these are not observables, comparison of observations to theory relies on ill-determined intermediate modeling steps. To avoid this step, we must propagate theoretical predictions forward to the observable level. This can be done by including galaxies, with all their observable properties, in simulations of structure formation. Initial attempts at this kind of prediction are provided by, for example, the GIF simulations [11].

We describe some example comparisons between SDSS observations and predictions of observables made using such simulations. While important details in the simulations remain to be checked, initial results are very encouraging. It seems likely that this method, comparing observations to theory at the observable level through structure formation simulations which include galaxies, will play an important role in the interpretation of precise new observations of the distribution of matter in the universe.

ACKNOWLEDGMENTS

Funding for the creation and distribution of the SDSS Archive has been provided by the Alfred P. Sloan Foundation, the Participating Institutions, the National Aeronautics and Space Administration, the National Science Foundation, the U.S. Department of Energy, the Japanese Monbukagakusho, and the Max Planck Society. The SDSS Web site is http://www.sdss.org/.

The SDSS is managed by the Astrophysical Research Consortium (ARC) for the Participating Institutions. The Participating Institutions are The University of Chicago,

Fermilab, the Institute for Advanced Study, the Japan Participation Group, The Johns Hopkins University, Los Alamos National Laboratory, the Max-Planck-Institute for Astronomy (MPIA), the Max-Planck-Institute for Astrophysics (MPA), New Mexico State University, University of Pittsburgh, Princeton University, the United States Naval Observatory, and the University of Washington.

REFERENCES

1. Jenkins, A. et al. 1998, Astrophysical Journal, 499, 20
2. Verde, L. et al. 2002, Monthly Notices of the Royal Astronomical Society, 335, 432
3. Benson, A. J., Cole, S., Frenk, C. S., Baugh, C. M., & Lacey, C. G. 2000, Monthly Notices of the Royal Astronomical Society, 311, 793
4. Berlind, A. A. & Weinberg, D. H. 2002, Astrophysical Journal, 575, 587
5. Stoughton, C. et al. 2002, Astronomical Journal, 123, 485
6. McKay, T., et al. 2001, astro-ph/0108013
7. McKay, T. A. et al. 2002, Astrophysical Journal Letters, 571, L85
8. Gladders, M. D. & Yee, H. K. C. 2000, Astronomical Journal, 120, 2148
9. Annis, J., et al. 2003, in preparation
10. Sheldon, E. S. et al. 2001, Astrophysical Journal, 554, 881
11. Kauffmann, G., Colberg, J. M., Diaferio, A., & White, S. D. M. 1999, Monthly Notices of the Royal Astronomical Society, 303, 188

The Dark Matter Distribution in Galaxy Cluster Cores

J.S. Arabadjis*, M.W. Bautz† and G. Arabadjis**

*Massachusetts Institute of Technology, Center for Space Research,
70 Vassar Street, Room 37-501, Cambridge, MA 02139
†Massachusetts Institute of Technology, Center for Space Research,
70 Vassar Street, Room 37-521, Cambridge, MA 02139
**Mitre Corporation, 202 Burlington Road, Bedford, MA 01730

Abstract.
Determining the structure of galaxy clusters is essential for an understanding of large scale structure in the universe, and may hold important clues to the identity and nature of dark matter particles. Moreover, the core dark matter distribution may offer insight into the structure formation process. Unfortunately, cluster cores also tend to be the site of complicated astrophysics. X-ray imaging spectroscopy of relaxed clusters, a standard technique for mapping their dark matter distributions, is often complicated by the presence of their putative "cooling flow" gas, and the dark matter profile one derives for a cluster is sensitive to assumptions made about the distribution of this gas. Here we present a statistical analysis of these assumptions and their effect on our understanding of dark matter in galaxy clusters.

Introduction

The cold dark matter (CDM) paradigm of modern cosmology has enjoyed spectacular success in describing the formation of large-scale structure in the universe [1, 2, 3, 4]. Galaxy-scale dark matter halos, however, exhibit several apparent inconsistencies with CDM, for example: the number of Milky Way satellites appears to be at least an order of magnitude lower than CDM predictions [5, 6, 7], and dark matter halos in dwarf and low surface brightness galaxies are much less cuspy than in CDM simulations [8, 9, 4]. Some reports [11, 12] even suggest that CDM fails on galaxy cluster scales for some clusters, but the latter are controversial [13, 14].

If CDM does indeed require alterations, there is no shortage of tailors. Proposed modifications include, though are not limited to, self-interacting dark matter [15, 16], warm dark matter [17], annihilating dark matter [18], scalar field dark matter [19, 20], and mirror matter [21], each of which is invoked to soften the core density profile. Many of these modifications will soften the core profile of galaxy clusters as well, although other astrophysical processes such as the adiabatic contraction of core baryons [22] may ameliorate this effect. Baryons, however, introduce a host of complications to CDM simulations; their effects will require a great deal of effort to disentangle [23].

In order to discriminate among CDM, its modifications, and other astrophysical influences, we have initiated a program to map the dark matter profiles of a sample of galaxy cluster cores. We use imaging spectroscopy from *Chandra X-ray Observatory* observations [24] and archival data [25] to extract the deprojected radial dependence of the baryon density and temperature of each cluster. In order to extract a dark matter profile from spatially resolved X-ray spectroscopy one usually assumes that the galaxy cluster is spherically symmetric and in hydrostatic equilibrium, and so for the most part we have restricted our sample to clusters for which these assumptions are most likely valid. Unfortunately, these clusters often contain "cooling flow" gas in their cores which complicates the spatial and spectral models. How one models the X-ray emission from this cool gas can significantly alter the resulting dark matter profile. If the model contains only a single emitting component (at temperature T and density ρ) at each radius, the inferred temperature profile will tend to dip significantly toward the center of a cooling flow cluster. If, however, gas in the the central few radial bins contains a second (cooler) component which is cospatial and isobaric with the first, the hot gas temperature profile remains flat into the core. The latter case tends to produce a larger central mass (see Figure 1A) and steeper density profile than the the former. Our problem, then, is the age-old exercise of choosing between a simple model 0^s and a complex model 0^c. Once we have done this, we will be better able to address the other issues listed above.

Models

In order to ascertain the importance of a second emission component we adopt a simplified geometry containing only two spherical shells (inner = 1, outer = 2). In both models (0^s and 0^c), shell 2 contains a (hot) thermal plasma at temperature T_{2h} and density ρ_{2h}. Model 0^s contains only one emission component in shell 1, characterized by a temperature T_{1h} and a density ρ_{1h}, whereas model 0^c contains a hot and a cool emission component in shell 1, described by T_{1h}, ρ_{1h}, T_{1c}, and ρ_{1c}. The X-ray emission from each component is modelled spectroscopically as using the MEKAL [26, 27, 28, 29] model as implemented in the XSPEC software package [30]. The best-fit parameter values of each model are calculated using a χ^2 minimization routine. Hereafter we will refer to the simple and complex model parameters using vectors θ^s and θ^c, respectively; i.e., $\theta^s = (T_{1h}, \rho_{1h}, T_{2h}, \rho_{2h})$ and $\theta^c = (T_{1h}, \rho_{1h}, T_{1c}, \rho_{1c}, T_{2h}, \rho_{2h})$.

At this point, tradition dictates that we employ a statistical test from the standard arsenals [31, 32, 33], such as the likelihood ratio test or the F-test, to choose between the two models. However, since θ^s lies on a boundary of θ^c (with $\rho_{1c} = 0$), these tests cannot be employed [34]. Instead, we construct an *empirical F-distribution* using Markov Chain Monte Carlo (MCMC) sampling, and gauge the significance of the complex model from the location of the F value of the data within that distribution [34].

Constructing an empirical F-distribution

Starting with the best-fit parameters θ_0 of model 0^s, we sample the 4D parameter space in its vicinity using MCMC sampling. This is done by running a Tcl script within XSPEC which calculates the probability distribution function P of a trial perturbation θ_1 about θ_0 given the observed data. The trial point is chosen using the trial distribu-

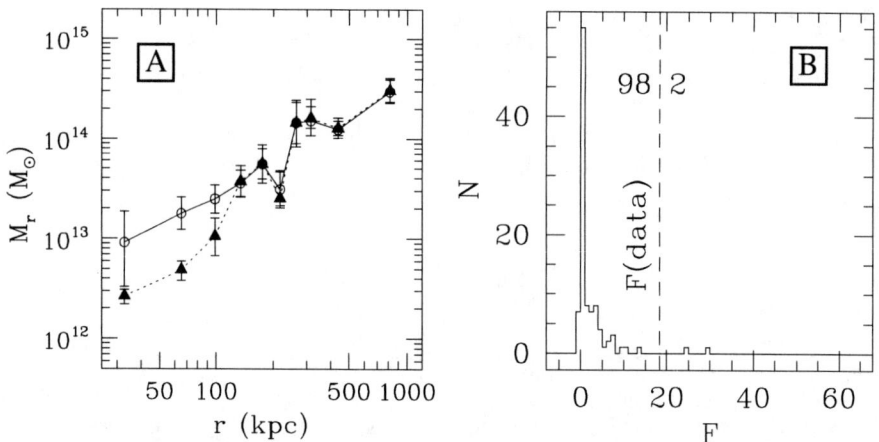

FIGURE 1. [A] One- and two-temperature mass profiles (dotted/triangles and solid/circles, respectively) of EMSS 1358+6245; [B] empirical F-distribution for models M^s and M^c of EMSS 1358+6245.

tion function $q(\theta_0, \theta_1)$. The choice of q is arbitrary; we restrict ourselves to functions which are symmetric in parameter space transitions, i.e. $q(\theta_i, \theta_j) = q(\theta_j, \theta_i)$ (this is the Metropolis algorithm – see ftp://ftp.cs.utoronto.ca/pub/radford/review.ps.Z; in this example we used a 4D gaussian deviate). This new parameter set is accepted if $P(\theta_1)/P(\theta_0)$ exceeds a random number on [0,1]. If not, the trial point is rejected and new one is selected. The sequence of accepted θ_i is a Markov Chain whose stationary distribution follows $P(\theta)$ [35]. We repeat this procedure until we have 100 values of θ for model 0 ^{s}.

For each of the parameter sets in the sample we simulate an X-ray spectrum, taking proper account of the instrument response and photon statistics. We then model each of the simulated data sets using both 0 ^{s} and 0 ^{c}, and tabulate the F-value of each data set:

$$F = \frac{\chi^2(\theta^s) - \chi^2(\theta^c)}{\chi^2(\theta^s)/\nu(M^s)} \qquad (1)$$

where $\nu(M^s)$ is the number of degrees of freedom of the simple model. (In practice, the normalization can be ignored.) The F-distribution for the cooling flow cluster EMSS 1358+6245 is shown in Figure 1B. The F-value of the original *Chandra* data set is indicated with a dashed line.

Conclusion

Of the 100 MCMC simulations that were run, only two resulted in an F-value which exceeded that of the data – that is, if M^s were the correct description, an F-value as large as that observed would occur with a probability of only 2% – meaning that the model with a separate, co-spatial cool component is preferred. The result is that a model with a steeper density profile and a larger central mass is favored. If this trend obtains for most cooling flow clusters, it may rule out several of the CDM modifications.

We have demonstrated a technique which provides a rigorous and quantitative procedure for deciding between emission models of cooling flow clusters. As we continue to model clusters in the *Chandra* archive, this method will faciliate comparisons with CDM simulations by helping to remove much of the uncertainty in the derivation of cluster core density profiles.

REFERENCES

1. Lahav, O, Bridle, S.L., Percival, W.J., Peacock, J.A., Efstathiou, G., Baugh, C.M., Bland-Hawthorn, J., Bridges, T., Cannon, R., Cole, S., Colless, M., Collins, C., Couch, W., Dalton, G., de Propris, R., Driver, S.P., Ellis, R.S., Frenk, C.S., Glazebrook, K., Jackson, C., Lewis, I., Lumsden, S., Maddox, S., Madgwick, D.S., Moody, S., Norberg, P., Peterson, B.A., Sutherland, W., & Taylor, K. 2001, *MNRAS*, **333**, 961
2. Peacock, J.A. *et al.* 2001, *Nature*, **410**, 169
3. Navarro, J.F., Frenk, C.S., and White, S.D.M. 1997, *ApJ*, **490**, 493
4. Moore, B., Quinn, T., Governato, F., Stadel, J. & Lake, G. 1999, *MNRAS*, **310**, 1147
5. Kauffman, G., White, S.D.M., & Guiderdoni, B. 1993, *MNRAS*, **264**, 201
6. Moore, B., Ghigna, S., Governato, F., Lake, G., Quinn, T., Stadel, J., & Tozzi, P. 1999a, *ApJL*, **524**, L19
7. Klypin, A.A., Kravtsov, A.V., Valenzuela, O., & Prada, F. 1999, *ApJ*, **k522**, 82
8. Burkert, A. 1995, *ApJ*, **447**, L25
9. McGaugh, S.S. & de Blok, W.J.G. 1998, *ApJ*, **499**, 41
10. Swaters, R.A., Madore, B.F. & Trewhella, M. 2000, *ApJL*, **531**, L107
11. Tyson, J.A., Kochanski, G.P. & Dell'Antonio, I.P. 1998, *ApJ*, **498**, L107
12. Smail, I., Ellis, R., Ritchett, M.J. & Edge, A.C. 1995, *MNRAS*, **273**, 277
13. Broadhurst, T., Huang, X., Frye, B., & Ellis, R. 2000, *ApJ*, **534**, 15
14. Shapiro, P.R. & Iliev, I.T., 2000, *ApJL*, **542**, L1
15. Spergel, D.N., and Steinhardt, P.J. 2000, *Phys. Rev. Lett.*, **84**, 17
16. Firmani, C., D'Onghia, E., Avila-Reese, V., Chincarini, G. & Hernández, X. 2000, *MNRAS*, **315**, L29
17. Hogan, C.J. & Dalcanton, J.J. 2000, *Phys. Rev. D*, **62**, 063511
18. Kaplinghat, M., Knox, L. & Turner, M.S. 2000, *Phys. Rev. Lett.*, **85**, 3335
19. Hu, W. & Peebles, P.J.E. 2000, *ApJ*, **528**, 61
20. Goodman, J. 2000, New Astron., **5**, 103
21. Mohapatra, R.N., Nussinov, S. & Teplitz, V.L. 2002, *Phys. Rev. D*, **66**, 063002
22. Hennawi, J.F. & Ostriker, J.P. 2002, *ApJ*, **572**, 41
23. Frenk, C.S. 2002, *Phi. Trans. Roy. Soc.*, **300**, 1277
24. Arabadjis, J.S., Bautz, M.W. & Garmire, G.P. 2002, *ApJ*, **572**, 66
25. Arabadjis, J.S. & Bautz, M.W. 2002, in preparation
26. Mewe, R., Gronenschild, E.H.B.M. & van den Oord, G.H.J. 1985, *A&AS*, **62**, 197
27. Mewe, R., Lemen, J.R. & van den Oord, G.H.J. 1986, *A&AS*, **65**, 511
28. Kaastra, J.S. 1992, *An X-Ray Spectral Code for Optically Thin Plasmas*, Internal SRON-Leiden Report, version 2.0.
29. Liedahl, D.A., Osterheld, A.L. & Goldstein, W.H. 1995, *ApJL*, **438**, L115
30. Arnaud, K.A. 1996, *Astronomical Data Analysis Software and Systems V*, George H. Jacoby & Jeannette Barnes, eds., *ASP Conf. Ser.*, **101**, 17
31. Bevington, P.R. 1969, *Data Reduction and Error Analysis for the Physical Sciences* (New York: McGraw-Hill)
32. Lampton, M., Margon, B. & Bowyer, S. 1976, *ApJ*, **208**, 177
33. Cash, W. 1979, *ApJ*, **228**, 939
34. Protassov, R., van Dyk, D.A., Connors, A., Kashyap, V.K. & Siemiginowska, A. 2002, *ApJ*, **571**, 545
35. Lewis, A. & Bridle, S. 2002, *Phys. Rev. D*, **66**, 103511

X-Ray Measurement of Dark Matter "Temperature" in Abell 1795

Yasushi Ikebe*[†], Hans Böhringer** and Tetsu Kitayama[‡]

*Code 661, NASA/Goddard Space Flight Center, Greenbelt Rd., Greenbelt, MD 20771, USA
[†]Joint Center for Astrophysics, University of Maryland, Baltimore County, 1000 Hilltop Circle, Baltimore, MD 21250, USA
**Max-Planck-Institut für extraterrestrische Physik, Postfach 1312, 85741 Garching, Germany
[‡]Department of Physics, Toho University, Miyama, Funabashi, Chiba 274-8510, Japan

Abstract. We establish a method from an X-ray observation of a galaxy cluster to measure the radial profile of the dark matter velocity dispersion, σ_{DM}, and to derive the dark matter "temperature" defined as $\mu m_p \sigma_{DM}^2$. The method is applied to the XMM-Newton observation of Abell 1795. The ratio between the specific energy of the dark matter and that of the intra cluster medium (ICM), which can be denoted as β_{DM} in analogy with β_{spec}, is found to be less than unity everywhere ranging $\sim 0.3 - 0.8$. In other words, the ICM temperature is higher than the dark matter "temperature", even in the central region where radiative cooling time is short and cooling flow phenomena is expected to be observed. A β_{DM} value smaller than unity can be most naturally explained by heating of the ICM in addition to gravitational heating.

INTRODUCTION

Early X-ray imaging observations with the *Einstein* observatory and *ROSAT* showed that in the central regions of clusters of galaxies the radiative cooling time is shorter than the age of the universe (e.g [1]). As a result, the intra cluster medium (ICM) should cool down to form a cold ($T < 10^6$K) gas phase inducing a global inflow of gas. This "cooling flow" (see [2] for a review) picture has been extensively discussed and formed a basic assumption in many arguments. The low resolution spectroscopy in 0.5-2 keV by *ROSAT* showed that in some clusters the ICM temperature actually decreases towards the center (e.g. [3, 4, 5]). Higher resolution spectroscopy in 0.5-10 keV with *ASCA*, however, can not be fully understood with the conventional cooling flow model. ASCA spectra of cooling flow clusters can be well explained by a two (hot and cool) phase plasma without significant excess absorption features (e.g. [6, 7]). A naive cooling flow model predicting a range of temperatures with intrinsic absorption could also fit the ASCA data but generally produce worse chi-square results (e.g. [8]). Most recently, very high resolution spectroscopy with XMM-Newton/RGS unambiguously show that there is very little X-ray emission from gas cooler than certain lower cut-off temperatures of $\sim 1 - 3$ keV ([9, 10, 11]). Unless a large amount of cooled gas or the metals in the cold gas are hidden ([12]), there must exist a heating mechanism that prevents the ICM from radiative cooling.

In order to shed some new light onto these "cooling flow phenomenon", we compare the temperature distribution of the ICM with the distribution of the velocity dispersion of the dark matter. A parameter, $\beta_{\text{spec}} \equiv \sigma_{\text{gal}}^2/(kT/\mu m_p)$, is often used as a measure of the average kinetic energy per unit mass in galaxies relative to that in the ICM. From observations of many clusters, the mean β_{spec} is ~ 1 with large scatter (e.g. [13]), indicating that the energy equipartition between galaxies and ICM is roughly achieved on average. In analogy with β_{spec}, we can introduce $\beta_{\text{DM}} \equiv \sigma_{\text{DM}}^2/(kT/\mu m_p)$ for comparison between the mean kinetic energy of the dark matter and that of the ICM, and define the dark matter "temperature" as $kT_{\text{DM}} \equiv \mu m_p \sigma_{\text{DM}}^2$. We, in this paper, obtain the radial profile of the β_{DM} value observationally for the first time.

METHOD OF MEASURING DARK MATTER VELOCITY DISPERSION

Under the assumptions of spherical symmetry and hydrostatic equilibrium, the ICM distribution is described by

$$\frac{GM(R)}{R} = -\frac{k_B T_g}{\mu m_p}\left(\frac{d\ln n_g}{d\ln R} + \frac{d\ln T_g}{d\ln R}\right), \quad (1)$$

where $M(R)$ is the total gravitating mass within a sphere of radius R, k_B is the Boltzmann constant, G is the Gravitational constant, n_g and T_g is the density and temperature of the ICM, respectively. From an X-ray observation, n_g and T_g are measured and $M(R)$ can be obtained via Eq. 1. When the dark matter particles tracing the same gravitational field are in steady state, they obey the Jeans equation

$$\frac{GM(R)}{R} = -\sigma_{\text{DM}}^2\left(\frac{d\ln \rho_{\text{DM}}}{d\ln R} + \frac{d\ln \sigma_{\text{DM}}^2}{d\ln R}\right), \quad (2)$$

where σ_{DM} is the radial velocity dispersion, and ρ_{DM} is the mass density, which is given as $\rho_{\text{DM}} = \frac{1}{4\pi R^2}\frac{dM}{dR} - \mu m_p n_g$. Therefore, once the total gravitating mass, M, and the gas density, n_g, profiles are obtained from an X-ray observation, Eq. 2 contains only one unknown parameter, σ_{DM}, in terms of which we can solve the equation under a given boundary condition.

DATA ANALYSIS

We applied the method to the XMM-Newton/EPIC-PN data of Abell 1795, which is a prototypical CF cluster (e.g. [14, 15]). From deprojected spectra in concentric shell regions, we derived the ICM temperature profile, which is illustrated in Fig. 1. The ICM density profile, n_g, can be obtained from the ICM temperature profile, $T(R)$, and the surface brightness profile, which may be substituted in Eq. (1) to derive the total mass profile, $M(R)$. We here instead start from a theoretically motivated analytical formula to model $M(R)$, which is then used in combination with $T(R)$ obtained from the X-ray spectra to predict the X-ray surface brightness profile of the cluster. By fitting the predicted brightness profile with the data, we determined an appropriate

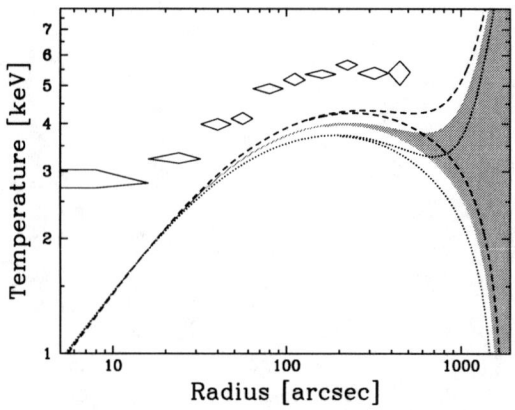

FIGURE 1. The diamonds show the ICM temperature profile, while the grey hatched region shows the dark matter "temperature" profile derived from the best-fit NFW model applied as the total mass profile. The dashed and dotted lines represent the extreme cases within errors. Since we assume H_0=70 km/s/Mpc, $\Omega_{m,0} = 0.3$, and $\Omega_{\Lambda,0} = 0.7$, 1 arcsec corresponds to 1.19 kpc at the cluster's redshift of 0.0616.

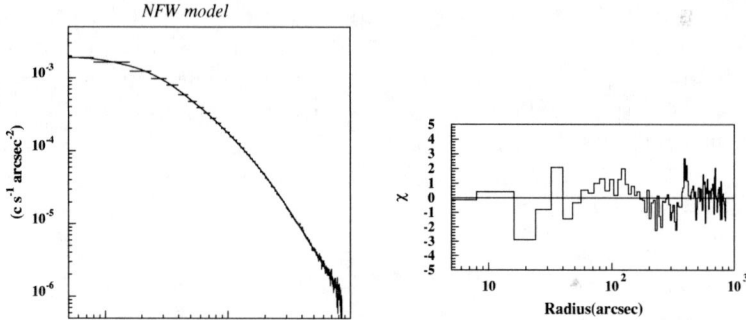

FIGURE 2. Left) The 0.5–10 keV X-ray count rate profile (crosses) is fitted with a model profile (solid line) predicted from a total mass profile given with the NFW model and from the observed ICM temperature profile. Right) Residuals for the best-fit model profile.

model to describe the total mass profile. Figure 2 shows the fit result, when the NFW model ([16, 17]) is applied to the mass profile, where the density is given as $\rho = \rho_0 (R/R_s)^{-1}[1+(R/R_s)^2]^{-2}$.

DARK MATTER "TEMPERATURE" PROFILE IN ABELL 1795

With the best-fit NFW model that well describes the data, we solved Eq. (2) in terms of σ_{DM} and obtained the dark matter "temperature" profile as shown in Fig. 1. The

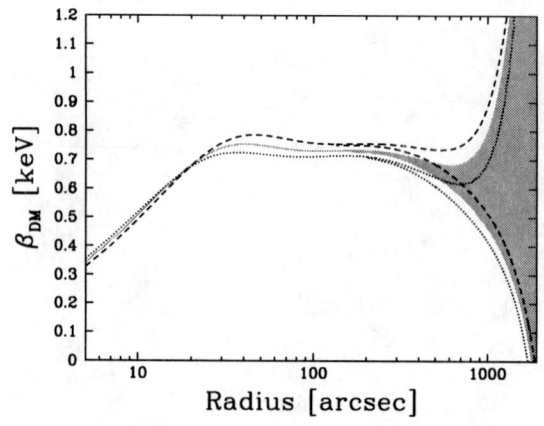

FIGURE 3. Radial profile of β_{DM}.

ICM temperature is found to be greater than the dark matter "temperature" everywhere, even in the central region where radiative cooling is expected to be most effective. In other words, the dark matter forms a temperature floor that limits the ICM temperature. The comparison of the temperatures can more directly be described by means of the β_{DM}, which ranges ~ 0.3-0.8 as shown in Fig. 3. From numerical simulation studies, $\beta_{DM} \sim 1 - 1.4$ is expected, if there is no cooling or additional heating (e.g. [18, 16, 19]). Therefore, our result on β_{DM} that is smaller than unity may be another piece of evidence for non-gravitational heating to prevent the ICM from cooling.

REFERENCES

1. Canizares, C. R., Stewart, G. C., & Fabian, A. C. 1983, ApJ, 272, 449
2. Fabian, A. C. 1994, ARA&A, 32, 277
3. Böhringer, H. et al. 1994, Nature, 368, 828
4. David, L. P., Jones, C., Forman, W. & Daines, S. 1994, ApJ, 428, 544
5. Allen, S. W. & Fabian, A. C. 1994, MNRAS, 269, 409
6. Ikebe, Y. et al. 1999, ApJ, 525, 58
7. Makishima, K. et al. 2001, PASJ, 53, 401
8. Allen, S. W. et al. 2001, MNRAS, 322, 589
9. Tamura, T. et al. 2001, A&A, 365, L87
10. Kaastra, J. S. et al. 2001, A&A, 365, L99
11. Peterson, J. R. et al. 2001, A&A, 365, L104
12. Fabian, A. C., Mushotzky, R. F., Nulsen, P. E. J., & Peterson, J. R. 2001a, MNRAS, 321, L20
13. Wu, X., Xue, Y., & Fang, L. 1999, ApJ, 524, 22
14. Edge, A. C., Stewart, G. C., & Fabian, A. C. 1992, MNRAS, 258, 177
15. Briel, U. G., & Henry, J. P. 1996, ApJ, 472, 131
16. Navarro, J. F., Frenk, C. S., & White, S. D. W. 1995, MNRAS, 275, 720
17. Navarro, J. F., Frenk, C. S., & White, S. D. M. 1996, ApJ, 462, 563
18. Metzler, C. A., & Evrard, A. E. 1994, ApJ, 437, 564
19. Bryan, G. L. & Norman, M. L. 1998, ApJ, 495, 80

Dark Matter, AGNs and Relativistic Jets

William K. Rose

Department of Astronomy, University of Maryland, College Park, MD 20742

Abstract. There is evidence that the Universe contains about ten times more nonluminous than luminous mass. It has been suggested that some or perhaps all of this dark matter is in the form of nonbaryonic mass. We discuss the possibility that nonbaryonic matter can be gravitationally captured by supermassive black holes and thereby affect the evolution of some quasars and AGNs. We argue that large amounts of nonbaryonic dark matter have not been captured by black holes emitting relativistic jets.

INTRODUCTION

It is known that massive black holes are routinely present at the centers of elliptical galaxies and spiral galaxy bulges. Recently Gephardt et al. 2000 have described a correlation between the estimated mass of a galaxy's central black hole and line-of-sight velocity dispersion. Richstone et al. 1998 have given a correlation between estimated black hole mass and bulge luminosity. The above results and other referenced measurements indicate that central black hole masses are approximately proportional to host galaxy spheroid masses and that the total mass density in black holes is crudely consistent with the mass equivalent energy density of quasar background radiation.

It has been suggested that galaxy formation occurs in the potential well of a dark matter halo. In this paper we will examine the possibility that significant amounts of the nonbaryonic mass in a dark matter halo may be captured by a supermassive black hole. Tidal torques are believed to be responsible for the production of galaxy angular momentum. The usual angular momentum parameter λ is proportional to the ratio of rotational velocity to virial velocity. It is given by the expression $\lambda = JE^{1/2}/GM^{5/2}$ where $J = MV_{rot} R$ and $E = GM^2/R$. For most systems λ is believed to be in the range $\lambda = 0.03 - .1$ but is expected to vary widely from galaxy to galaxy. We are interested in systems that have unusually low values of λ and therefore expect that large amounts of nonbaryonic mass capture will occur in a small percentage of galaxies.

There is evidence that highly relativistic jets are emitted from SMBHs at the centers of some quasars and radio galaxies. It has been argued (e.g. Phinney 1983, Rose et al. 1984) that a voltage drop can develop between the rotation axis and inner region of an accretion disk surrounding a Kerr (i.e. rotating) black hole. Such a voltage drop, which is likely to depend on magnetic field strength and geometry as well as black hole angular momentum, is a possible explanation for the production of highly relativistic jets.

The capture of dark matter in the form of massive neutrinos or other weakly interacting massive particles would increase the mass and therefore the Eddington luminosity of a black hole without causing significant radiation. In the discussion below we describe physical conditions required for the capture of nonbaryonic dark matter whose infall rate onto a black hole is not limited by the usual Eddington limit.

RESULTS

There is evidence that SMBHs exist at the center of massive galaxies early in their evolution. It is natural to ask if capture of mass from a surrounding dark matter halo can occur after the formation of a SMBH. For definiteness we assume that the nonluminous mass consists of a spherical halo of nonbaryonic dark matter and estimate how massive a central black hole must be to capture appreciable amounts of halo mass. According to the standard big bang scenario for the evolution of the universe cold dark matter candidates such as massive neutrinos are nonrelativistic when they freeze out at a temperature of about 3-10 x 10^{10}K and become noninteracting particles. After freeze out their number density scales as $(1+Z)^3$ and their temperature scales as $(1+Z)$. Peebles (1993) gives an expression for their number density at freeze out and estimates that if massive neutrinos are the dominant form of mass in the Milky Way then their local number density at freeze out is about $n \sim 1/m(GeV)$ particles cm^{-3}.

The cross section for gravitational capture of particles infalling onto a black hole is (Lightman et al. 1975)

$$\sigma = 16\pi \frac{G^2 M^2}{c^2 v^2} \quad (1)$$

where v is the infall velocity. The present (i.e. Z = 0) predicted temperature of hypothetical massive neutrinos is about 1K. In order that the relic mass density of such hypothetical particles not be so great as to lead to a Universe younger than 10^{10} years neutrino mass must either be less than 92 h^2 eV or greater than $\sim$ 3GeV (Peebles 1993). The $\sim$ 3GeV limit is known as the Lee-Weinberg bound. It exists because particles that become nonrelativistic before they decouple from baryonic mass must end up with much lower number densities than known neutrino species. The possible presence of a massive neutrino or other exotic particle (Kolb and Turner 1990) in the early universe is theoretically appealing because density fluctuations caused by such a particle might help explain the formation and spatial distribution of galaxies. From our point of view cold dark matter particles are interesting because of their possible capture by SMBHs.

If hypothetical massive neutrino masses equal 3 GeV then their thermal velocities are $\sim .1$ kms^{-1}. We assume that such dark matter particles fall from a radial distance of 10 kpc toward the central black hole of a massive galaxy. Their initial transverse velocities are assumed equal to their thermal velocities. Each noninteracting massive neutrino moves in its own independent orbit in the mean gravitational field of all other mass (nonbaryonic and baryonic), and conservation of particle angular momentum and energy holds. This implies that for the initial conditions we consider the cross section in equation (1) becomes $6 \times 10^{44} M_{10}^2$ with $M_{10} = M/10^{10}$ M$\odot$ and appreciable gravitational capture occurs only if black hole mass is $> 1.3 \times 10^{10}$ M$\odot$, which is much larger than black hole masses estimated for the Milky Way, M31 and Seyfert nuclei (Gephardt et al. 2000). Therefore, even a nonrotating cold dark matter halo around such galaxies should not be captured by the central black hole. The black hole mass estimate for M87 (Virgo A) is $\sim 3 \times 10^9$ M$_\odot$ and therefore some gravitational capture may occur. Figure 1 of Richstone et al. 1998 lists three black hole mass estimates that exceed 10^{10} M$_\odot$. Such black holes are sufficiently massive that nonbaryonic particle capture might become appreciable.

Accretion of nonbaryonic mass would produce little power output but could lead to a bimodal distribution of SMBH masses because black holes with masses as high as $\sim 10^{11}$ M$_\odot$ might be formed. Since gravitational capture adds little angular momentum to a central black hole the arguments given in the Introduction suggest that very massive black holes that are formed primarily from nonbaryonic matter would not produce highly relativistic jets. The baryonic mass accretion rate required for a 10^{11} M$_\odot$ black hole to radiate at its Eddington luminosity ($\sim 2 \times 10^{49}$ ergs^{-1}) is so large that it is unlikely that radiation pressure driven winds analogous to those of the stellar mass object SS433 would be formed.

The Eddington luminosity does not apply if emission is primarily in the form of gamma rays because the Klein-Nishina cross section replaces the Thomson cross section. Recent observations of gamma ray emitting AGNs show that Seyfert and radio galaxies represent one class of gamma ray AGN whereas blazars constitute another class (Dermer and Gehrels 1995). Detected Seyfert and radio galaxies are relatively nearby whereas blazars are observed to redshifts $Z \geq 2$ presumably because emission is from a relativistic jet pointing along the line of sight. Sources with the greatest estimated radio luminosities are observed to have the highest > 100 MeV gamma ray emission. In a number of blazars gamma ray emission dominated other forms of radiation. Several AGNs such as PKS 0528 + 134 have gamma ray luminosities of $\sim 10^{49}$ ergs^{-1} if radiation is assumed isotropic. Even with the Klein-Nishina correction to the Eddington luminosity included the estimated black hole mass of PKS 0528+134 is $\sim 2 \times 10^{10}$ M$_\odot$ if it is an isotropic radiator (Dermer and Gehrels 1995). As discussed above the existence of black holes this massive would make gravitational capture of nonbaryonic particles more probable.

Our principal conclusions are as follows. Black holes are initially formed from baryonic matter. Those that become sufficiently massive may accrete substantial amounts of hypothetical nonbaryonic matter. Giant elliptical galaxies are known to have little angular momentum even when their shapes are nonspherical. Radio quiet giant elliptical galaxies are probably the most likely objects to contain black holes formed primarily from massive, nonbaryonic particles.

REFERENCES

1. Dermer, C. D., and Gehrels, N. 1995, ApJ, 447, 103.
2. Gephardt, K. et al. 2000, ApJ, 539, L16.
3. Kolb, E. W., and Turner, M. S. 1990, The Early Universe (Addison-Wesley: Redwood City).
4. Peebles, P.J.E. 1993, Principles of Physical Cosmology (Princeton University: Princeton).
5. Phinney, E. S. 1983 in Astrophysical Jets, Eds. A. Ferrari, A. G. Pacholczyk (Reidel: Dordrecht).
6. Richstone, D. et al. 1998, Nature, 395, A14.
7. Rose, W. K., Guillory, J., Beall, J. H. and Kainer, S. 1984, Ap. J., 280, 500.

Cosmological Simulations of the Formation of Primordial Gas Clouds

Naoki Yoshida

Harvard-Smithsonian Center for Astrophysics, 60 Garden Street, Cambridge MA 02138

Abstract. We study the formation of primordial star-forming clouds in a ΛCDM universe using large N-body/SPH simulations. The simulations include non-equilibrium 9 species (e^-, H, H$^+$, He, He$^+$, He^{++}, H$_2$, H$_2^+$, H$^-$) chemistry and hydrogen molecular cooling. We determine that the critical halo mass to form primordial star-forming clouds is $5 \times 10^5 h^{-1} M_\odot$, with a weak dependence on redshift in a range $z > 16$.

INTRODUCTION

In currently favored Cold Dark Matter models, first star formation occurs in small mass ($\sim 10^6 M_\odot$) dark halos at redshift around 30, within which the primordial gas can cool and condense via hydrogen molecular cooling process. Recent detailed numerical studies indeed showed the first structure formation – the formation of primordial gas clouds and the first stars – at epochs as early as $z = 30$ (Abel et al. 2001; Bromm et al. 2001). In these simulations, dense, cold gas clouds form in the central parts of small halos, become self-gravitating, and contract further to form a core with mass about 100 $M_\odot$, which will eventually become a star. These previous works successfully explored how the very first star is formed, and it appears that the next step will be to study how and when *a large population of first stars* formed in the early universe. Surprisingly, there have not been many systematic studies on statistical properties of the early small-scale structure in a proper cosmological context.

Analytic modelling of the first baryonic structure formation is often based on an assumption that primordial gas can cool within halos with mass greater than a critical mass, so that the gas temperature, assumed to be close to the virial temperature of the host halo, is high enough to make the system reactive to produce hydrogen molecules that acts as a major coolant. The key element here is the relation between dark halos and star-forming clouds. Using a spherical collapse model, Tegmark et al. (1997) estimated a critical H$_2$ mass fraction needed for cooling of the primordial gas and a mass scale of halos within which they collapse. We use large, high-resolution N-body/SPH simulations to address this issue.

TABLE 1. Simulation parameters

Run	N_{tot}	Box size (h^{-1}kpc)	m_{gas} ($h^{-1}M_\odot$)	l_s (h^{-1}pc)
A	2×288^3	600	100.0	54
B	2×216^3	300	29.6	36
C	2×144^3	300	100.0	54

MINIMUM COLLAPSE MASS

We use parallel N-body/SPH solver GADGET (Springel, Yoshida & White 2001). Our simulations follow the non-equilibrium evolution of nine chemical species (e$^-$, H, H$^+$, He, He$^+$, He^{++}, H$_2$, H$_2^+$, H$^-$) using the method of Abel et al. (1997). We use the cooling rate of Galli & Palla (1998) for molecular hydrogen cooling. The largest of our simulations employs 48 million particles in a cosmological box of $600h^{-1}$kpc on a side (see Table 1). We work with the standard ΛCDM model ($\Omega_0 = 0.3$, $\Omega_b = 0.04$, $\Omega_\Lambda = 0.7$, h=0.7, and σ_8=0.9). Our primary goal in this section is to determine the minimum mass scale for the first structure formation. We locate dark matter halos in the simulation outputs and measure the mass of gas which is cold ($T < 0.5T_{vir}$) and dense ($\rho_{gas} > 10^{10} M_\odot \text{kpc}^{-3}$). Hereafter we refer to such cold, dense gas clumps as "gas clouds". In Figure 1 we plot the minimum mass of the halos that host gas clouds against output redshift. It shows the evolution of the minimum mass scale of the star-forming systems. In the figure we also show the mass trajectory of the most massive halo in each run. Figure 1 clearly shows that the minimum collapse mass scale is set at $5 \times 10^5 h^{-1} M_\odot$, with only a weak dependence on redshift in the range plotted. An excellent agreement is seen between the high resolution Run B (dot-dashed line) and low resolution Run C1 (filled circles). Our result seems to be converged on mass scales which our simulations probe. The relatively early formation epoch of the first bound object in Run A is due to the volume effect. Run A simulates an 8 times larger volume and thus contains a higher-σ density fluctuation than in Run C1 and C2. Run C1 and C2 differ in the assigned phase informations in the initial random Gaussian fields.

An important quantity which determines the onset of gas cooling is the number fraction of hydrogen molecules f_{H_2}. In the right panel of Figure 1 we plot f_{H_2} against the virial temperature for the halos at $z = 17$. Filled circles represent the halos hosting cold dense gas clouds, while open circles are for the other halos. Solid line is an analytical estimate of the H$_2$ fraction needed to cool the gas, which we compute *a là* Tegmark et al. (1997). In the figure halos appear clearly separated into two populations; those in which the gases have produced enough molecular hydrogen to cool, and the others that have not. Our analytic estimate of the critical H$_2$ fraction indeed agrees very well with the distribution of the gas in the f_{H_2} - T plane. Although the H$_2$ fraction primarily determines whether the gas in halos can cool or not, there are some halos with sufficiently high virial temperature within which gas clouds have not been formed (open circles in $T_{vir} > T_{cr}$). At $z = 17$, about 30-40% among the massive ($M > 5 \times 10^5 h^{-1} M_\odot$) halos are such 'deficient' halos. Yoshida et al. (2003, in preparation) argue that in these halos dynamical heating due to mass accretion and mergers prevents the gas cooling.

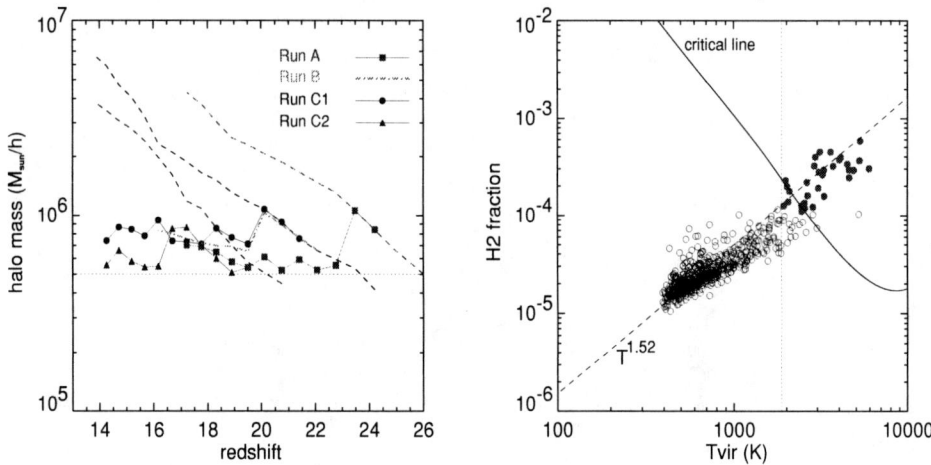

FIGURE 1. (Left panel) The minimum mass of the halos that host cold gas clumps. The solid lines with symbols indicate the minimum mass at each output redshift, and the dashed lines are the mass of the most massive halo in each run. (Right panel) The "Tegmark et al. 97" plot. We plot the gas mass weighted H_2 fraction for the halos which host gas clouds (filled circles) and for those which do not (open circles) for Run A at $z = 17$. The solid curve is the H_2 fraction needed to cool the gas at given temperature and the dashed line is the asymptotic H_2 fraction.

EFFECT OF PHOTO-DISSOCIATING RADIATION

We have also studied the effect of soft ultra-violet radiation on the formation of the primordial gas clouds. During our Run A, we turn on a uniform background radiation in the Lyman-Werner band (11.18-13.6 eV). We compare the effect of the radiation in two extreme cases; for one we assume the gas is optically thin, and for the other we take the gas self-shielding effect into account in a direct, but approximate manner. We compute the molecular hydrogen column density N_{H_2} around each simulation particle by summing up contributions from neighbouring particles out to 100 pc (typical virial radius for $\sim 10^6 M_\odot$ halos), and compute the effective shielding factor following Drain & Bertoldi (1996) to examine the possible maximum effect of gas self-shielding. Figure 2 shows the minimum collapse mass for three cases: (a) no radiation, (b) LW background radiation with $J = 10^{-23}$ erg s^{-1} cm^{-2} Hz^{-1} str^{-1}, and (c) LW background radiation with $J = 10^{-23}$ erg s^{-1} cm^{-2} Hz^{-1} str^{-1} with gas self-shielding.

The primordial gas cooling is suppressed by SUV radiation due to photo-dissociation of H_2. Consequently, the background radiation raises the minimum collapse mass scale (Machacek et al. 2001). Figure 2 shows that the minimum collapse mass is, approximately, $5 \times 10^5, 6.2 \times 10^5, 1 \times 10^6 h^{-1} M_\odot$, for case (a), (b), and (c), respectively. We also found that, for a radiation with an order of magnitude higher flux, the molecular fraction drops below the critical fraction (see Figure 1) for all the halos and the primordial gas cooling is nearly entirely prevented.

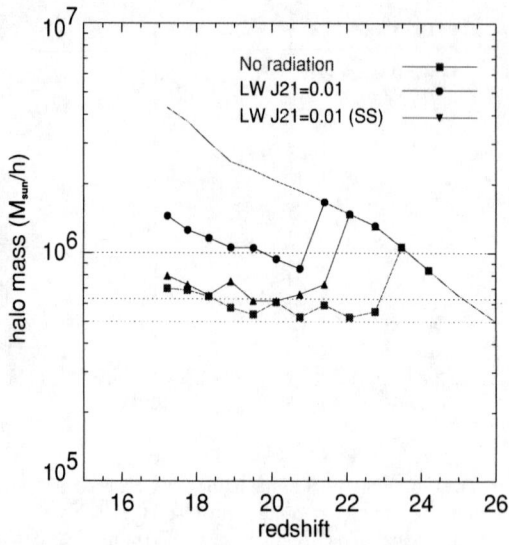

FIGURE 2. The minimum mass of the halos that host cold gas clumps. The solid lines with symbols indicate the minimum mass at each output redshift, and the dashed lines are the mass of the most massive halo in each run.

NY thanks Jonathan Tan and Paul Shapiro for fruitful discussions.

REFERENCES

1. Abel, T., Annimos, P., Norman, M. L., and Zhang, Y., New Astronomy, **2**, 181 (1997)
2. Abel, T., Bryan, G. L., and Norman, M. L., *Science*, **295**, 93, (2001)
3. Bromm, V., Coppi, P. S., and Larson, R. B., ApJ, **564**, 23 (2001)
4. Galli, D. and Palla, F., Astron. Astrophys, **335**, 403 (1998)
5. Haiman, Z., Abel, T., and Rees, M. J., ApJ, **534**, 11 (2000)
6. Machacek, M. E., Bryan, G. L., and Abel, T., ApJ, **548**, 509, (2001)
7. Springel, V., Yoshida, N. and White, S. D. M. 2001, New Astronomy, **6**, 79 (2001)
8. Tegmark, M., Silk, J., Rees, M., Blanchard, A., Abel, T., & Palla, F., ApJ, **474**, 1 (1997)

Halo Substructure and the Power Spectrum

Andrew R. Zentner* and James S. Bullock[†]

*Department of Physics, The Ohio State University, 174 W. 18th Ave., Columbus, OH 43210-1173
[†]Harvard-Smithsonian Center for Astrophysics, 60 Garden St., Cambridge, MA 02138

Abstract. In this proceeding, we present the results of a semi-analytic study of CDM substructure as a function of the primordial power spectrum. We show that the mass fraction of halos in substructure is not a strong function of spectral tilt and so it will be difficult to constrain tilt with strong lensing measurements of substructure. We also discuss the dwarf satellite issue in a cosmological context. We quantify the cosmology-dependence of the map between the observed central velocity dispersions of the Milky Way satellites and the circular velocities of their host halos. Finally, we point out that models with significantly tilted primordial power spectra may underpredict the number of Milky Way satellites with $V_{\mathrm{max}} \gtrsim 40$ km s^{-1}.

1. INTRODUCTION

The ΛCDM model of a flat Universe dominated by cold, collisionless dark matter (CDM), and a cosmological constant (Λ) has emerged as the standard framework for the growth of cosmic structure. With $\Omega_{\mathrm{M}} \approx 0.3$, $h \approx 0.7$, and a nearly scale-invariant primordial spectrum of adiabatic density perturbations ($P(k) \propto k^n$, $n \approx 1$), ΛCDM is remarkably successful at reproducing large scale observations. In contrast, this paradigm faces several challenges on galactic and sub-galactic scales [1, 2]. In Zentner & Bullock [3] (ZB), we emphasized that inflation does not predict *exactly* scale-invariant (*i.e.*, $n = 1$) primordial spectra. Many models of inflation predict "tilted" spectra ($n \neq 1$), spectral index "running" ($\mathrm{d}n/\mathrm{d}\ln k \neq 0$), or other deviations from scale-invariance that have dramatic consequences on small scales. We showed that spectra with tilts of $n \sim 0.9$ and/or running and fixed by COBE on large scales can greatly reduce the predicted central densities of dark matter halos, alleviating the "central density problem" plaguing ΛCDM. Further, the neighborhood of $\sigma_8 \sim 0.75$ implied by these tilts is provocatively close to many recent estimates of "low" σ_8 values [4].

In this proceeding, we report on results from follow-up work to ZB. We study the dependence of CDM halo substructure on the primordial power spectrum (PPS). Our models of the PPS are the same as those in ZB. We COBE normalize all spectra and we assume a cosmological model with $\Omega_{\mathrm{M}} = 1 - \Omega_{\Lambda} = 0.3$, $\Omega_{\mathrm{B}} h^2 = 0.02$, and $h = 0.72$. The important characteristics of each input spectrum are summarized in Table 1. Numerical simulations cannot have both the resolution and the statistics needed to study substructure so we model substructure semi-analytically using host halo merger histories [5] and a scheme for approximating subhalo orbits and tidal mass loss. Our model expands on previous work by Bullock *et al.* [6] and Taylor and Babul [7]. We calibrated our model against available data from N-body simulations; nevertheless, our results *must* be regarded as preliminary estimates to be verified by extensive N-body

TABLE 1. Initial power spectra from the inflationary models discussed in ZB.

Model Description	Model Name	$n(k_{COBE})$	$dn(k_{COBE})/d\ln k$	σ_8
Scale-invariant	$n=1$	$\equiv 1$	$\equiv 0$	$\simeq 0.95$
Inverted Power Law	IPL4	$\simeq 0.94$	$\simeq -0.001$	$\simeq 0.83$
Running-mass model I	RM I	$\simeq 0.84$	$\simeq -0.004$	$\simeq 0.65$
Running-mass model II	RM II	$\simeq 0.90$	$\simeq -0.001$	$\simeq 0.75$
Running-mass model III	RM III	$\simeq 1.1$	$\simeq -0.001$	$\simeq 1.21$
Broken scale-invariant	BSI	$= 1.0$	$= 0$	$\simeq 0.97$

work. We present results based on 100 merger tree realizations. We give a detailed description of our model and further results in a forthcoming paper.

2. SUBSTRUCTURE MASS FRACTIONS

Efforts have been made to use flux ratios in multiply-imaged quasars to detect substructure in galactic halos and to use these measurements to constrain cosmology. In particular, Dalal and Kochanek [8] (DK) considered bounds on the PPS. As such, it is important to understand the theoretical predictions for halo substructure as a function of the PPS and, more generally, substructure distributions as a function of cosmology.

Our results on the substructure mass fraction and the PPS are summarized in Figure 1. DK took a typical lens mass of 3×10^{12} $M_\odot$ and the lenses in their sample have a median redshift of $z_\ell \approx 0.6$, so we present results for a 3×10^{12} $M_\odot$ halo at $z = 0.6$; however, our results do not change appreciably as a function of mass or redshift. Lensing measurements are sensitive to the mass fraction in substructure projected onto the plane of the lens at a halo-centric distance of order the Einstein radius, $R_E \sim 5$ kpc. Consequently, we show in Fig. 1 the mass fraction in substructure for the entire halo *and* the mass fraction in substructure in a 2D projection of radius $R = 10$ kpc.

Notice that the mass fraction is not a strong function of tilt and/or running. In tilted models, halos accrete their substructure later, compensating for the fact that the subhalos less dense and are more easily destroyed by tides. It will be difficult to use substructure measurements to constrain these parameters. Only the BSI model, with a sharp drop in power at $\sim 10^{10}$ $M_\odot$, shows deviation from the $n = 1$ model that is significant compared to the scatter. It may be possible to constrain models with such an abrupt break (*e.g.*, warm dark matter). DK found the halo mass fraction in substructures of mass $M_{sat} \lesssim 10^9$ $M_\odot$ to be $0.006 \lesssim f_{sat} \lesssim 0.07$. All of our models are consistent with this bound.

3. THE DWARF SATELLITES

The "dwarf satellite problem", namely that ΛCDM predicts roughly an order of magnitude more halos with $V_{max} \lesssim 40$ km s^{-1} than observed Milky Way (MW) satellites, is an often-discussed challenge to ΛCDM [1]. Stoehr *et al.* [9] (S02) and Hayashi *et al.* [10] (H02) proposed that substructure halos may be significantly less concentrated

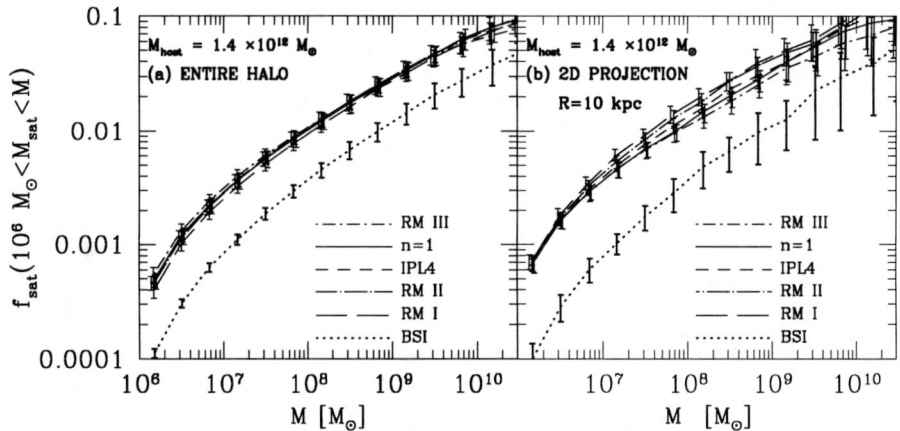

FIGURE 1. The fraction of the host halo mass bound up in substructures of mass between $10^6 M_\odot$ and M as a function of M. (a) The mass fraction in substructure for the entire halo. (b) The mass fraction in a 2D halo-centric cylindrical projection of radius $R = 10$ kpc. The lines represent the average mass fractions and the errorbars show the dispersion among the 100 realizations. The models are labeled in each panel.

than comparable field halos due to tidal effects. This implies that the values of $V_{\max}$ that correspond to the observed central velocity dispersions, $\sigma_\star$, of the MW satellites are larger than the values inferred by other authors. One must be cautious. Results regarding subhalo the subhalo velocity function (VF) are quite sensitive to mass resolution (S02), initial subhalo concentrations, and subhalo accretion times (H02). Our model represents the one extreme; we do not allow for redistribution of mass within a subhalo's tidal limit. Using our model, we can also quantify the cosmology dependence of the mapping between $\sigma_\star$ and $V_{\max}$. We have assumed that CDM halos can be well described by NFW [11] profiles with a particular $V_{\max}$ and $R_{\max}$ (the radius at which $V_{\max}$ is attained) and calculated all combinations of $V_{\max}$ and $R_{\max}$ that lead to the observed values of $\sigma_\star$ for each of the MW satellites. We have assumed that the stars have isotropic dispersion tensors and that the stellar distributions are given by King profiles [12] with parameters given by Mateo [13].

Results for Carina are shown in Figure 2, along with a scatter plot of $V_{\max}$ vs. $R_{\max}$ for the surviving satellites in 10 realizations of a $M_{\text{host}} = 1.4 \times 10^{12}$ $M_\odot$ host halo at $z = 0$ for the $n = 1$ and RM I models. This plot shows how allowing for less concentrated halos helps to alleviate the dwarf satellite problem. Less concentrated halos (larger $R_{\max}$) require a larger $V_{\max}$ to match the observed values of $\sigma_\star$ and these larger halos are intrinsically scarcer objects. Feedback mechanisms (*e.g.*, [6]) then explain the dearth of smaller halos. Figure 2 also demonstrates that the mapping between $\sigma_\star$ and $V_{\max}$ is dependent upon the PPS and, more generally, upon cosmology and so the same observational data imply a *cosmology dependent* "observed" VF.

In the right panel of Fig. 2 we present the predicted VFs along with separate "observed" velocity functions for the $n = 1$ and RM I models. The RM I VF is a factor of ~ 2 lower than the $n = 1$ VF because fewer halos survive to $z = 0$ at a given mass *and* typical halos are less concentrated so that $V_{\max}$ is lower at a given mass. Also notice

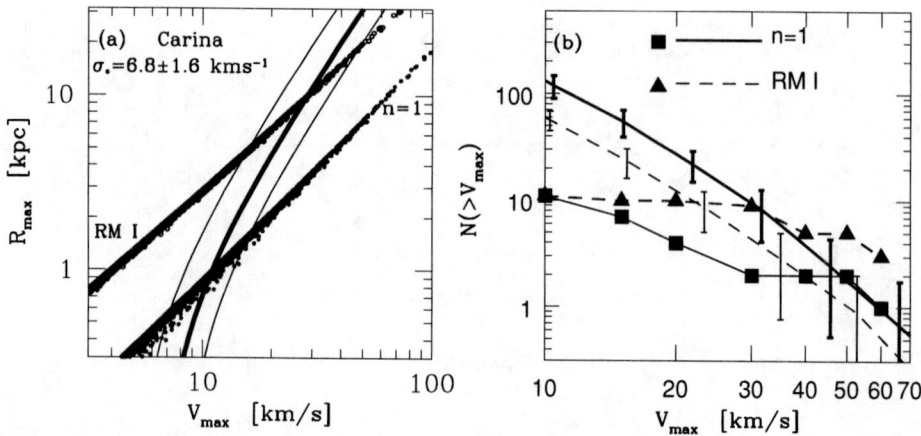

FIGURE 2. (a) The lower group of points represent a scatter plot of V_{max} vs. R_{max} for 10 realizations of the $n = 1$ model. The upper points correspond to RM I. The lines show the region that yields an observed value of $\sigma_\star = 6.8 \pm 1.6$ km s^{-1} for Carina. The thick solid line corresponds to the central value of $\sigma_\star$ while the thin solid lines correspond to the 1σ errors. Consistency demands that Carina resides in a halo that has structural properties that lie in the region of overlap between the thin solid lines and the scattered points for each cosmology. (b) The predicted VFs (lines) and scatter for the $n = 1$ and RM I models along with the "observed" VF (shapes) for each model inferred from the observed values of $\sigma_\star$.

that the "observed" VF is shifted significantly higher at high V_{max}. This suggests that significantly tilted power spectra $n \lesssim 0.9$ may *underpredict* the number of MW satellites at high V_{max}. Moreover, if subhalos are less centrally concentrated than field halos as in S02 and H02, this underprediction only becomes *more* pronounced. However, we recommend circumspection. Our results concerning VFs are sensitive to several assumptions such as the isotropy of the dispersion tensor.

REFERENCES

1. Moore, B. *et al.*, ApJ, **524**, L19-L22 (1999); Klypin, A. A. *et al.*, ApJ, **522**, 82 (1999).
2. Debattista, V. P. and Sellwood, J. A., ApJ, **543**, 704 (2000); Keeton, C. R., ApJ, **561**, 46 (2001); Alam, S. M. K. *et al.*, ApJ, **572**, 34 (2002); McGaugh, S. *et al.*, ApJ, **584**, in press, (2003).
3. Zentner, A. R. and Bullock, J. S., PRD, **66**, 043003 (2002), (ZB).
4. Viana, P. T. *et al.*, ApJ, **569**, 75; Bahcall, N. *et al.*, ApJ, **585**, in press (2002); Schuecker, P. *et al.*, A&A, in press (2002); Allen, S. W. *et al.*, astro-ph/0208394, (2002); Brown, M. L. *et al.*, astro-ph/0210213, (2002); Hamana, I. *et al.*, astro-ph/0210450 (2002); Peirpaoli, W. J. *et al.*, astro-ph/0210567, (2002).
5. Somerville, R. S. and Kolatt, T. S., MNRAS, **305**, 1, (1999).
6. Bullock, J. S., Kravtsov, A. V., and Weinberg, D. H., ApJ, **539**, 517, (2000).
7. Taylor, J. E., and Babul, A., ApJ, **559**, 716, (2001).
8. Dalal, N. and Kochanek, C., ApJ, **572**, 25, (2001); Dalal, N. and Kochanek, C., PRL, submitted, (2002).
9. Stoehr, F. *et al.*, MNRAS, **335**, L84, (2002).
10. Hayashi, E. *et al.*, ApJ, in press, astro-ph/0203004, (2002).
11. Navarro, J. F., Frenk, C. S., and White, S. D. M., ApJ, **490**, 493, (1997).
12. King, I., ApJ, **67**, 471, (1962).
13. Mateo, M., Ann. Rev. Astron. Astrophys., **36**, 435, (1998).

THE INTERGALACTIC
AND INTRACLUSTER MEDIUM

The Lyman-α Forest as a Cosmological Tool

David H. Weinberg*, Romeel Davé[†], Neal Katz** and Juna A. Kollmeier[‡]

*The Ohio State University, Dept. of Astronomy, Columbus, OH 43210
[†]University of Arizona, Dept. of Astronomy, Tucson, AZ 85721
**University of Massachusetts, Dept. of Physics and Astronomy, Amherst, MA, 91003
[‡]Ohio State University, Dept. of Astronomy, Columbus, OH 43210

Abstract. We review recent developments in the theory of the Lyα forest and their implications for the role of the forest as a test of cosmological models. Simulations predict a relatively tight correlation between the local Lyα optical depth and the local gas or dark matter density. Statistical properties of the transmitted flux can constrain the amplitude and shape of the matter power spectrum at high redshift, test the assumption of Gaussian initial conditions, and probe the evolution of dark energy by measuring the Hubble parameter $H(z)$. Simulations predict increased Lyα absorption in the vicinity of galaxies, but observations show a Lyα deficit within $\Delta_r \sim 0.5h^{-1}$ Mpc (comoving). We investigate idealized models of "winds" and find that they must eliminate neutral hydrogen out to comoving radii $\sim 1.5h^{-1}$ Mpc to marginally explain the data. Winds of this magnitude suppress the flux power spectrum by ~ 0.1 dex but have little effect on the distribution function or threshold crossing frequency. In light of the stringent demands on winds, we consider the alternative possibility that extended Lyα emission from target galaxies replaces absorbed flux, but we conclude that this explanation is unlikely. Taking full advantage of the data coming from large telescopes and from the Sloan Digital Sky Survey will require more complete understanding of the galaxy proximity effect, careful attention to continuum determination, and more accurate numerical predictions, with the goal of reaching $5-10\%$ precision on key cosmological quantities.

PHYSICS OF THE FOREST

The 1990s saw four epochal advances in our understanding of the Lyα forest. Spectra of quasar pairs showed coherence over scales of a hundred kpc and more, implying large sizes and thus low densities for the absorbing structures [1, 2, 3, 4]. Keck HIRES spectra of unprecedented resolution and signal-to-noise demonstrated the ubiquity of weakly fluctuating Lyα absorption in the high redshift universe [5], and they revealed the presence of metal lines associated with low column density hydrogen absorbers [6, 7]. Finally, and most directly relevant to this review, a combination of numerical simulations and related analytic models led to a compelling new physical picture of Lyα forest absorption [8, 9, 10, 11, 12, 13, 14, 15].

The basic numerical result is simple to summarize: given a cosmological scenario motivated by independent observations, 3-d simulations that incorporate gravity, gas dynamics, and photoionization by the UV background produce something very much like the observed Lyα forest, an outcome that requires no *ad hoc* adjustments to the model. The top three rows of Figure 1 illustrate this point, showing, respectively, the observed Lyα forest of the $z = 3.62$ quasar Q1422+231, expanded views of four selected regions of this spectrum, and simulated spectra of the same length along four randomly selected lines of sight through a cosmological simulation. The simulation uses smoothed particle hydrodynamics (SPH) with 128^3 dark matter particles and 128^3 gas particles in a periodic cube of comoving size $11.111h^{-1}$ Mpc (1422 km s^{-1} at $z = 3$). It assumes a ΛCDM model (inflationary cold dark matter with a cosmological constant), with $\Omega_m = 0.4$, $\Omega_\Lambda = 0.6$, $h = 0.65$, $\Omega_b = 0.02h^{-2} = 0.0473$, inflationary spectral tilt $n = 0.95$, and a power spectrum normalization that corresponds to $\sigma_8 = 0.8$ at $z = 0$. Spectra are extracted from the $z = 3$ simulation output using the methods of [12].

Studies of the Lyα forest have traditionally focused on "lines" identified by a decomposition procedure and the "clouds" or "absorbers" that produce them. Hydrodynamic simulations show that typical marginally saturated absorption features at $z \sim 3$ arise in filamentary structures, which are analogous to (but smaller in scale than) today's galaxy superclusters. However, in the simulations there is no sharp distinction between the "lines" and the "background," and one can also characterize the Lyα forest as "Gunn-Peterson" [16] absorption produced by a smoothly fluctuating intergalactic medium [12, 17, 18, 19].

What makes the fluctuating IGM perspective especially powerful is the tight relation between the density and temperature of low density cosmic gas, $T \approx T_0 (\rho/\bar{\rho})^\alpha$ with $\alpha \approx 0.6$, which emerges from the balance between photoionization heating and adiabatic cooling [20, 21]. The Lyα optical depth of photoionized gas is $\tau \propto n_{\rm HI} \propto \rho^2 T^{-0.7} \Gamma^{-1}$, where $T^{-0.7}$ accounts for the temperature dependence of the recombination rate and Γ is the rate at which neutral atoms are ionized by the cosmic UV background. The temperature-density relation then leads to the "Fluctuating Gunn-Peterson Approximation" (FGPA), $F \equiv \exp(-\tau) = \exp[-A(\rho/\bar{\rho})^\beta]$, which relates the continuum normalized flux F to the local gas overdensity [17, 22, 18, 19]. The index β lies in the range $1.6-2$ depending on the thermal history of the gas, and the proportionality constant A is itself proportional to $\Omega_b^2 h^3 T_0^{-0.7} \Gamma^{-1}$. Thus, one can think of a Lyα forest spectrum as providing a 1-dimensional, non-linear map of the gas overdensity $\rho/\bar{\rho}$ along the line of sight. The map is smoothed by thermal broadening and distorted by peculiar velocities, but these effects are small enough to leave a tight correlation between Lyα optical depth and local gas density even in redshift space [17, 23]. Furthermore, pressure gradients in the diffuse, photoionized IGM are usually weak compared to gravitational forces, so the gas overdensity traces the dark matter overdensity fairly well.

The FGPA provides a valuable way of thinking about the information content of the Lyα forest, and it can be a useful calculational tool if one has a way to produce realizations of cosmic density and velocity fields. The analytic models of the Lyα forest by [8, 24, 13, 14] essentially combine the FGPA with a log-normal or Zel'dovich approximation model for creating non-linear density and velocity fields. Alternatively, one can run an inexpensive, gravity-only N-body simulation and assume that the diffuse gas traces the dark matter (or add an approximate treatment of gas pressure [25, 26, 27]). The dotted lines in the third row of Figure 1 show spectra extracted from a particle-mesh (PM) N-body simulation, run from the same initial conditions as the SPH simulation, illustrating both the effectiveness and the limitations of this approach. The two spectra trace each other over most of their length, with the largest breakdowns occurring in regions where shock heating has pushed the gas above the $T \propto \rho^\alpha$ temperature-density relation, which makes the optical depth in the SPH simulation lower than the FGPA predicts.

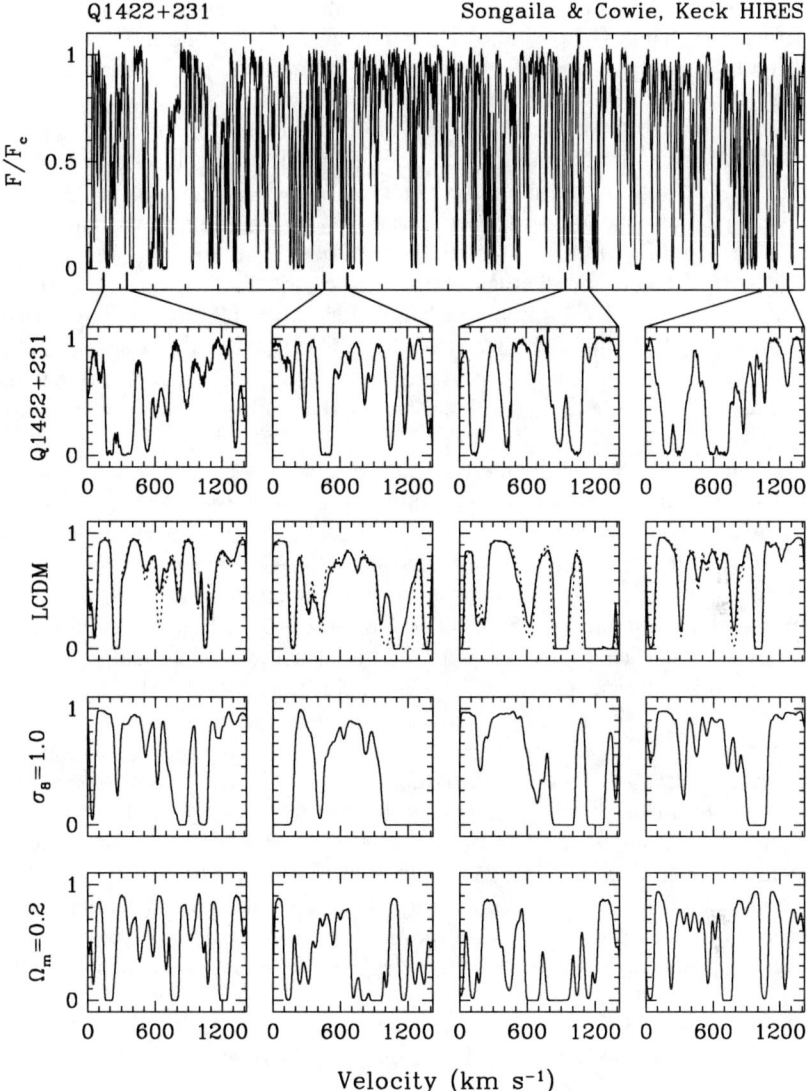

FIGURE 1. Simulating the Lyα forest. The top panel shows a continuum normalized, Keck HIRES spectrum of the Lyα forest region of the quasar Q1422+231 ($z = 3.62$), from [6]. The next row of panels shows blowups of four regions 1422 km s^{-1} in length. The third row shows simulated spectra extracted along four random lines of sight at $z = 3$ from an SPH simulation (solid) and a PM simulation (dotted) of the ΛCDM model, with $\Omega_m = 0.4$, $h = 0.65$, and $\sigma_8 = 0.8$ (at $z = 0$). The fourth and fifth rows show PM results for different cosmological parameter values: $\sigma_8 = 1.0$ and $\Omega_m = 0.2$, respectively. All simulated spectra are 1422 km s^{-1} in length.

The N-body+FGPA technique offers a convenient way to investigate the response of the Lyα forest to changes in cosmological parameters. Comparing the third and fourth rows of Figure 1 illustrates the effect of increasing the matter fluctuation amplitude by 20%, to $\sigma_8(z=0) = 1.0$, with the same Fourier phases in the initial conditions. The increased clustering of the high amplitude model is evident in the greater incidence of saturated absorption and in the merging of multiple small scale features into single larger scale features. The bottom row shows spectra from a model with $\sigma_8 = 0.8$ and the same linear power spectrum but a lower matter density, $\Omega_m = 0.2$. The reduction in Ω_m lowers the value of the Hubble parameter $H(z)$ at $z = 3$, and as a result features are more densely packed in redshift space, producing "choppier" spectra. In fact, our $11.111 h^{-1}$ Mpc simulation cube is only 1024 km s^{-1} in length for $\Omega_m = 0.2$, and we have "padded" each spectrum to 1422 km s^{-1} by replicating the first 400 km s^{-1}. Reducing Ω_m also raises the fluctuation amplitude at $z = 3$ (since σ_8 is held fixed at $z = 0$ and there is less late time growth for lower Ω_m), but this effect is less important than the effect on the Hubble parameter.

The relative simplicity of the underlying physics and the existence of superb data at redshifts that are only sparsely probed by other observables make the Lyα forest a potentially powerful tool for testing cosmological models. Given independent estimates of Γ (e.g., from the quasar luminosity function or the proximity effect), one can use the mean opacity of the forest to obtain a lower limit to the cosmic baryon density [18, 28]. However, the observational constraints on Γ are loose, so for higher precision applications one must treat Γ as a free parameter and adjust it to match a single observable, usually taken to be the mean opacity. This normalization also absorbs uncertainties in Ω_b, T_0, and h. We have followed this practice for each of the simulations in Figure 1 and will adopt it in all of our subsequent analyses. After normalizing to the mean opacity, the variation about the mean — i.e., the structure of the Lyα forest — is a prediction of the cosmological model, driven primarily by the structure in the underlying dark matter distribution. Ideally, one would like to calculate predictions for each cosmology using high resolution hydrodynamic simulations, but for some purposes other numerical or analytic approximations may be accurate enough. The adequacy of such approximations must be tested on a case-by-case basis, and the improving quality of the observational data places ever more stringent demands on the accuracy of the theoretical calculations.

CONTINUOUS FLUX STATISTICS

Because of the tight predicted correlation between the local Lyα optical depth and the local gas density, the most natural statistics for characterizing the Lyα forest are those that treat each quasar spectrum as a continuous 1-dimensional field, rather than a collection of discrete "lines." We have reviewed this approach elsewhere [19, 23], and here we provide a brief recap with different examples and emphasis.

Figure 2 shows the power spectrum of continuum normalized flux, converted from 1-d to 3-d as described by [22], and multiplied by k^3 to yield the variance per $\ln k$. The curves are noisy because they are based on a single $11.111 h^{-1}$ Mpc comoving cube, but we use the same phases in each simulation, so the relative behavior should be minimally

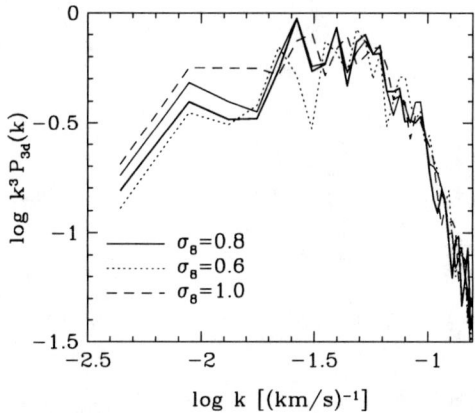

FIGURE 2. The flux power spectrum. Heavy and light solid curves show the 3-d flux power spectrum from, respectively, the SPH simulation and a PM simulation with the same initial conditions. Dotted and dashed curves show results with lower and higher matter fluctuation amplitudes, corresponding to $\sigma_8 = 0.6$ and 1.0 at $z = 0$. All results are from a single $11.111h^{-1}$ Mpc cube realized with the same Fourier phases.

affected by noise. At scales $k > 0.03$ (km s^{-1})$^{-1}$, the power spectra turn over because of the combined effects of non-linearity, peculiar velocities, and thermal broadening. The predictions in this regime are also affected by the finite resolution of the simulations. At larger scales, the SPH and N-body+FGPA methods give similar but not identical results for the same cosmological model, and the amplitude of the flux power spectrum increases with the amplitude of the matter power spectrum, as expected based on the FGPA and on Figure 1.

On large scales, the shape of the 3-d flux power spectrum is similar to that of the linear matter power spectrum $P_m(k)$ [22, 29]. The close connection between the shape and amplitude of the flux power spectrum and the shape and amplitude of $P_m(k)$ is the basis of Croft et al.'s [22] method for recovering $P_m(k)$ from Lyα forest data. Applying this method to a sample of 30 Keck HIRES spectra and 23 Keck LRIS spectra yields a matter power spectrum in remarkably (or, perhaps, disappointingly) good agreement with the "concordance" ΛCDM model favored by CMB, supernova, weak lensing, and low-z large scale structure data ([29]; see [30] for an independent analysis of the same flux power spectrum and [31] for an independent comparison to other cosmological constraints). McDonald et al. [32] reach similar conclusions from a "forward" comparison of hydrodynamic simulation predictions to the flux power spectrum measured from eight HIRES spectra. The Lyα forest power spectrum tests the ΛCDM model in a previously unexplored regime of redshift and lengthscale. It confirms one of the scenario's key predictions, a linear power spectrum that bends from the primeval k^n towards k^{n-4} on small scales. The implied constraints on cosmological parameter combinations complement those from the CMB and other data. The Sloan Digital Sky Survey (SDSS) has already obtained moderate resolution spectra of several thousand high redshift quasars, and these will soon provide measurements of the flux power spectrum with much greater precision on large scales.

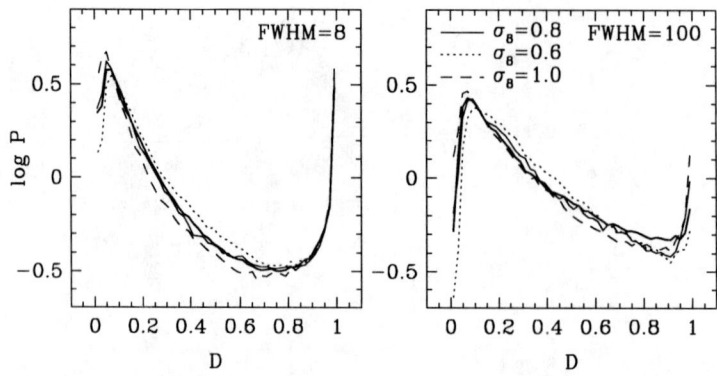

FIGURE 3. PDF of the flux decrement $D = 1 - e^{-\tau}$, measured from simulated spectra smoothed with a Gaussian of FWHM=8 km s^{-1} (left) or 100 km s^{-1} (right). Line types as in Fig. 2.

Figure 3 shows a different statistic, the probability distribution function (PDF) of the flux decrement $D = 1 - F = 1 - e^{-\tau}$, for the same set of simulations. The measurements in the left hand panel are from simulated spectra smoothed with a Gaussian of FWHM = 8 km s^{-1}, comparable to the resolution of Keck HIRES or VLT UVES spectra. The SPH and N-body+FGPA predictions agree well for $\sigma_8 = 0.8$. Models with higher fluctuation amplitude have a broader distribution of densities ρ and a correspondingly broader distribution of flux decrements D, with more saturated and low-opacity pixels and fewer pixels of intermediate opacity. The differences in the predicted PDFs for $D \sim 0.3 - 0.7$ are $\Delta \log P \sim 0.1$ — not enormous, but readily measurable at high statistical significance with reasonable observational samples. The fraction of saturated pixels (defined here by $D > 0.96$) increases from 7.2% to 8.3% to 9.1% as σ_8 goes from 0.6 to 0.8 to 1.0. There are larger differences in the number of nearly transparent pixels, but continuum fitting uncertainties make it difficult to measure the PDF accurately near $D = 0$. One can investigate the scale dependence of matter clustering by smoothing the spectra (or by observing them at lower spectral resolution), analogous to studying galaxy counts in cells of increasing size. Figure 3 (right) shows results for 100 km s^{-1} smoothing. The predicted PDFs are narrower, since smoothing drives pixel values towards the mean, but the dependence on σ_8 is similar. The saturated pixel fraction doubles, from 1.9% to 3.8%, as σ_8 rises from 0.6 to 1.

For cosmological models with Gaussian initial conditions, the shape of the flux PDF depends mainly on the amplitude of mass fluctuations, once one has chosen Γ to match the mean opacity, [33, 23]. The effective physical scale of this amplitude measurement is determined by the spectral smoothing or, for spectra that fully resolve the observed absorption features, by a combination of thermal broadening and gas pressure effects. Models with non-Gaussian initial conditions predict significantly different flux PDFs [23]. Observational tests to date show good agreement with models that have Gaussian initial conditions and a $P_m(k)$ amplitude compatible with ΛCDM predictions [18, 23, 32]. The uncertainties in these tests are comparable to the model differences in Figure 3. Reducing them requires careful attention to the effects of continuum fitting and noise and

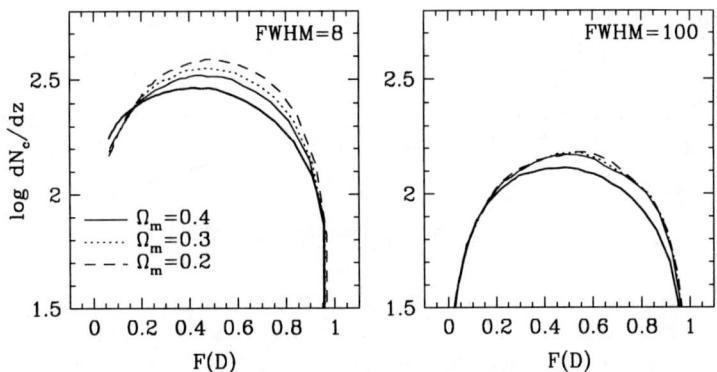

FIGURE 4. Threshold crossing frequency in spectra with FWHM=8 km s^{-1} (left) and 100 km s^{-1} (right). Curves show the number of times per unit redshift that the spectrum crosses a threshold of flux decrement D as a function of the filling factor, the fraction $F(D)$ of pixels with flux decrement less than D. Heavy and light solid curves show the SPH and PM results, respectively, for $\Omega_m = 0.4$. Dotted and dashed lines show PM results for $\Omega_m = 0.3$ and 0.2, respectively.

to the accuracy of the theoretical predictions, but it should be possible to obtain tight constraints on the normalization of the matter power spectrum, and some information on its shape by studying different smoothing scales. Consistency between results from the flux power spectrum and the flux PDF can be a sensitive diagnostic for primordial non-Gaussianity.

Figure 4 shows the threshold crossing frequency, the number of times per unit redshift that the absorption spectrum crosses a decrement threshold D. Following [34], we plot dN_c/dz as a function of filling factor, the fraction $F(D)$ of pixels with flux decrement less than D, which cleanly separates the information in dN_c/dz from the information in the PDF and makes the model predictions nearly independent of the photoionization rate Γ. As one would expect from Figure 1, the threshold crossing frequency increases as Ω_m decreases because a lower Hubble parameter $H(z)$ "squeezes" redshift separations relative to comoving distances. The threshold crossing frequency drops slightly as the amplitude of $P_m(k)$ increases and gravitationally driven merging "smooths" structure [23], but the effect is small over the range $\sigma_8 = 0.6 - 1$. The crossing frequency also drops for redder matter power spectra, since these also lead to smoother structure [23]. For high spectral resolution, dN_c/dz also depends on the gas temperature, which determines the level of thermal broadening.

The great promise of the threshold crossing statistic is its potential for constraining the Hubble parameter at high redshift. In spatially flat models with a cosmological constant, the ratio $H(z)/H_0$ is determined by Ω_m. Alternatively, if Ω_m is known independently, the ratio $H(z)/H_0$ can constrain the equation of state of dark energy [35, 36]. Unfortunately, dN_c/dz also depends on other cosmological and IGM parameters, and it is numerically difficult to predict with high accuracy even when the model is fully specified. The difference between the PM and SPH predictions in Figure 4 is comparable to the Ω_m effects themselves, and the SPH result is still affected by the finite numerical resolution. On the observational side, accurate measurement of dN_c/dz requires high signal-to-noise spectra. Furthermore, while moderate resolution spectra can provide useful diagnostics for the shape and amplitude of the matter power spectrum [23], the $H(z)$ application demands high spectral resolution, since smoothing over a scale that is fixed in redshift units erases the sensitivity to $H(z)$ (see Fig. 4, right). Exploiting dN_c/dz as a probe of dark energy thus represents a theoretical and observational challenge.

THE GALAXY PROXIMITY EFFECT

One can also use the observable correlations between Lyman Break Galaxies (LBGs) and the Lyα forest to study the environments of high redshift galaxies. Several groups have investigated this issue theoretically using simulations [37, 38, 39, 40], and Adelberger et al. have carried out an observational study using an LBG survey in fields probed by seven quasar lines of sight ([41], hereafter ASSP). On large scales — cubes of comoving size $\sim 13h^{-1}$ Mpc — ASSP find a clear correlation of galaxy overdensity with Lyα flux decrement, the expected signature of galaxy formation in overdense environments [37, 39]. However, the observed Lyα decrement *decreases* within $\Delta_r \approx 1h^{-1}$ Mpc of LBGs (comoving, redshift-space separation), where the simulations predict that absorption should be strongest [38, 39, 40].

Figure 5 illustrates this conflict. In the upper left panel, triangles show the predicted mean decrement in bins of Δ_r, while filled circles show the ASSP data points. The predictions come from an SPH simulation of a $22.222h^{-1}$ Mpc cube, and we select the 40 galaxies with the highest star formation rates to approximate the magnitude limit of ASSP's spectroscopic survey (see [39] for details). Plausible random errors in galaxy redshifts can reduce the discrepancy at $\Delta_r \approx 1-2h^{-1}$ Mpc, but they cannot explain the results at the smallest scales, especially the innermost data point at $\Delta_r = 0-0.5h^{-1}$ Mpc, $\langle D \rangle = 0.11$ [38, 39, 42]. The right column of Figure 5 shows simulated spectra along 12 lines of sight that pass within $0.5h^{-1}$ Mpc of a target galaxy, centered on the galaxy redshift space position. All but one of these spectra have $D \geq 0.6$ at the galaxy redshift, and half have $D \geq 0.9$.

The obvious conclusion is that feedback from the observed galaxies reduces neutral hydrogen in their immediate surroundings. Local photoionization by the galaxies' stars or active nuclei proves insufficient; this argument can be cast in general terms that seem difficult to escape [39]. The natural alternative is some form of supernova or AGN-driven wind [41, 38, 40]. Here (and in [42]) we have investigated highly idealized "wind" models in which we simply eliminate all neutral hydrogen in a sphere of comoving radius R_w around each target galaxy. Squares, pentagons, and hexagons in Figure 5 show results for $R_w = 0.75$, 1.0, and $1.5h^{-1}$ Mpc. Stars show a model where we place winds around all 641 resolved galaxies in the simulation volume (instead of the top 40 that constitute the "observed" sample) and scale the wind volume in proportion to the galaxy baryon mass, with a normalization $R_w = 1h^{-1}$ Mpc around the 40th-ranked galaxy. In the right column, dotted curves show spectra for the $R_w = 1.5h^{-1}$ Mpc model.

The most important lesson from Figure 5 is that eliminating neutral gas to a *real space* distance R_w does not eliminate absorption within a *redshift space* separation R_w (e.g., to $\Delta V = \pm 200$ km s^{-1} in the spectrum plots). Peculiar infall velocities allow gas

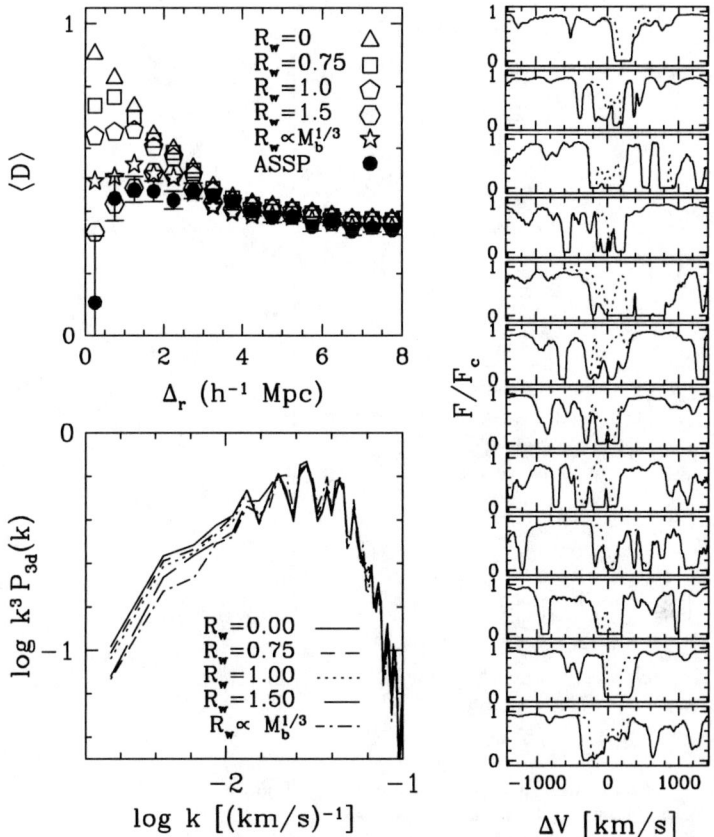

FIGURE 5. Influence of idealized "winds" on the conditional mean flux decrement and the flux power spectrum, based on an SPH simulation of a $22.222h^{-1}$ Mpc (comoving) cube. Triangles in the upper left panel show the mean flux decrement in pixels that lie at redshift-space separation Δ_r from one of the 40 brightest galaxies in the cube. Results are averaged over $0.5h^{-1}$ Mpc (comoving) bins of Δ_r. Squares, pentagons, and hexagons show the effect of removing all neutral hydrogen in spheres of comoving radius 0.75, 1.0, and $1.5h^{-1}$ Mpc around these galaxies before extracting spectra, and stars show a model with sphere volume proportional to galaxy baryon mass. Filled circles show the observational estimates of ASSP. Sample spectra on the right illustrate the effect qualitatively. Solid curves show spectra along 12 lines of sight selected to pass within $0.5h^{-1}$ Mpc (24") of a target galaxy at redshift space position $\Delta V = 0$. Dotted curves show the corresponding spectra for $R_w = 1.5h^{-1}$ Mpc. The lower left panel shows the flux power spectrum for the unmodified SPH simulation (solid) and the various "wind" models (as marked).

at larger distances to produce absorption near the galaxy redshift, making the energetic requirements on any wind explanation of the observed Lyα deficit much more stringent. Only the mass-scaled and $R_w = 1.5h^{-1}$ Mpc models come close to matching the ASSP data. Winds that fully ionize or perfectly entrain gas to such large distances are not a natural outcome of hydrodynamic simulations with stellar feedback [43, 40], and even reaching $1.5h^{-1}$ Mpc in ~ 1 Gyr requires a sustained propagation speed ~ 600 km s^{-1}.

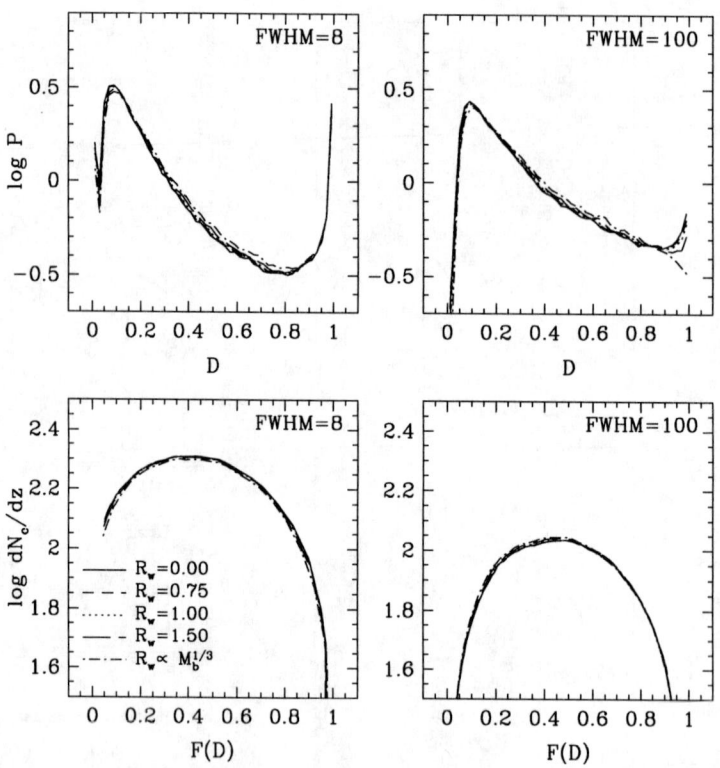

FIGURE 6. Influence of winds on the flux decrement PDF (top, as in Fig. 3) and the threshold crossing frequency (bottom, as in Fig. 4). Results are shown for the $22.222h^{-1}$ Mpc SPH simulation with no modification (solid) and with the "wind" models illustrated in Fig. 5.

Given the challenges facing the wind explanation, it is worth considering the alternative possibility that extended Lyα *emission* from the target galaxies is "filling in" the corresponding region of the absorption spectrum. Steidel et al. [44] observed two extended Lyα "blobs" apparently associated with LBGs, with angular extents ~ 15" and AB apparent magnitudes 21.02 and 21.14 in the Lyα band. Cooling radiation from gas settling into massive galaxies at $z = 3$ naturally produces Lyα fluxes of this order [45, 46]. The three quasars that contribute to ASSP's innermost data point have G-band AB magnitudes of 20.1, 21.6, and 23.4, so if *all* of a target galaxy's Lyα cooling radiation went down the slit it could potentially replace the quasar flux absorbed by the surrounding IGM. However, a 1.4" slit at an angular separation $\Delta\theta \sim 15-20$" from a galaxy should intercept at most $\sim 1.4/2\Delta\theta \sim 0.03-0.05$ of the galaxy's extended Lyα flux, at least on average, so this explanation seems to fail by $1-2$ orders of magnitude. Furthermore, a fourth pair involving the $G = 17.8$ quasar Q0302-0019, shows no sign of absorption near the galaxy redshift, and in this case the quasar is clearly too bright for galaxy emission to compete with it. (This pair and two others are dropped from ASSP's $\langle D \rangle$ calculation because of possible Lyβ contamination.) At this point, the Lyα emission explanation seems unlikely; it can be conclusively ruled out by observing more close pairs involving bright quasars or by obtaining symmetrically placed spectra away from the observed quasars to search for galaxy emission.

Assuming for now that winds are the correct explanation for the observed Lyα deficit, one can ask whether they completely spoil the picture painted in §§1 and 2. The lower left panel of Figure 5 shows the flux power spectrum for the various "wind" models. Eliminating neutral hydrogen to distances $R_w = 0.75$ or $1h^{-1}$ Mpc around bright galaxies has only a small impact on the flux power spectrum, because the filling factor of the "bubbles" is small and absorption close to the galaxies remains nearly saturated in any case. However, these models also do not explain the ASSP results. In the two more extreme models, winds suppress the flux power spectrum on large scales by $0.1 - 0.2$ dex, comparable to the 0.1-dex 1σ uncertainty that [29] quote for the normalization of the matter power spectrum. Thus, winds of this magnitude could have systematic effects on $P_m(k)$ determinations that are significant relative to the present observational uncertainties. The influence of winds on the flux PDF and threshold crossing frequency is smaller, as shown in Figure 6. Even the more extreme models have a negligible impact on dN_c/dz, and they only slightly alter the shape of the PDF. The most significant effect is on the fraction of saturated pixels for 100 km s^{-1} smoothing, which drops from 2.5% for the no-wind case to 1.9% for $R_w = 1.5h^{-1}$ Mpc and 1.3% for the mass-scaled model.

PROSPECTS AND CHALLENGES

From the above discussion, it is clear that one immediate challenge is to better understand the galaxy proximity effect. Figures 5 and 6 show that winds can reach substantial distances from bright galaxies without having much impact on the global statistics of the Lyα forest (see also [43, 40]), but even the more extreme wind models considered in §3 do not reproduce the ASSP results very well. Simulations with more realistic wind physics [38, 43, 40] can shed further light on this problem, but major progress will have to await further observational studies, since current inferences rest crucially on a handful of galaxy-quasar pairs.

Studies with Keck HIRES and VLT UVES have provided a (still growing) trove of high quality data on the Lyα forest at $z \sim 2 - 4$. The fluctuating IGM perspective, furthermore, shows that large samples of moderate resolution spectra can be a powerful resource for studying structure on large scales, since continuous flux statistics do not require resolution of individual "lines." Such samples can be assembled quickly with large telescopes, and the SDSS is producing an enormous sample at resolution $R \sim 2000$ in the course of its normal operations. One of the observational frontiers is the use of correlations across multiple lines of sight. Quasar pair studies provided the first decisive evidence for large coherence scales of absorbing structures [1, 2, 3, 4], but larger samples allow more ambitious goals, such as using the Alcock-Paczynski test [47] to measure spacetime geometry [48, 49, 50], improving measurements of the flux power spectrum with cross correlations [51], and mapping large scale 3-dimensional structure at high redshift [52, 53].

Analyses of existing data have already led to an important cosmological conclusion, namely that models with matter clustering similar to that of "concordance" ΛCDM at $z \sim 3$ are consistent with the observed Lyα forest while models with substantially different clustering amplitudes or suppression of small scale power are not [18, 54, 55, 23, 32, 56, 57, 58, 26, 29, 30]. They have also provided constraints on the temperature of the diffuse IGM [59, 60, 61] and indications of helium reionization at $z \sim 3.2$ [61, 62, 63, 64]. However, to do justice to the quality and quantity of data and keep pace with the tightening observational constraints from other observables, we must play for higher stakes. It looks possible in principle to achieve precision of $5 - 10\%$ on quantities like the matter fluctuation amplitude and $H(z)$, but even without the potential complications of galaxy feedback, inferences at this level require more extensive theoretical modeling. Many effects that are unimportant at the 25% level — differences between approximate methods and full hydrodynamics, numerical resolution and box size limitations, spatial fluctuations in IGM temperature, details of continuum determination — may become critical at the $5 - 10\%$ level. Despite these challenges, the Lyα forest is the most promising tool we have for precision cosmological measurements at $z \sim 2 - 4$. These measurements might, in the end, simply confirm the cosmological model favored by other data, but complementary constraints have the potential to break parameter degeneracies and thereby reveal subtle quantitative discrepancies. These in turn could yield insight into the nature of dark energy, the mechanisms of inflation, or some other fundamental aspect of our universe.

REFERENCES

1. Bechtold, J., Crotts, A. P. S., Duncan, R. C., & Fang, Y. 1994, ApJ, 437, L83
2. Dinshaw, N., Impey, C. D., Foltz, C. B., Weymann, R. J., & Chaffee, F. H. 1994, ApJ, 437, L87
3. Dinshaw, N., Foltz, C. B., Impey, C. D., Weymann, R. J., & Morris, S. L. 1995, Nature, 373, 223
4. Crotts, A. P. S., & Fang, Y. 1998, ApJ, 502, 16
5. Hu, E.M., Kim, T.S., Cowie, L.L., Songaila, A., & Rauch, M. 1995, AJ, 110, 1526
6. Songaila, A. & Cowie, L.L. 1996, AJ, 112, 335
7. Ellison, S. L., Lewis, G. F., Pettini, M., Chaffee, F. H., & Irwin, M. J. 1999, ApJ, 520, 456
8. Bi, H.G., 1993, ApJ, 405, 479
9. Cen, R., Miralda-Escudé, J., Ostriker, J.P., & Rauch, M. 1994, ApJ, 437, L9
10. Petitjean, P., Mücket, J. P., & Kates, R. E. 1995, A&A, 295, L9
11. Zhang, Y., Anninos, P., & Norman, M.L. 1995, ApJ, 453, L57
12. Hernquist L., Katz, N., Weinberg, D.H., & Miralda-Escudé, J. 1996, ApJ, 457, L51
13. Bi, H.G., & Davidsen, A. 1997, ApJ, 479, 523
14. Hui, L., Gnedin, N., & Zhang, Y. 1997, ApJ, 486, 599
15. Theuns, T., Leonard, A., Efstathiou, G., Pearce, F. R., & Thomas, P. A. 1998, MNRAS, 301, 478
16. Gunn, J.E., & Peterson, B.A. 1965, ApJ, 142, 1633
17. Croft, R.A.C., Weinberg, D.H., Katz, N., Hernquist, L., 1997, ApJ, 488, 532
18. Rauch, M., Miralda-Escudé, J., Sargent, W. L. W., Barlow, T. A., Weinberg, D. H., Hernquist, L., Katz, N., Cen, R., & Ostriker, J. P., 1997, ApJ, 489, 7
19. Weinberg, D. H., Katz, N., & Hernquist, L. 1998, in ASP Conference Series 148, Origins, eds. C. E. Woodward, J. M. Shull, & H. Thronson, (ASP: San Francisco), 21, astro-ph/9708213
20. Hui, L., & Gnedin, N. 1997, MNRAS, 292, 27
21. Weinberg, D. H., Hernquist, L., & Katz, N. 1997, ApJ, 477, 8
22. Croft, R. A. C., Weinberg, D. H., Katz, N., & Hernquist, L. 1998, ApJ, 495, 44
23. Weinberg, D. H., et al. 1999, in Evolution of Large Scale Structure: From Recombination to Garching, eds. A.J. Banday, R. K. Sheth, & L. N. Da Costa, (Twin Press: Vledder NL), 346 astro-ph/9810142
24. Bi, H., Ge, J., & Fang, L.-Z. 1995, ApJ, 452, 90

25. Gnedin, N. Y., & Hui, L. 1998, MNRAS, 296, 44
26. Zaldarriaga, M., Hui, L., & Tegmark, M. 2001, ApJ, 557, 519
27. Viel, M., Matarrese, S., Mo, H. J., Theuns, T., & Haehnelt, M. G. 2002, MNRAS, 336, 685
28. Weinberg, D.H., Miralda-Escudé, J., Hernquist, L., & Katz, N., 1997, ApJ, 490, 564
29. Croft, R. A. C., Weinberg, D. H., Bolte, M., Burles, S., Hernquist, L., Katz, N., Kirkman, D., Tytler, D. 2002, ApJ, 581, 20
30. Gnedin, N. Y. & Hamilton, A. J. S. 2002, MNRAS, 334, 107
31. Tegmark, M. & Zaldarriaga, M. 2002, PRD, 66, 103508
32. McDonald, P., Miralda-Escudé, J., Rauch, M., Sargent, W. L. W., Barlow, T. A., Cen, R., & Ostriker, J. P. 2000, ApJ, 543, 1
33. Cen, R. 1997, ApJ, 479, L85
34. Miralda-Escudé J., Cen R., Ostriker, J.P., & Rauch, M. 1996, ApJ, 471, 582
35. Kujat, J., Linn, A. M., Scherrer, R. J., & Weinberg, D. H. 2002, ApJ, 572, 1
36. Viel, M., Matarrese, S., Theuns, T., Munshi, D., & Wang, Y. 2002, MNRAS, submitted, astro-ph/0212241
37. McDonald, P., Miralda-Escudé, J., & Cen, R. 2002, ApJ, 580, 42
38. Croft, R. A. C., Hernquist, L., Springel, V., Westover, M., White, M. 2002, ApJ, 580, 634
39. Kollmeier, J.A., Weinberg, D.H., Davé, R., Katz, N., 2002, ApJ, submitted, astro-ph/0209563
40. Bruscoli, M., Ferrara, A., Marri, S., Schneider, R., Maselli, A., Rollinde, E., Aracil, B., MNRAS, submitted, astro-ph/0212126
41. Adelberger, K.L., Steidel, C.C., Shapley, A.E., Pettini, M. 2002, ApJ, in press, astro-ph/0210314
42. Kollmeier, J.A., Weinberg, D.H., Davé, R., & Katz, N., these proceedings
43. Theuns, T., Viel, M., Kay, S., Schaye, J., Carswell, R. F., & Tzanavaris, P. 2002, ApJ, 578, L5
44. Steidel, C. C., Adelberger, K. L., Shapley, A. E., Pettini, M., Dickinson, M., & Giavalisco, M. 2000, ApJ, 532, 170
45. Haiman, Z., Spaans, M., & Quataert, E. 2000, ApJ, 537, L5
46. Fardal, M. A., Katz, N., Gardner, J. P., Hernquist, L., Weinberg, D. H. & Davé, R. 2001, ApJ, 562, 605
47. Alcock, C., & Paczyński, B. 1979, Nature, 281, 358
48. Hui, L., Stebbins, A., & Burles, S. 1999, ApJ, 511, 5
49. McDonald, P. & Miralda-Escudé, J. 1999, ApJ, 518, 24
50. McDonald, P. 2001, ApJ, submitted, astro-ph/0108064
51. Viel, M., Matarrese, S., Mo, H. J., Haehnelt, M. G., & Theuns, T. 2002, MNRAS, 329, 848
52. Liske, J., Webb, J. K., Williger, G. M., Fernández-Soto, A., & Carswell, R. F. 2000, MNRAS, 311, 657
53. Rollinde, E., Petitjean, P., Pichon, C., Colombi, S., Aracil, B., D'Odorico, V., & Haehnelt, M. G. 2002, MNRAS, submitted
54. Croft, R. A. C., Weinberg, D. H., Pettini, M., Katz, N., & Hernquist, L. 1999, ApJ, 520, 1
55. Theuns, T., Leonard, A., Schaye, J., & Efstathiou, G. 1999, MNRAS, 303, L58
56. Narayanan, V. K., Spergel, D. N., Davé, R., & Ma, C. 2000, ApJ, 543, L103
57. Theuns, T., Schaye, J., & Haehnelt, M. G. 2000, MNRAS, 315, 600
58. Meiksin, A., Bryan, G., & Machacek, M. 2001, MNRAS, 327, 296
59. Ricotti, M., Gnedin, N. Y. & Shull, J. M. 2000, ApJ, 534, 41
60. McDonald, P., Miralda-Escudé, J., Rauch, M., Sargent, W. L. W., Barlow, T. A., & Cen, R. 2000, ApJ, 562, 52
61. Schaye, J., Theuns, T., Rauch, M., Efstathiou, G. & Sargent, W. L. W. 2000, MNRAS, 318, 817
62. Theuns, T., Zaroubi, S., Kim, T., Tzanavaris, P., & Carswell, R. F. 2002, MNRAS, 332, 367
63. Theuns, T., Bernardi, M., Frieman, J., Hewett, P., Schaye, J., Sheth, R. K., & Subbarao, M. 2002, ApJ, 574, L111
64. Bernardi, M., et al. 2003, AJ, 125, 32

The Entropy in Groups—A Clue to Galaxy Formation

Richard Mushotzky

*Laboratory for High Energy Astrophysics
NASA Goddard Space Flight Center, Code 662
Greenbelt, Maryland US*

Abstract. The entropy in the hot x-ray gas in groups of galaxies is a fossil of the process of galaxy formation. The amount of entropy in these low mass systems considerably exceeds that predicted from structure formation models. To explain these results requires "extra" energy which is a relic of the process of star formation and active galaxy heating. We present new XMM results on the entropy and entropy profiles. These results are inconsistent with pre-heating scenarios which have been developed to explain the entropy floor in groups but are broadly consistent with models of structure formation which include the effects of heating and/or the cooling of the gas. The total entropy in these systems provides a strong constraint on all models of galaxy and group formation, and on the poorly defined feedback process which controls the transformation of gas into stars and thus the formation of structure in the universe.

I INTRODUCTION

The standard theory of the formation of structure by the evolution of dark matter halos in a cold dark matter scenario has been remarkably successful as shown by several of the talks at this meeting (cf., Frenk, 2002; Peacock, 2002). But it has several crucial "missing pieces" and problems. While the evolution of dark matter halos is well understood, it is not known by what detailed processes gas becomes galaxies, clusters, and groups. It is well known that if one does not include processes other than gravity, the cooling of gas in dark matter halos does not form the galaxies that we see today. This set of processes has been given the generic name of "feedback," since it is thought to be directly related to the formation of stars and black holes, which feed back energy into the gas as it cools and collapses into the dark matter halos. However, it is not at all clear what is the origin of the "feedback" process and how it controls the efficiency of the conversion of gas into stars and thus governs the star formation rate in the universe and the optical, UV, and infrared properties of galaxies. The interaction between star formation via cooling and the energy released by massive stars and supernova is quite complex and, given the absence of a fundamental theory of star formation, is rather difficult to model. In addition to the direct production of stars, one also needs to predict the chemical evolution of galaxies and how it is connected with their formation, mass, and age.

The largest uncertainties in galaxy formation models are related to the relative efficiency of cooling, feedback, and star formation; what is the origin of the energy in the feedback process and in what form the energy is produced (heat, mass motion, etc). We also need to understand how these processes regulate what fraction of baryons is converted into luminous matter, what fraction is ejected out of dark halos by the feedback process, what fraction remains in the hot gas, and where and how the metals are distributed.

The stars and hot x-ray emitting gas in groups are the fossils of this process and contain the missing information.

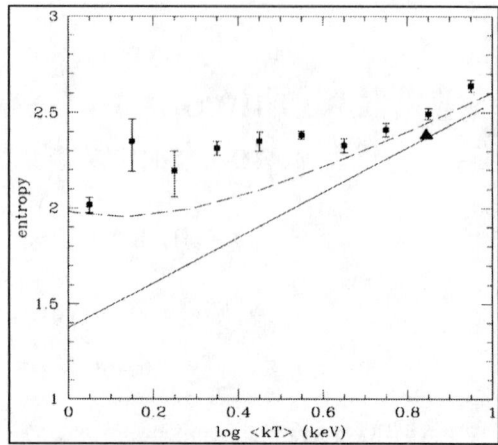

Figure 1: the entropy of a set of groups and clusters vs the x-ray temperatures. The solid line is the entropy expected from shocks. Notice that at low temperatures, the observed entropy considerably exceeds that predicted from shocks.

The well-tested cold, dark matter model makes strong predictions on the formation of groups and clusters including their mass profiles, their evolution with cosmic time, and the number of objects per unit mass per unit volume. One of the more robust predictions is how entropy is produced in the shocks in the infalling gas (Eke et al., 1998), with most of the entropy in the x-ray emitting gas predicted to be from the accretion shock of gas falling into the dark matter potential. Recent ROSAT and ASCA results (Ponman et al., 1999; Finoguenov et al., 2002) show that the gas in groups has much more entropy than can be provided by shocks, indicating the importance of some process other than gravity (Figure 1).

II. WHY ARE GROUPS INTERESTING AND IMPORTANT

Most of the matter in virialized systems is in groups (Fukugita Hogan and Peebles 1998) and at least 60% of the galaxies, thus making groups the "average" place in the universe.

The lower mass of groups compared to clusters makes them much more sensitive to "additional" physics than clusters. In massive clusters, gravity is dominant and the additional physics that we are searching for is much more difficult to discern.

In addition, massive clusters are special objects, representing the highest density regions of the universe, and it is likely that the history of structure formation in them has been different from lower mass systems. If one can believe the recent numerical simulations (Davé et al., 2001), a very large fraction of all baryons at z < 1 are in the hot phase with $T > 10^6$ K. These baryons are enriched in metals and have a large fraction of all the metals ever created. Thus these "missing" baryons hold the key to the understanding of the origin of structure formation.

How can we observe this process of structure formation and metal enrichment?

Groups and cluster of galaxies and the gas that they contain are a relic of the formation of both structure and metal formation. Groups form over a wide redshift range with more or less equal efficiency per unit redshift (Mo and White, 2002) and thus are relics of the epoch at which they formed.

III. SYMPTOMS OF THE "PROBLEM"

It has been known for over 15 years (Kaiser, 1986) that the x-ray emission from clusters and groups does not follow many of the scaling predictions of CDM models without additional physics. There are a large number of such "problems":

1) The deviation of the observed relation between temperature and luminosity $L \sim T^3$ from the $L \sim T^2$ expected has been known for many years (Figure 2, Horner et al., 2002). This indicates that some process other than gravity has modified the observed luminosity or temperature or both.
2) The lack of strong evolution in the x-ray luminosity function compared to that expected from simple scaling arguments (Evrard and Henry, 1989) also indicates that there is additional physics required.
3) The surface brightness profiles of low mass systems are systematically flatter than high mass systems (White, 1991), which is not expected in CDM models.
4) The deviation of the normalization of the total cluster mass/ temperature relation differs by $\sim 40\%$ from simulations in which only gravity is included (Horner et al., 1999).
5) While the total baryonic content of clusters is close to that expected from nucleosynthesis and a low Ω universe, the baryonic fractions in spiral and elliptical galaxies are considerably lower, indicating counter to expectations, that these systems have lost baryons. Most assumptions are that galaxies represent baryonic concentrations (Silk, 2002).
6) There is also a theoretical problem with overcooling, in the dense universe at early times when cooling is initially very efficient. This leads to an unrealistically large fraction of the halo baryon content cooling and forming stars, producing many more stars and galaxies than observed (van den Bosch, 2002).

All of these are symptoms of the lack of inclusion of "feedback" in the theoretical predictions. In the last few years there have been many theoretical efforts to include such processes (e.g., Thomas et al., 2002; Davé et al., 2002; Balogh et al., 1999), which have produced results more consistent with the observations but which are, as yet, relatively unconstrained by observations.

IV. THE ENTROPY PROBLEM IN GROUPS

It was first noted by Ponman et al., 1999, based on ROSAT data, that groups of galaxies had systematically more entropy at $R \sim 0.1\,R_{vir}$ than could be produced by shocks in the standard CDM model (Eke et al., 1998). This was confirmed by analysis of the ASCA data (Lowenstein, 2001; Finoguenov et al., 2002) and is now detected at other scales as well. This "excess" entropy seems to enter at mass scales less than 5×10^{14} M but may also be present in clusters (Ikebe et al., 2002) at a lower percentage level.

The "extra" entropy can be produced by a variety of physical process. Most of the theoretical effort has gone into models where the gas is heated before it falls into the cluster, e.g., heating the IGM before the collapse of the group (Tozzi and Norman 2001; Babul et al., 2002; Borgani et al., 2002). Other models have discussed heating the gas in the group/cluster relying on the physical feedback from star formation and AGN (Lowenstein, 2000; Brighenti and Matthews, 2001), removing the low entropy gas via cooling, which increases the entropy in the remaining gas and forms galaxies (Thomas et al., 2002; Muangwong et al., 2002; Davé et al., 2002; Voit et al., 2002).

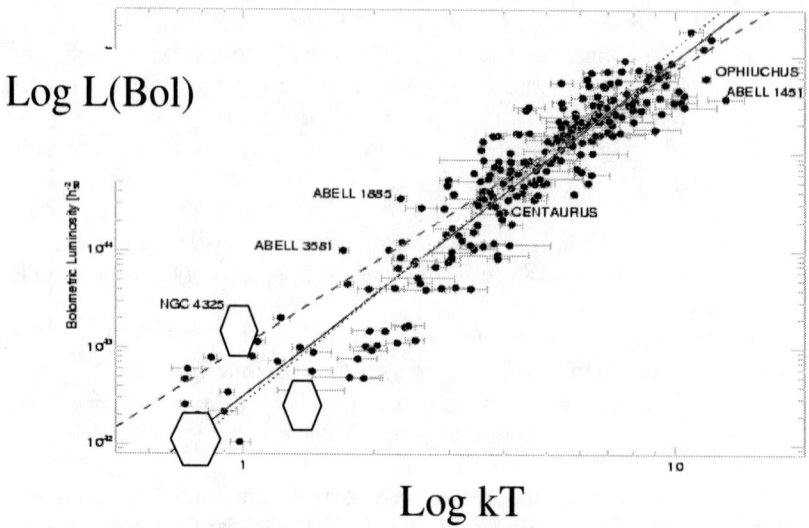

Figure 2: The X-ray luminosity temperature relation for 270 clusters (Horner 2001). The solid line is the best-fit $L \sim T^3$. The open symbols are the groups discussed later.

If the extra entropy is due to heat deposited inside the group, it would require more than 1 keV per particle and represent a major event in the universe. The only sources of extra heat with sufficient energy are supernova explosions that produce the metals seen in the cluster gas (the feedback mechanism in most theories of galaxy formation) or mechanical energy from active galaxies. The amount of energy in the preheating scenario depends on the epoch at which the heat is delivered to the gas and can be considerably less if the heating is at high redshift.

How can we test these scenarios? The implications for the production of energy in the universe and galaxy formation are critical and control the formation of stars and galaxies.

Most authors have chosen to test their models using the L(x) vs T relation (cf., the references above), the total amount of extra entropy at R ~ 0.1 R_{vir}, and the mass-T relation. With suitable fine tuning, all three classes of models (pre-heating, internal heating, and cooling) can fit these observational data but with different free parameters and problems.

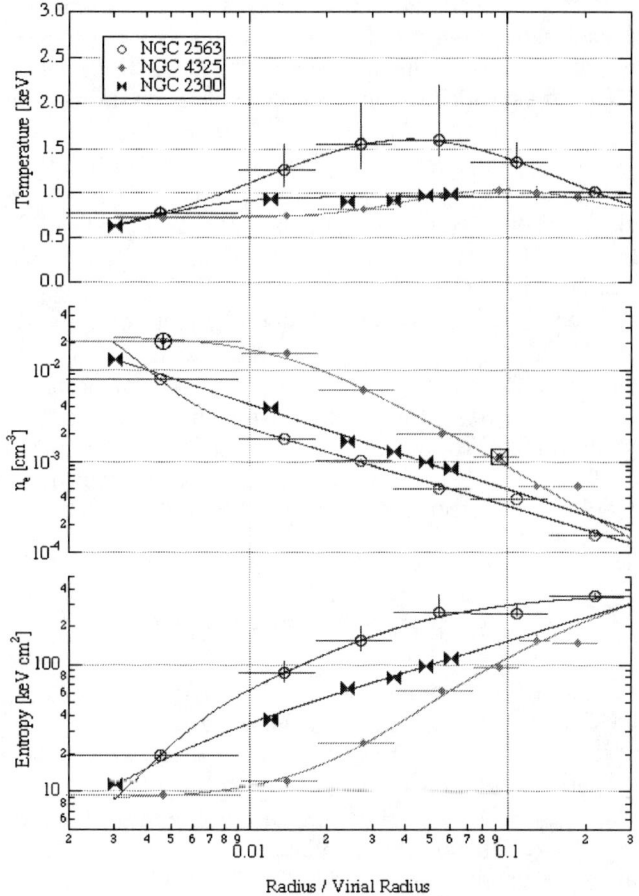

Figure 3: The temperature (top), density (middle) and entropy (bottom) profiles for NGC4325, NGC2300, and NGC2563 groups derived from XMM data vs the scaled radius in virial units.

The pre-heating models require a very large amount of energy injected at high redshifts and do not specify a source of the energy. The SN heating models require very efficient thermalization of the available energy in supernova, or an increase in the amount of energy per supernova. The cooling models overpredict the observed mass in stars. The only theoretical calculation which directly connects the semi-analytic models of star formation to the group entropy (Bower et al., 2001) are also able to account for the observed L-T relations with the same free parameters that successfully produce the galaxy luminosity function (Frenk, this symposium) but do not discuss the physics of the process.

The shape of the entropy profile in individual objects is a strong discriminator between these otherwise successful models. If the gas is strongly heated before it falls into the cluster, e.g., heating the IGM, low mass systems should have a roughly isentropic (constant entropy) profile (Babul et al., 2002). The two other scenarios: heating the gas after it has fallen in, or removing the low entropy gas via cooling, can have variable entropy profiles and values depending on history and amount of heating/cooling and thus determining the entropy profile fixes the free parameters of the models. However, at present the models disagree as to the magnitude of these effects, since in the models they depend greatly on the numerical resolution of the codes. In particular, SPH codes have difficulty with localized heating as well, since the SN explode in high density regions where the code "wants" to radiate away the energy instead of producing heat.

Another check on these models is the spatial distribution of the gas compared to the dark matter. The same process that produces the "extra" entropy puffs up the gas compared to that in a pure gravitational collapse model and thus changes the fraction of the mass in gaseous baryons in the innermost regions. This effect is larger at lower masses and smaller radii. At present it is not clear if the internal heating model is capable of actually expelling baryons into the IGM, but it seems rather likely. The pre-heating model should affect the entire IGM, while the pure cooling model does not affect the IGM.

V. THE ENTROPY PROFILES

Before XMM there were serious problems in deriving the entropy profiles in groups. While ROSAT had sufficient angular resolution, it had poor spectral resolution, and it was not clear that the derived temperatures were accurate. ASCA had sufficient energy resolution to obtain high quality spectra to derive accurate temperatures but had relatively poor spatial resolution and a complex energy dependent point spread function that requires extensive modeling. There was also controversy about the interpretation of the ASCA spectral data due to the possible spatial mixing of temperature gradients (Buote, 2000) and the possible existence of multi-phase gas. The solution is to derive the temperature profile from groups with XMM and Chandra, which have much better spatial resolution and compare these profiles to the models. XMM is the instrument of choice because of its large field of view and excellent signal to noise. We have observed three groups of galaxies, all of which have a flat temperature profile over a wide range in radius and rising entropy profile ruling out external heating and arguing strongly for either internal heating by SN or AGN or a combination of cooling and heating in the cluster.

The temperature profiles for the three groups are subtly different with respect to the virial radius but are roughly isothermal at $R > 0.1$ R_{virial}. Of the three systems, only NGC4325 shows a decline in the temperature profile shown in most models of cluster formation (Loken et al., 2002). The entropy profiles are different and have factors of 3 range in entropy value at the same virial radius. There is no entropy floor nor is the gas isentropic. The values of the entropy agree well with previous results from ROSAT and ASCA ~ 100 kev-cm^2 at 0.1 R_{virial}. **The fact that the values of entropy and the entropy profiles are all different and are not isentropic rules out the pre-heating models.**

The same data set also allows a determination of the gas mass, binding (dark) mass and the baryonic mass fractions vs radius (Figure 4).

The mass profiles for the three groups are very similar outside the central $R < 0.01$ R_{virial}, with the fraction of the mass that is in gas being similar at larger radii. However, at small radii the baryonic fraction is very different in the three systems.

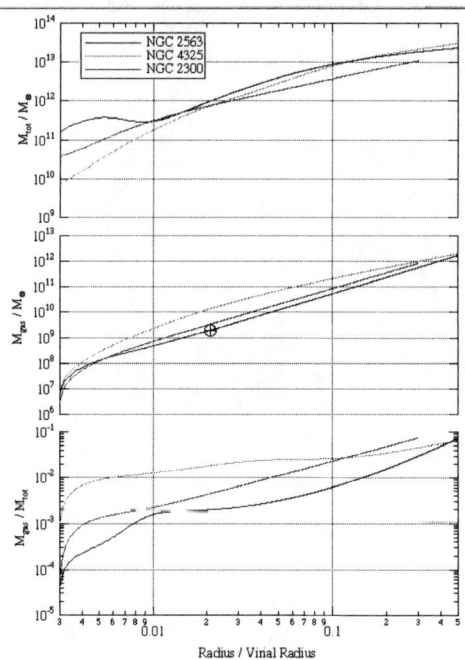

Figure 4. The total mass (top), gas mass (middle) and gas mass fraction (bottom) for the same 3 groups vs radius in virial units.

All three groups have much smaller M_{gas}/M_{tot} than rich clusters consistent with the internal heating scenarios in which the gas is puffed up. Even at r ~ 0.3 R_{virial}, the groups have M_{gas}/M_{tot} 0.06 compared to clusters 0.16 $h_{50}^{-3/2}$ (Allen ,Schmidt and Fabian 2002).

We suspect that cooling probably does not dominate the extra entropy since the stellar masses are very similar (Heldson and Ponman, 2002) and the ratio of stellar mass with respect to the total mass are similar (Mulchaey et al., 2002). However, the stellar-to-gas mass ratio is not constant in these systems and the NGC4325 system has a lower value of stellar to gas mass. This maybe an indication that cooling has had a role (higher stellar fraction in higher entropy systems) but the effect is not linear.

The lower baryonic fraction in groups indicates that a significant fraction of their baryons have been expelled, probably in a group wind which enrich the IGM in metals and energy. The variation in stellar to binding mass at exactly the same mass scale is also unexpected if, as seems to be the case, most of the stellar mass is in very old stars.

The XMM chemical abundances in NGC4325 and NGC2563 are very similar (~ 0.35 solar in NGC4325; 0.27 solar in NGC2563) but NGC2300 is only 0.1 solar.

But with rather different gas masses the total mass in metals differs by ~ 3 at $R \sim 0$. R_{virial}. This confirms the pattern seen in Finoguenov et al. (2002), in which the extra entropy does not track the present day abundance in the gas.

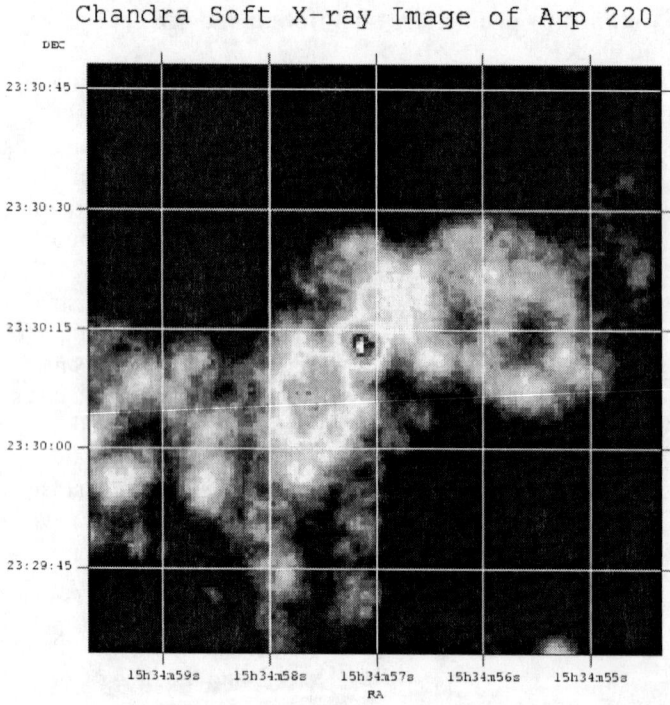

Figure 5: Chandra image of the galactic wind in the starburst galaxy Arp220. The image is about 20kpc across.

Further XMM observations will see if this entropy pattern is true for all groups.

If true, it indicates that the "extra" entropy is created inside the group and is not due to heating of the intergalactic medium at high redshift. Correlation of the XMM entropy profile with the dynamical state of the group, metallicity of the gas, and presence of a radio source or active nucleus will help determine the origin of the "extra" entropy.

VI. WHAT IS THE ORIGIN OF THE EXTRA HEAT?

We now know from direct x-ray imaging spectroscopy that starburst galaxies winds can escape their galaxy and are metal enriched (Figure 5, 6, Martin et al., 2002). The data on galactic winds (Strickland et al., 2002) shows that most of the energy from supernova goes into the wind and that the mass loss rate in the wind is ~ the total star formation rate. Thus the supernova energy is not radiated away and sufficient energy exists to produce the observed large values of "extra" heat. This process also happens at high redshift (Pettini et al., 2002; Steidel, these proceedings) and most of the high z rapidly star forming galaxies drive winds with v ~ 500 km/sec. The effect of active galaxies is not clear. Chandra observations of clusters show that the effects of AGN and jets on the hot gas do not obviously heat the gas. However, the amount of potential energy is large and conditions might be rather different in groups and in the early universe.

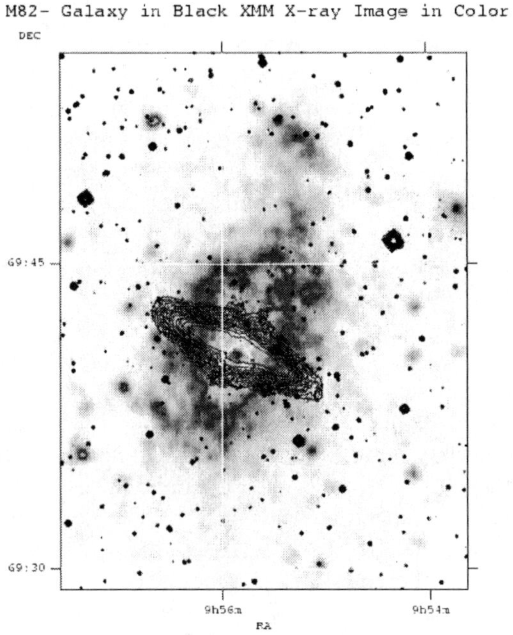

Figure 6. XMM image of the x-ray emission from M82. The black contours are the optical emission. The bright spot to the north is ~ 10 kpc from the center. There is x-ray line emission from O and Fe in the spot.

VII. WHAT HAVE WE LEARNED AND WHERE DO WE GO FROM HERE

The entropy in groups is indeed higher than predicted by pure gravitational formation.

The entropy profiles are not consistent with pre-heating models, ruling out massive heating of the IGM at high redshift. The observed level of entropy and the profiles are consistent with the "extra" entropy being due to both heating and cooling models.

Pure cooling models apparently overpredict the fraction of mass in stars, but the effects of cooling are important. Heating models using supernova correctly predict the entropy profiles, but would naturally predict a correlation of metallicity with entropy which is not detected. If all the heat comes from SN, one needs very high thermalization efficiencies if the standard 10^{51} ergs/ type II SN is used. But such efficiencies apparently are necessary in starburst galaxies. Heating models using AGN have to be very finely tuned to produce the narrow range in entropy at a fixed mass scale.

Detailed comparison of mass in stars, mass in gas, and entropy will strongly constrain cooling scenarios (Mulchaey et al., in prep). More XMM observations of low z groups will derive the full range of entropy parameters and provide correlations with abundance, baryon fraction, AGN activity, local density, and a large number of other parameters.

Astro-E2 observations will measure the entropy at large radii not possible with XMM or Chandra because of their high internal background.

To go to higher redshift, need Constellation-X spectra combined with Chandra images to examine the evolution of entropy with time and mass. Groups at $z = 0.5$ have been detected by Chandra and are ~ 1' in size in size and have fluxes that are easily within Constellation-X's sensitivity capability. We also need more detailed theory to understand the range of entropies and entropy profiles and directly connect to galaxy formation models and the nature of the "feedback" process.

Acknowledgements: this work has drawn extensively on analysis of XMM data with E. Figueroa-Feliciano, M. Loewenstein, and S. L. Snowden. I would like to thank the XMM-Newton project team for their efforts which have made this work possible. Special thanks are due to Pat Tyler who assisted greatly with the manuscript.

REFERENCES

Allen, S. W.; Schmidt, R. W.; Fabian, A. C 2002MNRAS.334L..11A
Babul, A., Balogh, M. L., Lewis, G. F., & Poole, G. B. 2002, MNRAS, 330, 329
Balogh, M. L., Babul, A., & Patton, D. R. 1999, MNRAS, 307, 463
Borgani S., Governato F., Wadsley J. N. P., Quinn, T., Stadel, J., & Lake, G. 2002 MNRAS, in press
Brighenti, F. & Mathews, W. G. 2001, ApJ, 553, 103
Bower, R. G.; Benson, A. J.; Lacey, C. G.; Baugh, C. M.; Cole, S.; Frenk, C. S 2001MNRAS.325,497
Buote, D. A. 2000, MNRAS, 311, 176
Davé, R Cen, R Ostriker, J. P.; Bryan, Greg L.; Hernquist, Lars; Katz, N.; Weinberg, D. H.; Norman, M. L.; O'Shea, B 2001ApJ...552..473
Davé, R., Katz, N., & Weinberg, D. H. 2002, ApJ, submitted.
Eke, V., Navaro, J., & Frenk, C. S. 1998, ApJ, 503, 569

Evrard, G. & Henry, J. P. 1991, ApJ, 383, 95
Finoguenov, A., Jones, C., Boehringer, H., & Ponman, T. J. 2002, ApJ, 578, in press
Fukugita, M.; Hogan, C. J.; Peebles, P. J. E 1998ApJ...503
Helsdon, S. F., & Ponman, Trevor, J. 2002 astro-ph/0212046
Horner, Donald J.; Mushotzky, Richard F.; Scharf, Caleb A 1999ApJ...520...
Horner, D. 2001, PhD Thesis, University of Maryland
Ikebe,Y. 2002 A&A submitted
Kaiser, N. 1986, MNRAS, 222, 323
Loewenstein, M. 2000, ApJ, 532, 17
Loewenstein, M. 2001, ApJ, 557, 553
Loewenstein M., & Mushotzky, R. 2002, ApJ, submitted
Loken, C., Norman, M., Nelson, E., Burns, J., Bryan, G., & Motl, P. 2002, astrop-ph 0207095
Martin , Crystal Kobulnicky Henry Heckman Timothy astro-ph/0203513
Mo, H. J.; White, S. D. 2002MNRAS.336
Mulchaey, J. S., et al. 2002b, in preparation
Pettini, Max; Rix, Samantha A.; Steidel, Chuck C.; Hunt, Matthew P.; Shapley, Alice E.; Adelberger, Kurt L 2002Ap&SS.281..
Ponman, T. J., Cannon, D. B., & Navarro, J. F. 1999, Nature, 397, 135
Silk: : Joseph MNRAS, submitted astro-ph/0212068
Strickland, D. 2001, astro-ph0107116
Thomas, Peter A.; Muanwong, Orrarujee; Kay, Scott T.; Liddle, Andrew R 2002MNRAS.330L
Tozzi, P.; Norman, C. 2001, ApJ, 546, 63
Trinchieri, G., Fabbiano, G., & Kim, D.-W. 1997, A&A, 318, 361
van den Bosch, F. C. 2002, MNRAS, 332, 456
Voit, G., Mark Bryan, G. L., Balogh M. L, Bower, R. G. 2002, ApJ, in press
White, Raymond E., III 1991ApJ...367...

Nonthermal Particles and Radiation Produced by Cluster Merger Shocks

Robert C. Berrington[*] and Charles D. Dermer[†]

[*]*ASEE Postdoctoral Fellow, Naval Research Laboratory, Code 7653, Washington, DC 20375-5352*
[†]*Naval Research Laboratory, Code 7653, Washington, DC 20375-5352*

Abstract. We have developed a numerical model for the temporal evolution of particle and photon spectra resulting from nonthermal processes at the shock fronts formed in merging clusters of galaxies. Fermi acceleration is approximated by injecting power-law distributions of particles during a merger event, subject to constraints on maximum particle energies. We consider synchrotron, bremsstrahlung, Compton, and Coulomb processes for the electrons, nuclear, photomeson, and Coulomb processes for the protons, and knock-on electron production. Broadband radio through γ-ray light curves radiated by nonthermal protons and primary and secondary electrons are calculated both during and after the merger event. Using ROSAT observations to establish typical parameters for the matter density profile of clusters of galaxies, we find that merger shocks are weak and accelerate particles with relatively soft spectra. Our results suggest that only a minor contribution to the diffuse extragalactic γ-ray background can originate from cluster merger shocks.

INTRODUCTION

In the hierarchical merging cluster scenario, clusters grow in mass by accreting nearby clusters. Approximately 30–40% of galaxy clusters show evidence of substructure in both the optical [1] and X-ray wavelengths [2]. Velocity differences between the observed structures is ≈ 1000–2000 km s^{-1}. With gravitational forces driving the interaction between the two systems, cluster mergers are consistent with highly-parabolic orbits. Typical sound speeds within the intracluster medium (ICM) are ≈ 1000 km s^{-1}, so shocks will form at the interaction boundary of the two systems.

Shock fronts that form in the ICM as a result of a cluster merger event are thought to be associated with the cluster *radio relics*, which are diffuse emission found on the cluster periphery with no known optical counterpart. The shock compression will orient any existing cluster magnetic field into the plane of the shock. Radio relics are characterized by highly organized magnetic fields with field strengths in the ~ 1 μG range with linearly polarized field lines in the vicinity of the shock [3]. The shock front will accelerate a fraction of the thermal particles within the ICM by first-order Fermi acceleration.

It has been proposed that cluster mergers are the dominant contributor to the diffuse γ-ray background [4]. Some unidentified EGRET sources are claimed to be associated with γ-ray emission from galaxy clusters [5]. Excess EUV emission from Coma can be explained by nonthermal electrons accelerated at merger shocks [6]. Variations in radio surface brightness will result from superposition of cluster emissions [7].

MODELS

We present the results of a computer code designed to calculate the time-dependent particle distribution functions evolving through adiabatic and radiative losses for electrons and protons accelerated by the first-order Fermi process at the cluster merger shock [8]. The model calculates the shock speed from the shock formed in a cluster merger event by the interaction speeds expected from two point masses interacting under their mutual gravity. The point masses are assumed to be on elliptical orbits whose onset is initiated by the collapse and merger of primordial density fluctuations. The collision is a result of the two bodies "falling out" of the Hubble flow and onto a nearby cluster. Expected total orbital energies are $E_{\text{tot}} \sim 10^{63}$–$10^{64}$ ergs. Expected collision velocities between a $10^{15} M_\odot$ and a $10^{14} M_\odot$ mass cluster range from $\sim$1800–3500 km s^{-1}. The electron and proton distribution functions originate from a momentum power-law injection spectrum. In terms of particle kinetic energy $E_{e,p} = (\gamma_{e,p} - 1) m_{e,p} c^2$, the injection function is

$$Q_{e,p}(E_{e,p},t) = Q^0_{e,p} c^{s(t)-1} \left[E_{e,p} \left(E_{e,p} + 2 m_{e,p} c^2 \right) \right]^{-\frac{s(t)+1}{2}} \left(E_{e,p} + m_{e,p} c^2 \right) e^{-\frac{E_{e,p}}{E_{\max}(t)}}, \quad (1)$$

where an exponential cutoff $E_{\max}$ has been applied and is determined by the maximum energy associated with the available time since the beginning of the merger event, by a comparison of the Larmor radius with the size scale of the system, and by a comparison of the energy-gain rate through first-order Fermi acceleration with the energy-loss rate due to adiabatic, synchrotron and Compton processes. The injected particle function is normalized through the normalization constant $Q^0_{e,p}$ which is determined by

$$\int_{E_{\min}}^{E_{\max}} dE \, E_{e,p} Q_{e,p}(E,t) = \frac{\eta_{e,p}}{2} A_s \eta^e_{\text{He}} \langle n \rangle m_p v_s^3 . \quad (2)$$

where $\eta_{e,p}$ is an efficiency factor, v_s is the shock speed A_s is the area of the shock front; $E_{\max}$ and $E_{\min}$ are the maximum and minimum particle energies; η^e_{He} is an enhancement factor to account for the presence of ions heavier than Hydrogen. We assume an efficiency factor $\eta_{e,p} = 5\%$ for both protons and electrons. With this method the total energy deposited into nonthermal particle production is $\eta_{e,p} E_{\text{tot}}$. The particle density $\langle n \rangle$ is calculated according the β model, and averaged over the shock front.

The time evolving particle spectrum is determined by solving the Fokker-Planck equation in energy space for a spatially homogeneous ICM, given by

$$\frac{\partial N(E,t)}{\partial t} = \frac{1}{2} \frac{\partial^2}{\partial E^2} [D(E,t) N(E,t)] - \frac{\partial}{\partial E} [\dot{E}_{\text{tot}}(E,t) N(E,t)] + Q(E,t) - \sum_{i=\pi, p\gamma, d} \frac{N(E,t)}{\tau_i(E,t)} . \quad (3)$$

The quantity $\dot{E}_{\text{tot}}(E,t)$ represents the total synchrotron, Compton, Coulomb, and adiabatic energy-loss rate for electrons, and the sum of the Coulomb and adiabatic energy-loss rates for protons. Both protons and electrons are subject to diffusion in energy space by Coulomb interactions. The protons experience catastrophic losses due to proton-proton collisions on the timescale τ_π, proton-γ collisions on the timescale $\tau_{p\gamma}$, and spatial diffusion from the cluster on timescale τ_d. Spectra of secondary electrons and positrons

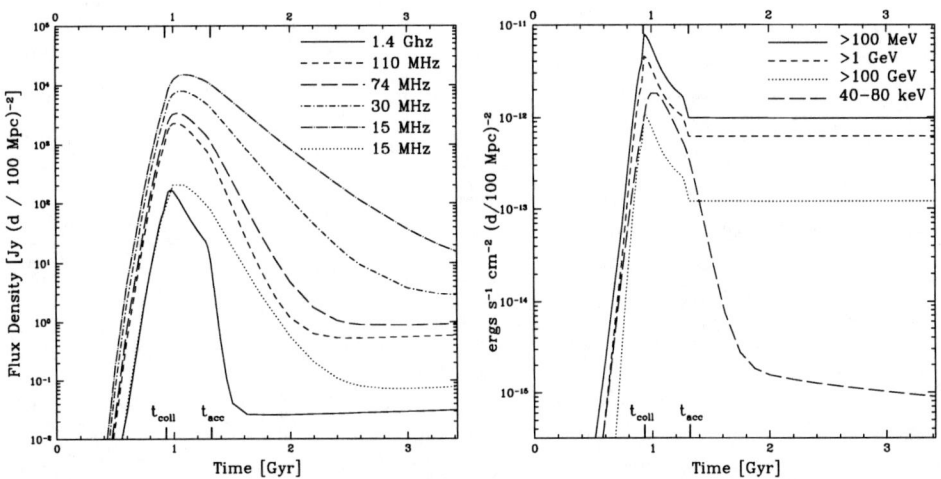

FIGURE 1. Light curves at various observing frequencies produced by a shock formed in a merger between $10^{14} M_\odot$ and $10^{15} M_\odot$ clusters that begins at $z_i=0.3$ and is evolved to the present epoch (t=3.42 Gyr). All light curves are for a magnetic field strength of B=1.0 μG except the 15 MHz light curve is also calculated with a magnetic field strength of B=0.1 μG (dotted curve). Radio light curves in Jansky units are given at 15 MHz, 30 MHz, 74 MHz, 110 MHz, and 1.4 GHz on the left panel, and light curves in energy flux units are given at 40-80 keV, >100 MeV, >1 GeV and >100 GeV in the right panel.

are calculated from pion-decay products, and are subject to the same physical processes as the primary electrons. $Q(E,t)$ is the particle injection function.

The synchrotron, Compton, bremsstrahlung, and pion-decay γ-ray spectral components are calculated from the particle spectra following the methods described by [8]. We use a standard parameter set with a mean ICM number density $n_0 = 10^{-3}$ cm^{-3} and a uniform cluster magnetic field of $B = 1.0$ μG. Thus, $V(t) = A(t)v_s t$.

RESULTS

Fig. 1 shows the light curves at various energies for a cluster merger shock between a $10^{14} M_\odot$ and $10^{15} M_\odot$ with radii of 1.5 Mpc and 0.75 Mpc, respectively, with a central gas number density of $n_0=10^{-3}$, and a magnetic field of B=1 μG is assumed. The assumed core radius for the dominant cluster profile is 250 kpc, and β=0.75. We assume a cosmology defined by $H_0 = 70$ km s^{-1} Mpc^{-1}, and $(\Omega_0,\Omega_R,\Omega_\Lambda)=(0.3,0.0,0.7)$. We assume the shock forms at a redshift of $z_i=0.3$. The light curves exhibit a similar characteristic independent of frequency with the peak luminosity occurring when the centers of mass of the two clusters pass. The emission slowly decay and approach a plateau at times $t > t_{acc}$ when particle injection has stopped. The rate of decay slows with decreasing energy with the lowest energies exhibiting the slowest decay rates. The late time plateaus are from π^0 γ-rays for photons with E>1 GeV, and secondary electrons for the radio energies. With these expected fluxes, it is unlikely that the more than a

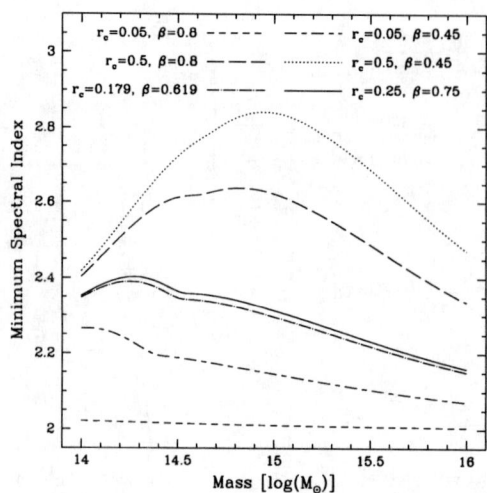

FIGURE 2. Calculations of the hardest particle injection spectral indices s_{min} formed in cluster merger shocks as a function of the dominant cluster mass, for various values of r_c and β. The mass of the merging cluster is assumed to be $10^{14} M_\odot$. Our adopted values are plotted by the solid line, and dot-dashed line shows s_{min} plotted against the dominant cluster mass for the average values of r_c and β obtained by ROSAT observations of 45 Abell clusters [9].

few of the isotropic unidentified EGRET sources can be attributed to the radiation of nonthermal particles produced in cluster merger shocks.

The diffuse extragalactic γ-ray background is a featureless power law with photon index of 2.10($\pm$0.03). Fig. 2 shows the hardest particle injection spectral index, s_{min}, for a cluster merger shock as function of the mass of the dominant cluster with varying values of core radii, r_c, and β's. The particle spectral indices expected from the average values of r_c and β obtained from ROSAT observations of 45 Abell clusters [9] are 2.2-2.4. Only the dark matter profiles with strong central peaks produce particle indices <2.1. This discrepancy in the particle indices suggests that nonthermal radiation from cluster merger shocks can make only a minor contribution to the diffuse extragalactic γ-ray background unless dark matter halos contain strong central peaks.

REFERENCES

1. Beers, T. C., Geller, M. J., & Huchra, J. P., *ApJ*, **257**, 23 (1982).
2. Forman, W., *et al.*, *ApJ*, **243**, L133 (1981).
3. Enßlin, T. A., *et al.*, *Astron. & Astrophys.*, **332**, 395.
4. Loeb, A. and Waxman, E., *Nature*, **405**, 156 (2000).
5. Totani, T. and Kitayama, T., *ApJ*, **545**, 572 (2000).
6. Atoyan, A. M. and Völk, H. J., *ApJ*, **535**, 45 (2000).
7. Waxman, E. and Loeb, A., *ApJ*, **545**, L11 (2000).
8. Berrington, R. C. and Dermer, C. D., *ApJ*, *in press* (2002) (astro-ph/0209436).
9. Wu, X. and Xue, Y., *ApJ*, **542**, 578 (2000).

Intermittency and Large-Scale Structure in the Universe: A New Window for the Study of the Nonlinear Regime of Structure Formation

Priya Jamkhedkar*[†] and Li-Zhi Fang*

*Department of Physics, University of Arizona, Tucson
[†]Currently with Laboratory for Physical Sciences, and Department of Electrical and Computer Engineering, University of Maryland, College Park; email: priyaj@lps.umd.edu

Abstract. We study the weak nonlinear regime of structure formation using high resolution and high signal-to-noise ratio (S/N) samples of Quasi Stellar Objects' (QSOs) Lyα transmission spectra. Using a space-scale decomposition, the Discrete Wavelet Transform (DWT), we show that the field traced by Lyα transmission flux is intermittent on scales less than 2000 km/s. The distribution of the local power of fluctuations is spiky with almost no power between the spikes. This spike-gap-spike feature gets more pronounced on smaller scales (128 − 16 km/s). We show that the structure functions and the intermittent exponent are not only able to quantitatively differentiate between different dark matter models but also qualitatively describe the nature of non-Gaussianity. Structure functions and the intermittent exponent are powerful tools for describing an intermittent field.

An intermittent field is a random field characterized by strong enhancements (peaks or spikes) with almost no field value between the spikes. The probability distribution function (PDF) of field fluctuations on a scale r may be long-tailed with respect to a Gaussian field and have a higher probability for low value events than for a Gaussian field

Mathematically, the ratio of the high to the low order moments of an intermittent field, $\rho(x)$, diverges as the scale decreases, i.e.,

$$\frac{S_r^{2n}}{(S_r^2)^n} = \frac{\langle [\rho(x+r)-\rho(x)]^{2n} \rangle}{[\langle [\rho(x+r)-\rho(x)]^2 \rangle]^n} \simeq \left(\frac{r}{L}\right)^{\zeta(n)},$$

where S_r^{2n} is the nth order structure functions on a scale r, L the sample size, $\langle ... \rangle$ is for an ensemble average, and the intermittent exponent $\zeta(n)$ is negative. The field becomes spikier on smaller scales.

The intermittent exponent: The intermittent exponent, $\zeta(n)$, as a function of the order, n, can qualitatively distinguish between different types of fields:

Gaussian field	$\zeta(n)$ =	0
Self-similar field	$\zeta(n)$ =	0
Lognormal field	$\zeta(n)$ ∝	$-(n^2-n)$
Monofractal field	$\zeta(n)$ ∝	$-(n-1)$

Samples and Units: The samples used in this study are Keck QSO Lyα transmission

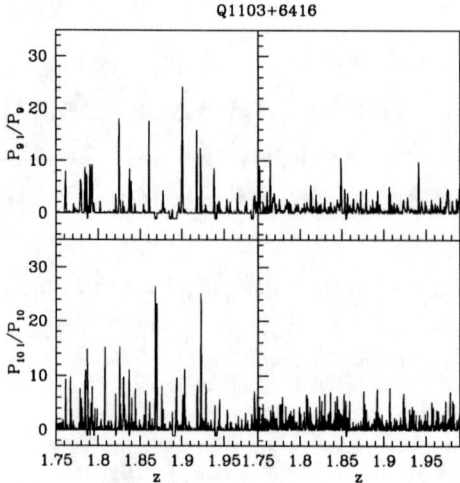

FIGURE 1. Distribution of local powers of the flux field on the scales 64 km/s (top panels), and 32 km/s (bottom panels). The suffix l is for position in space. The left panels are for real fields and right are for their PR counterparts.

spectra, and pseudo-hydro simulation samples for Low density Cold Dark Matter model (LCDM) with $\Omega_m = 0.3$ and $\Omega_\lambda = 0.7$, and the Warm Dark Matter (WDM) model with particle masses $m = 300, 600, 800, 1000$ eV. We use the DWT for this analysis. The suffices, 8, 9, 10 & 11, represent the scales 128, 64, 32 & 16 km/s, respectively.

Local power spectrum of fluctuations: Fig. 1 shows the distribution of local powers on the scales 64 and 32 km/s, for a Keck QSO (left hand panels). The right hand panels represent the local powers of a phase randomized (PR) field. The PR field is obtained from the original field by randomizing the phases of the Fourier transform of the original field. Thus the mean powers on the left and right panels are the same. The real field is spikier than its PR counterpart. The spikiness increases as the scale decreases.

Structure functions of the samples: Fig. 2 shows the dependence of the structure functions on the scale. All the samples are intermittent because the ratio of the high to low order structure functions diverges as the scale decreases. Moreover the real samples are less intermittent than the pseudo-hydro simulation samples.

Fig. 3 shows the dependence of the intermittent exponent on the order n for Keck samples. The graph is fitted with lines of the form $n^\alpha(n-1)$. For a lognormal field, $\alpha = 1.0$, while for a monofractal field $\alpha = 0.0$. For the Keck samples we see that on large scales α is close to that of a monofractal field but increases as the scale decreases. The dotted line is for a Gaussian field.

Figs. 4(a) and 4(b) show similar dependence of ζ on the order n for the LCDM and the WDM models. Both these samples are close to a monofractal field even on small scales. The intermittent exponent can describe the type of intermittency.

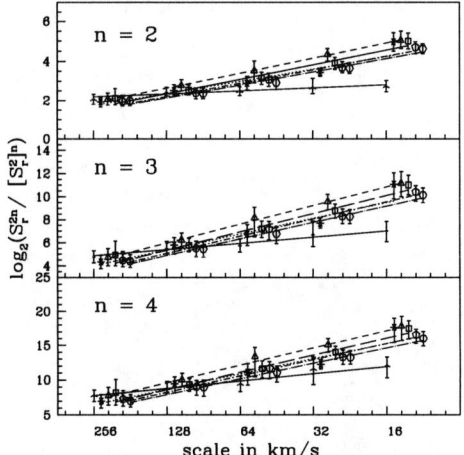

FIGURE 2. $\log_2[S_r^{2n}/(S_r^2)^n]$ vs. scale for Keck data (vertex with three legs) and the models of the LCDM (star) and WDM $m_W =$ 1000 (hexagon), 800 (pentagon), 600 (square) and 300 (triangle) eV. The error bars are given by the maximum and minimum of bootstrap resampling. For clarity, the points for simulation samples are shifted horizontally to the right from the corresponding scale.

Conclusions: Intermittency is important in the study of the nonlinear evolution of structure in the universe. The cosmic mass field is highly intermittent. Using structure functions and the intermittent exponent we find that pseudo-hydro samples of the LCDM and WDM models differ from the real field both quantitatively and qualitatively.

Acknowledgments: We thank Hu Zhan (University of Arizona, Tucson), Long-Long Feng (University of Science and Technology of China, PRC), Hongguang Bi (Rare Medium Group, Inc.), Jesus Pando (DePaul University, Chicago), Wei Zheng (the Johns Hopkins University, Baltimore), and David Tytler and David Kirkman (University of California, San Diego) for their contributions to this study.

REFERENCES

1. P. Jamkhedkar, "Intermittency and Large Scale Structures in the Universe," Ph.D. thesis, University of Arizona, Tucson, 2002 (available from http://www.ece.umd.edu/~priyaj).
2. J. Pando, L.-L. Feng, P. Jamkhedkar, W. Zheng, L. Z. Fang, "Non-Gaussian features of the transmitted flux of QSO's Lyα absorptions: the intermittent exponent," Astrophysical Journal, vol. 574, pp. 575-589, 2002.
3. P. Jamkhedkar, H. Zhan, L. Z. Fang, "The intermittent behavior of the cosmic mass field revealed by a QSO's Lyα forest," Astrophysical Journal, vol. 543, pp. L1-L4, 2000.

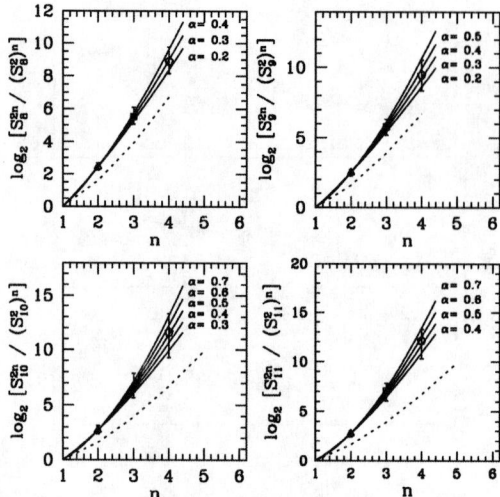

FIGURE 3. $\log_2[S_r^{2n}/(S_r^2)^n]$ vs. n for 128, 64, 32 and 16 km/s for the Keck data. The error bars are given by bootstrap resampling. The fitting curves are $n^\alpha(n-1)$. The dotted curves are for a Gaussian field, i.e., $\log_2[S_r^{2n}/(S_r^2)^n] = \log_2(2n-1)!!$.

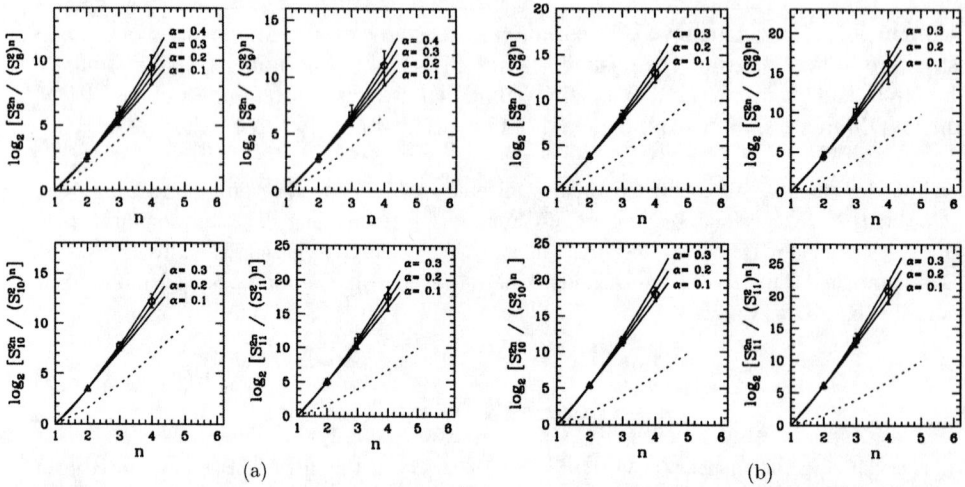

FIGURE 4. (a) Same as Fig 3 but for the LCDM samples. (b) Same as Fig 3 but for the WDM ($m = 300$ eV) samples.

The Galaxy Proximity Effect in the Lyα Forest

Juna A. Kollmeier*, David H. Weinberg*, Romeel Davé† and Neal Katz**

The Ohio State University, Dept. of Astronomy, Columbus, OH 43210
†*University of Arizona, Dept. of Astronomy, Tucson, AZ 85721*
**University of Massachusetts, Dept. of Physics and Astronomy, Amherst, MA, 91003*

Abstract. Hydrodynamic cosmological simulations predict that the average opacity of the Lyα forest should increase in the neighborhood of galaxies because galaxies form in dense environments. Recent observations (Adelberger et al. [1]) confirm this expectation at large scales, but they show a *decrease* of absorption at comoving separations $\Delta_r \lesssim 1h^{-1}$ Mpc. We show that this discrepancy is statistically significant, especially for the innermost data point at $\Delta_r \lesssim 0.5h^{-1}$, even though this data point rests on three galaxy-quasar pairs. Galaxy redshift errors of the expected magnitude are insufficient to resolve the conflict. Peculiar velocities allow gas at comoving distances $\gtrsim 1h^{-1}$ Mpc to produce saturated absorption at the galaxy redshift, putting stringent requirements on any "feedback" solution. Local photoionization is insufficient, even if we allow for recurrent AGN activity that keeps the neutral hydrogen fraction below its equilibrium value. A simple "wind" model that eliminates all neutral hydrogen in spheres around the observed galaxies can marginally explain the data, but only if the winds extend to comoving radii $\sim 1.5h^{-1}$ Mpc.

BASIC PREDICTIONS

In a recent paper [2], we discuss a variety of predictions for galaxy-Lyα forest correlations from hydrodynamic simulations. In this proceeding, we focus on the "galaxy proximity" effect on small scales ($\lesssim 2h^{-1}$ Mpc comoving). Using smoothed particle hydrodynamics simulations (SPH) of a ΛCDM universe ($\Omega_m = 0.4, \Omega_\Lambda = 0.6, h = 0.65, \Omega_b = 0.02h^{-2} = 0.0473$, $\sigma_8 = 0.80$), we generate synthetic Lyα forest spectra from the temperature, gas density, and velocity at each spatial location in skewers through the simulation box. Once we have created synthetic spectra, we use the known positions of the simulated galaxies to compute the mean flux decrement, $\langle D \rangle = \langle 1 - e^{-\tau} \rangle$, as a function of comoving separation from the galaxy, Δ_r.

Figure 1a shows the basic predictions for the mean flux decrement, computed for the 150 galaxies with the highest star formation rates within a simulation $50h^{-1}$ Mpc on a side. We see a clear trend of increasing decrement (absorption) with decreasing Δ_r, a signature of the dense environments these galaxies occupy. The observed points, from [1], show a similar trend at large scales, but the trend flattens at $\Delta_r \lesssim 2.5h^{-1}$ Mpc, and absorption *decreases* for $\Delta_r \lesssim 1h^{-1}$ Mpc. The innermost point, $0 \leq \Delta_r \leq 0.5h^{-1}$ Mpc is in especially severe conflict with the theoretical prediction. We investigate several possibilities for resolving this conflict below. Similar investigations have been carried out by [3], [4], and our results are compatible with theirs to the extent that they overlap.

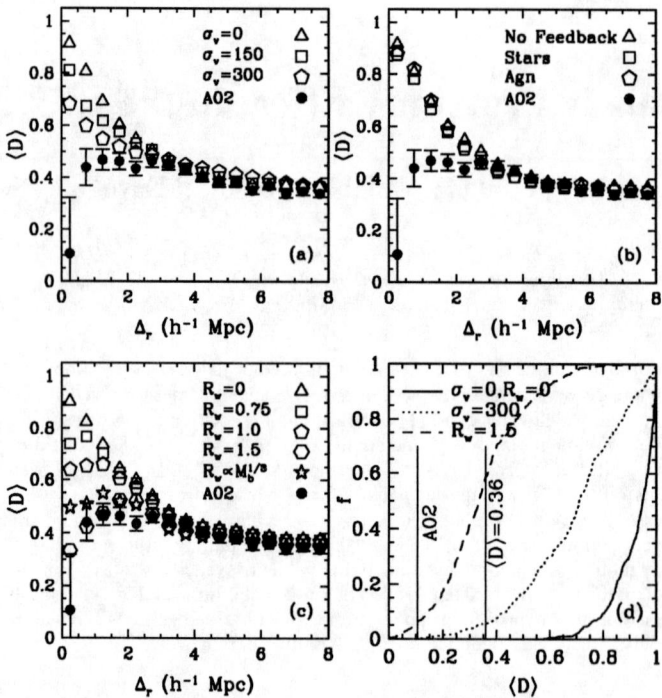

FIGURE 1. The Conditional Mean Flux Decrement. (a) The conditional mean decrement compared with the baseline calculation, and calculations with redshift errors as indicated in panel and described in text (b) Effect of photoionization feedback from stars or AGN on the numerical predictions, (c) Effect of spherical winds/gas removal on the predictions, (d) Cumulative distribution of 3-tuples at $\Delta_r = 0.25$ Mpc as a function of $\langle D \rangle$. Vertical lines at $\langle D \rangle = 0.11$ and 0.36 correspond to the Adelberger et al. [1] value for this separation and the global mean respectively. Symbols are as indicated in the panel.

REDSHIFT ERRORS

Since we expect galaxies to occupy peaks in the density distribution, if the redshift of the galaxy is incorrectly estimated, then one may compute an artificially low value of $\langle D \rangle$ at small Δ_r because one samples a region of lower density than the region around the true galaxy redshift. We estimate the size of this effect by adding a redshift error, drawn from a Gaussian distribution, to each galaxy redshift. Squares and pentagons in Figure 1a show the result for rms redshift errors $\sigma_v = 150 \text{km s}^{-1}$ and $\sigma_v = 300 \text{km s}^{-1}$, respectively. The sign of the effect is as expected — increasing the redshift errors does decrease the values of $\langle D \rangle$ at small Δ_r, pushing them towards the global mean. It is clear, however, that even with significant redshift errors (Adelberger et al. estimate $\sim 150 \text{km s}^{-1}$) the difference between the theoretical curves and the observations remains substantial.

LOCAL PHOTOIONIZATION

There is some evidence that a significant amount of Lyman continuum radiation is leaking from the interstellar media of Lyman Break Galaxies (LBGs) [5]. It is then plausible that these galaxies may affect their immediate surroundings in the form of photoionization from the stars within them. There is also evidence indicating that $\sim 3\%$ of LBGs host AGN [6]. If the timescale between AGN outbursts is sufficiently short, then, in contrast to the stellar case, the gas surrounding the galaxies can remain out of photoionization equilibrium between outbursts, and the neutral hydrogen fractions around galaxies that have recently hosted AGN may be further suppressed. We have incorporated simple models for these two scenarios within the simulations by including the non-uniform ionizing background near galaxies in each case, as well as the additional effect of non-equilibrium neutral fractions in the AGN case. For details of these models see [2]. Figure 1b compares the results of these two calculations to the original, no-feedback, calculation. Photoionization has minimal impact on the mean decrement even at small Δ_r. It is tempting to think that increasing either the AGN luminosities or the escape fraction of ionizing photons could produce a larger effect, but this is not the case because the total output of the sources cannot exceed the UV background, which is itself constrained by the mean (unconditional) flux decrement. The models we have presented are close to maximal, with the observed galaxies or their AGN assumed to produce 50% of the entire UV background.

WINDS

Outflows have been detected in LBGs by looking at the difference between absorption and emission features within these systems [7]. Strong winds from supernovae are a generic property of starburst galaxies and have the effect of shocking and sweeping up the material in their wake, both of which lead to decreased absorption inside the "sphere of influence" of the wind. We have constructed very simple "wind" models in which we eliminate all neutral hydrogen in a spherical region of radius R_{wind} around each galaxy in the simulation sample. We note that this model is highly optimistic since it assumes either *perfect* entrainment of the material in the volume out to R_{wind} or sufficient energy injection to completely ionize hydrogen within this radius. Neither condition is necessarily expected to hold for realistic winds ([3], [4]).

Figure 1c shows the result of the wind model for the 40 galaxies with the highest star formation rates in a simulation box of side $22.22h^{-1}$ Mpc (comoving). Here the squares, pentagons, and hexagons correspond to models with constant comoving radii $R_{wind} = 0.75, 1.0,$ and $1.5h^{-1}$ Mpc respectively. Stars show a model in which the volume of the wind around a galaxy is proportional to the baryonic mass of the galaxy, normalized such that the $40th$ brightest galaxy has a wind radius of $1h^{-1}$ Mpc and including winds around all 641 resolved galaxies in the box. The largest winds in this model extend to $R_{wind} \sim 2h^{-1}$ Mpc, requiring an average propagation speed $V \sim 750 \text{km s}^{-1}(1\text{Gyr}/t)$ for a wind duration t and $h = 0.65$. Only the most extreme wind models come close to matching the observational results. Note, in particular that $1h^{-1}$ Mpc winds do not eliminate, or even drastically reduce, absorption at $\Delta_r \lesssim 1h^{-1}$ Mpc because much of the absorption at these separations in *redshift space* comes from infalling gas that is further than $1h^{-1}$ Mpc in *real space*. These peculiar velocity effects are also the reason that photoionization has such a tiny impact; even for near maximal models, the ionization does not strongly affect gas at such large distances.

STATISTICAL FLUKE?

Since the innermost data point comes from only three galaxy-los pairs, we must also ask whether it could just be an anomalous statistical fluctuation. We have done a Monte Carlo calculation in which we draw 500 sets of 3 galaxies from our population of $z = 3$ simulated galaxies and compute the value of $\langle D \rangle$ at $\Delta_r \leq 0.5h^{-1}$ Mpc for each 3-tuple. Figure 1d shows the cumulative distribution of the flux decrement from these samples. We see that our baseline calculation can virtually never get to decrements as low as those observed. Even with a redshift error of 300km s^{-1}, one sees decrements below 0.36 only $\sim 5\%$ of the time. For the most extreme wind model, we find decrements as low as the observed one, $\langle D \rangle = 0.11$, $\sim 5\%$ of the time. Despite the limited size of the current data set, the observed "LBG proximity effect" stands as a striking result, not easily explained.

REFERENCES

1. Adelberger, K.L., Steidel, C.C., Shapley, A.E., Pettini, M. 2002, ApJ, in press, astro-ph/0210314
2. Kollmeier, J.A., Weinberg, D.H., Davé, R., Katz, N., 2002, ApJ, submitted astro-ph/0209563
3. Croft, R., Hernquist, L., Springel, V., Westover, M., White, M., 2002, ApJ, **580**, 634
4. Bruscoli, M., Ferrara, A., Marri, S., Schneider, R., Maselli, A., Rollinde, E., Aracil, B., astro-ph/0212126 (2002)
5. Steidel, C. C., Pettini, M., & Adelberger, K. L. 2001, ApJ, **546**, 665
6. Steidel, C., Hunt, M., Shapley, A., Adelberger, K., Pettini, M., Dickinson, M., & Giavalisco, M. 2002, ApJ, **576**, 653
7. Pettini, M., Rix, S. A., Steidel, C. C., Adelberger, K. L., Hunt, M. P., & Shapley, A. E. 2002, ApJ, **569**, 742

Keck Absorption-Line Spectroscopy of Galactic Winds in Massive Infrared-Luminous Galaxies[1]

David S. Rupke*, Sylvain Veilleux*[†] and D. B. Sanders**[‡]

Department of Astronomy, University of Maryland, College Park, MD 20742
[†]*Cottrell Scholar of Research Corporation*
**Institute for Astronomy, University of Hawaii, Honolulu, HI 96822*
[‡]*Max-Planck-Institut für extraterrestrische Physik, Garching, German*

Abstract. A proposed mechanism for the seeding of intergalactic space with metals and energy is outflows (winds) of gas from galaxies, powered by star formation and/or AGN activity. Using moderately-high resolution spectroscopy from Keck II, we observe blueshifted absorption in the Na I I D doublet in 8 of 11 starburst-dominated ultraluminous infrared galaxies at $\langle z \rangle = 0.14$. We interpret these blueshifts as outflows of warm neutral gas. Mass outflow rates for each source, when normalized to the corresponding global star formation rate, are 25% on average. The average escape fraction of the gas could be as high as 40 – 50%. Most effort to date has focused on the escape of mass, metals, and energy from small galaxies; if our results are confirmed, then very massive galaxies could play an important role in energizing and enriching the IGM.

INTRODUCTION

Large-scale galactic outflows, energized by stellar winds and supernovae ejecta or a central AGN, are ubiquitous in the local universe and at high redshift [1, 2, 3] (see also contribution from C. Steidel, this volume). These outflows (winds) may play a role in galaxy formation and evolution; they provide feedback to star formation; they are likely a significant source of metals and energy for the intergalactic and intracluster media; they could be responsible for the mass-metallicity relation of ellipticals; and they may contribute to formation of Lyα-absorbing clouds and reionization. Since much effort has focused on winds from galaxies of small-to-moderate mass, we are here concerned with outflows in high-mass galaxies. We wish to answer the following: (1) Are there winds of warm neutral gas in *massive* star-forming galaxies? (2) What are their properties? (3) How do winds from massive galaxies impact the IGM, and thus the chemical and thermal evolution of the universe?

Observers usually look for spatially resolved, optical-line-emitting warm gas or x-ray-emitting hot gas to probe the existence and properties of superwinds. For largely unresolved sources, where wind emission is faint and spatially blended with the galaxy, spectral features like interstellar absorption lines blueshifted from systemic are excellent

---

[1] Based on observations at the W. M. Keck Observatory, which is operated as a scientific partnership among the California Institute of Technology, the University of California, and the National Aeronautics and Space Administration. The Observatory was made possible by the generous financial support of the W. M. Keck Foundation.

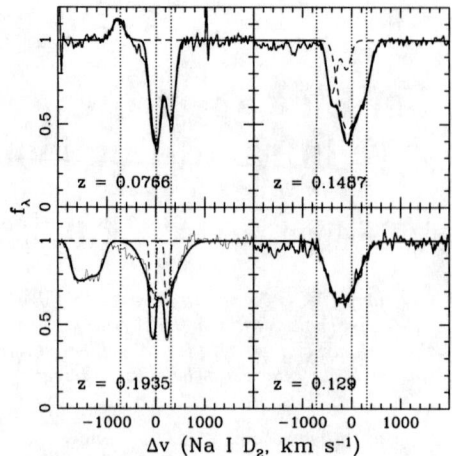

FIGURE 1. The Na I D doublet in 4 ultraluminous infrared galaxies. The solid lines are the data and fits. Dashed lines indicate the components of the fit. The vertical dotted lines locate Na I D and He I $\lambda 5876$ in the rest frame of each galaxy.

indicators of an outflow. Absorption-line spectroscopy of the Na I D doublet has been successful in detecting outflows in local galaxies [4, 5], and UV lines have been used at high redshift [1, 2, 3]. To begin filling the gap between high- and low-z observations, we present Na I D absorption-line observations of a sample of galaxies at $z = 0.04 - 0.27$ [6].

DATA, ANALYSIS, RESULTS

Our sample [6] consists of 11 starburst-dominated ultraluminous infrared galaxies (ULIGs; $\log[L_{IR}/L_\odot] \geq 12$) from the 1 Jy sample [7]. ULIGs are massive galaxies with high star formation/AGN activity. The optical spectral types of our targets indicate that they are mostly powered by star formation. The average redshift and star formation rate for our sample are $\langle z \rangle = 0.14$ and $\langle SFR \rangle = 280 M_\odot \, yr^{-1}$.

Using the Echellette Spectrograph and Imager (ESI) on Keck II, we observed the spectral range $3900 - 10900$ Å at a resolution of $R \sim 4600$, or $\Delta v \sim 65 \, km \, s^{-1}$. We fit $1 - 3$ components to each Na I D profile (Fig. 1) using SPECFIT [8] and measure the following: outflow velocity relative to systemic, Δv; Doppler width, b; covering fraction, C_f; and column density of Na. Using a simple mass-conserving free wind model, the mass outflow rate for each object is

$$\dot{M}(H) = 21 \left(\frac{\Omega}{4\pi} \right) C_f \left(\frac{r_\star}{1 \, kpc} \right) \left(\frac{N(H)}{10^{21} \, cm^{-2}} \right) \left(\frac{\Delta v}{200 \, km \, s^{-1}} \right) M_\odot \, yr^{-1}$$

summed over all components (Ω is the opening angle of the wind and $r_\star$ its inner radius). We compute global star formation rates for each object from L_{IR} (after correcting for

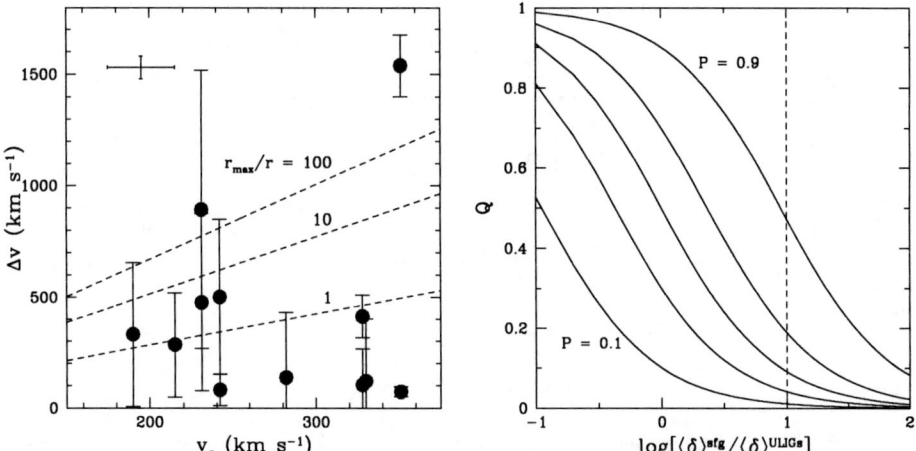

FIGURE 2. [Left] Outflow velocity for each blueshifted component vs. estimated circular velocity of the host galaxy. The vertical bars represent the FWHM of the fitted profile. The dotted lines are v_{esc} for a singular isothermal sphere, truncated at r_{max}, for various values of r_{max}/r. Material with $v > v_{esc}$ may represent escaping gas. [Right] Fraction of gas injected into the IGM that originates in ULIGs, Q, as a function of the ratio of the ejection efficiency of all other star-forming galaxies to that in ULIGs. The lines represent different values of P, the fractional contribution of ULIGs to the total star formation rate density of the universe at a given z. The vertical dotted line is $\langle\delta\rangle^{sfg}/\langle\delta\rangle^{ULIGs} = 10$, which matches current measurements if $f_{esc}^{sfg} = f_{esc}^{ULIGs}$.

AGN activity) and estimate $\dot{M}_{esc}$, the amount of gas mass escaping each galaxy, by comparing the outflow velocities to the estimated escape velocity (Fig. 2). The reheating efficiency, escape fraction, and "ejection efficiency" are

$$\eta \equiv \frac{\dot{M}_{tot}}{SFR}, \quad f_{esc} \equiv \frac{\dot{M}_{esc}}{\dot{M}_{tot}}, \quad \text{and} \quad \delta \equiv \frac{\dot{M}_{esc}}{SFR}.$$

We find components blueshifted by more than 70 km s^{-1} in 8 of 11 objects (73%). For those galaxies hosting a wind, we measure the following properties:
- *component velocities:* $\Delta v_{max} = 120 - 1538$ km s^{-1}, average $\langle\Delta v_{max}\rangle = 527$ km s^{-1} (median 372 km s^{-1})
- *mass outflow rate:* $\dot{M} = 13 - 133 M_\odot$ yr^{-1}, average $\langle\dot{M}\rangle = 56 M_\odot$ yr^{-1}
- *reheating efficiency:* $\eta = 0.1 - 0.7$, average $\langle\eta\rangle = 0.31$
- *gas escape fraction:* $\langle f_{esc}\rangle = 0.4 - 0.5$ for $r_{max}/r = 10$ (see Fig. 2), which gives an "ejection efficiency" of $\langle\delta\rangle = 0.1$ ($\langle f_{esc}\rangle$ and $\langle\delta\rangle$ are averaged over the whole sample)

CONCLUSIONS

ULIGs are surprisingly efficient at powering outflows, considering their large dynamical masses ($\sim 10^{11} M_\odot$; [9]) and concentrated molecular gas content ($\sim 10^{10} M_\odot$; [10]).

Based on a similar-size sample, dwarfs and spirals are somewhat more efficient, with $\eta \sim 1-5$ [11]. We measure a large gas escape fraction (though this number is uncertain). The amount of hot, enriched gas that escapes is likely to be even higher. Current models suggest that dwarf and/or normal-size galaxies expel metals into the intergalactic medium and could enrich it to observed levels [12, 13]; however, if ULIGs are a significant source of star formation in the universe, they could also be a significant source of metals and energy in the IGM (Fig. 2).

This work was supported by a Cottrell Scholarship from the Research Corporation, NASA/LTSA grant NAG 56547, NSF/CAREER grant AST-9874973, and NASA/JPL contract 961566.

REFERENCES

1. Pettini, M., Shapley, A. E., Steidel, C. C., Cuby, J.-G., Dickinson, M., Moorwood, A. F. M., Adelberger, K. L., and Giavalisco, M., *Ap. J.* **554**, 981-1000 (2001).
2. Frye, B., Broadhurst, T., and Benítez, N., *Ap. J.* **568**, 558-575 (2002).
3. Adelberger, K. L., Steidel, C. C., Shapley, A. E., and Pettini, M. *Ap. J.* in press (2002), astro-ph/0210314.
4. Phillips, A. C., *A. J.* **105**, 486-498 (1993).
5. Heckman, T. M., Lehnert, M. D., Strickland, D. K., and Armus, L., *Ap. J. Supp.* **129**, 493-516 (2000).
6. Rupke, D. S., Veilleux, S., and Sanders, D. B., *Ap. J.* **570**, 588-609 (2002).
7. Kim, D.-C., and Sanders, D. B., *Ap. J. Supp.* **119**, 41-58 (1998).
8. Kriss, G. A., "Fitting Models to UV and Optical Spectra," in *ADASS III*, edited by D. R. Crabtree, R. J. Hanisch, and J. Barnes, ASP Conference Series 61, ASP, San Francisco, 1994, pp.437ff.
9. Tacconi, L. J., Genzel, R., Lutz, D., Rigopoulou, D., Baker, A. J., Iserlohe, C., and Tecza, M. *Ap. J.* **580**, 73-87 (2002).
10. Downes, D., and Solomon, P.M., *Ap. J.* **507**, 615-654 (1998).
11. Martin, C. L., *Ap. J.* **513**, 156-160 (1999).
12. Silich, S., and Tenorio-Tagle, G., *Ap. J.* **552**, 91-98 (2001).
13. Aguirre, A., Lernquist, L., Schaye, J., Weinberg, D. H., Katz, N., & Gardner, J., *Ap. J.* **560**, 599-605 (2001).

A Filament Between Galaxy Clusters A3391 & A3395

Eric Tittley* and Mark Henriksen*

Joint Center for Astrophysics, Physics, UMBC, Baltimore MD 21250

Abstract. Filamentary gas spanning the region between the galaxy clusters Abell 3391 and Abell 3395 has been detected using *ASCA* and *ROSAT* archived data. The gas has a minimum flux of 1.3×10^{-12} erg cm^{-2} s^{-1} (0.8 - 10 keV). Within this filament resides a galaxy group for which a flux of $(2.0 \pm 0.3) \times 10^{13}$ erg cm^{-2} s^{-1} (0.8 - 10 keV) is determined. An analysis using raytracing determines light scattered into the filamentary region contributes 13% of the count-rate. The structure in which the filamentary gas resides is postulated to be a filament aligned nearly lengthwise with the line of sight. Identification of this structure as a filament is based on the angular and redshift distribution of surrounding galaxies and clusters. The distribution is compatible with the structure being a quasi-linear structure tilted to the line-of-sight such that the ratio of the depth to the tangential distance is between 6:1 and 18:1. The filamentary gas contains on the order of 10^{13} M$_\odot$ of gas which is close to 2% of the total mass of the system.

THE ENVIRONMENT OF A (POSSIBLY) ALIGNED FILAMENT.

First, we will review the evidence that the galaxy clusters A3391 and A3395 are located in a filament. The evidence is provided by both the linear arrangement on the sky and the similar redshifts of the constituent structures.

The x-ray sky brightness (Fig. 1, left) reveals a swath of cataloged objects. The *ROSAT* All-Sky-Survey events (RASS) data for the region surrounding the filament were used to find event flux by determining the radii encompassing a fixed number of events (100). There is a clear north-north-westerly trend.

The galaxy distribution (Fig. 1, right) paints the same picture as the x-ray emission. The galaxy sky surface density ($m_R < 15.5$) was determined from Digitized Sky Survey-II data using SExtractor to separate galaxies from stars. The contours are at levels of 2, 4, 6, ... σ above the field density. The dashed contour corresponds to the -2σ level and encircles the bright star, Canopus, which masks faint objects.

The clusters and groups are all located at a similar redshift. The redshift distribution for the A3391-A3395 system is illustrated in Fig. 2. The mean local redshift of galaxies with available redshifts from NED (all from [1]) is given by the colour background image. Note that 1000 km s^{-1} corresponds to $10h^{-1}$ Mpc while 1° corresponds to $2.6h^{-1}$ Mpc (at 15 000 km s^{-1}). The contours outline the galaxy spatial distribution and are derived from the same list of galaxies. The white circles are centred on objects listed in the Table 1. The large black circle corresponds to the ASCA GIS field of view displayed in the figure on the next page.

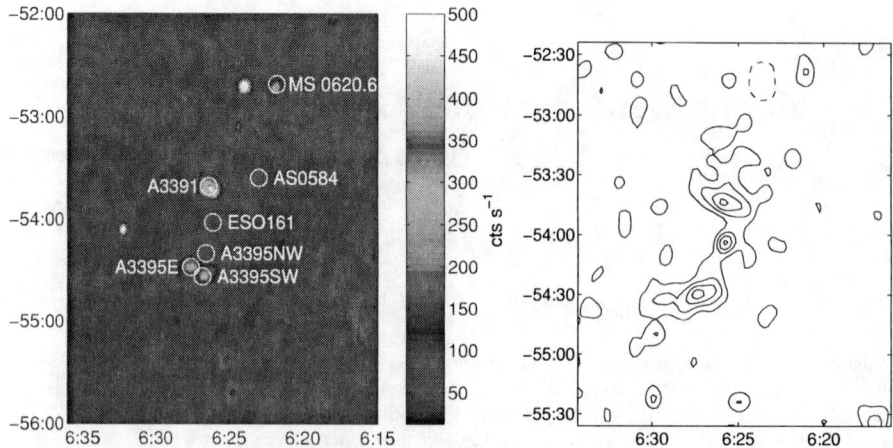

FIGURE 1. The structure field from the RASS (left). The galaxy sky surface density (right).

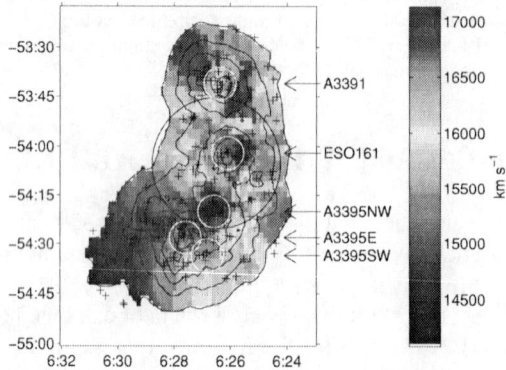

FIGURE 2. The galaxy spatial and redshift distribution.

INTER-CLUSTER GAS IN THE ALIGNED FILAMENT?

This section will 1) summarize the evidence for an excess of emission from the region between A3391 and A3395 and 2) rule out the excess can be entirely explained by scattering of x-ray photons from A3391 and A3395 by *ASCA*'s optics.

The presence of the excess emission is illustrated in Fig. 3. For the figure, the *ASCA* GIS events data were converted to an image via adaptive smoothing kernel. This preliminary image was convolved with a Gaussian with a HWHM of $1''$ (*ASCA* HPD=$2.9''$). A background count rate derived from the database of source-removed exposures was subtracted. The colour scale is in units of σ above the background.

Raytracing was used to model the scattering of light. An x-ray intensity map of the region extracted from the RASS was used as input to the *ASCA* raytracing program,

TABLE 1. Members of the supercluster/filament. The members are listed from northern-most to southern-most. Data from [2, 3, 4, 5, 1]

Obj	RA	Dec	L_X [10^{43} erg s^{-1}]	T [keV]	V_r [km s^{-1}]
A3380	$6^h06^m58^s$	$-49°29'$			16 700
MS 0620.6-5239	$6^h21^m44^s$	$-52°41'$	2.6 ± 0.6		15 300
AS0584	$6^h22^m58^s$	$-53°36'$	<0.01		14 200
A3391	$6^h26^m15^s$	$-53°41'$	16 ± 2	4–5	16 500
ESO161 -IG 006	$6^h26^m06^s$	$-54°02'$	0.12 ± 0.01		15 600
A3395 NW	$6^h26^m35^s$	$-54°20'$			14 500
A3395 E	$6^h27^m37^s$	$-54°27'$			15 000
A3395 SW	$6^h26^m49^s$	$-54°33'$			15 400

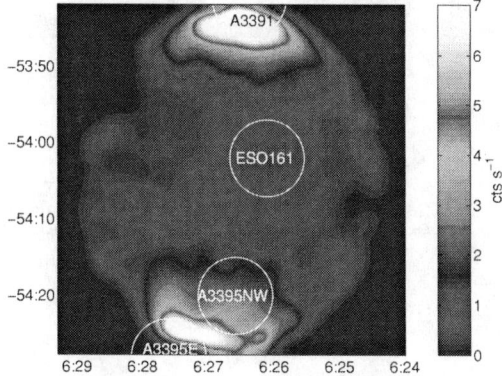

FIGURE 3. The *ASCA* GIS image.

trace_asca[6]. To determine the significance of scattered light in the inter-cluster region, the inter-cluster region was 'blanked' with a constant count-rate equal to the sky background and the raytracing was performed again. The input and results are illustrated in Fig. 4. If scattered photons were the the dominant source of intercluster flux, the two output images would have similar flux in the intercluster region. The extended emission found between A3391 and A3395 is clearly not an artefact of light scattering from the brighter sources just beyond the field of view of the x-ray telescope. There is, however, still a contribution (13%) of light to the field.

The spectral information from the *ASCA* GIS, SIS, and *ROSAT* PSPC instruments were used to estimate the fluxes in the intercluster region (minus the contribution from a region around the galaxy group and the regions around A3391 and A3395). Simultaneous fitting did not work due to normalization discrepancies between *ROSAT* and *ASCA*. Computed flux rates for the intercluster emission are given in Table 2. The rate for the GIS observation, at a distance of $150h^{-1}$ Mpc, corresponds to a luminosity, $L = 6.0 \times 10^{42} h^{-2}$ erg s^{-1} [0.8 – 9.0 keV]. The errors include the deviation in the

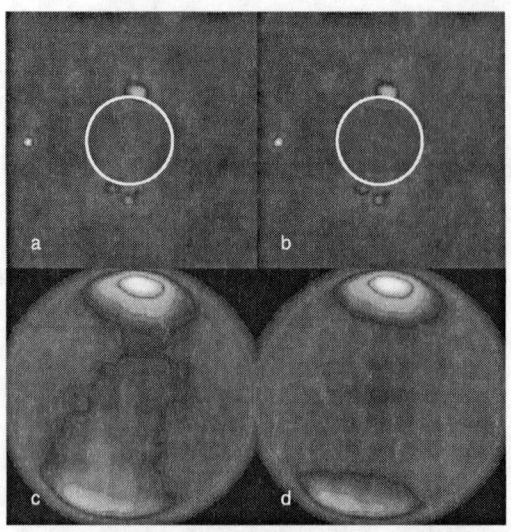

FIGURE 4. Input to trace_asca from the RASS (top). Raytraced images of the *ASCA* GIS FOV (bottom).

TABLE 2. Computed flux rates for the intercluster emission.

Instrument	Energy Window [keV]	X-ray flux $[10^{-13}\,\mathrm{erg\,cm^{-2}\,s^{-1}}]$
ASCA GIS	0.8 - 9.0	13 ± 1
ASCA SIS	0.5 - 6.0	9 ± 1
ROSAT PSPC	0.4 - 2.4	6.0 ± 0.6

background rate.

Please see Tittley and Henriksen [7] for further details.

REFERENCES

1. Teague, P. F., Carter, D., and Gray, P. M., *ApJS*, **72**, 715–753 (1990).
2. De Grandi, S., Bohringer, H., Guzzo, L., Molendi, S., Chincarini, G., Collins, C., Cruddace, R., Neumann, D., Schindler, S., Schuecker, P., and Voges, W., *ApJ*, **514**, 148 (1999).
3. Gioia, I. M., Maccacaro, T., Schild, R. E., Stocke, J. T., Liebert, J. W., Danziger, I. J., Kunth, D., and Lub, J., *ApJ*, **283**, 495–511 (1984).
4. Henriksen, M., and Jones, C., *ApJ*, **465**, 666 (1996).
5. Stocke, J. T., Morris, S. L., Gioia, I. M., Maccacaro, T., Schild, R., Wolter, A., Fleming, T. A., and Henry, J. P., *ApJS*, **76**, 813–874 (1991).
6. Ptak, A. F., *"X-Ray Constraints on Accretion and Starburst Processes in Galactic Nuclei"*, Ph.D. thesis, University of Maryland College Park (1997).
7. Tittley, E. R., and Henriksen, M., *ApJ*, **563**, 673–686 (2001).

HIGH-REDSHIFT STRUCTURE

Active Galaxies from $Z = 0$ to $Z = 6$

Amy J. Barger**

Department of Astronomy, University of Wisconsin-Madison 53706
Department of Physics and Astronomy, University of Hawaii 96822
Institute for Astronomy, University of Hawaii, Honolulu, Hawaii 96822

Abstract. The high angular resolution and sensitivity of the *Chandra X-ray Observatory* has yielded large numbers of faint X-ray sources with measured redshifts in the soft (0.5 − 2 keV) and hard (2 − 8 keV) energy bands. Many of these sources show few obvious optical signatures of active galactic nuclei (AGN). We use *Chandra* observations of the Hubble Deep Field-North region, together with other shallower surveys, to show that the number density of sources in the 10^{42} ergs s^{-1} to 10^{44} ergs s^{-1} luminosity range is rising, or is at least constant, with decreasing redshift. Broad-line AGN are the dominant population at higher luminosities, and these sources show the well-known rapid positive evolution with increasing redshift to $z \sim 3$. We argue that the dominant supermassive black hole formation has occurred at recent times in objects with low accretion mass flow rates rather than at earlier times in more X-ray luminous objects with high accretion mass flow rates. We also argue that there are too few moderate luminosity AGN at $z = 5 - 6.5$ to ionize the intergalactic medium.

INTRODUCTION

A primary objective in current observational research is to map the growth of supermassive black holes (SMBHs). A challenge in achieving this goal is that we need a complete census of SMBHs in the Universe, including sources that are obscured by gas and dust. Fortunately, high energy X-rays can penetrate extremely large column densities of gas and dust. Thus, hard X-ray surveys can provide an accurate accounting of all but the most highly obscured SMBHs. Early analyses of deep X-ray surveys with the *Chandra X-ray Observatory* found that optical selections of active galactic nuclei (AGN) had missed large numbers of accreting SMBHs. Ref. [1] showed that the redshift history of SMBH growth is more strongly peaked to low redshifts than had been inferred from optically-selected samples. With ultradeep *Chandra* observations, we can resolve the X-ray background, determine the evolution of SMBHs with time, and search for the first SMBHs at high redshifts.

THE XRB

The two deepest X-ray images of the sky are the 2 Ms exposure of the *Chandra* Deep Field-North (CDF-N) [2] and the 1 Ms exposure of the *Chandra* Deep Field-South (CDF-S) [3]. These exposures have resolved $> 80 - 90\%$ of the $2 - 8$ keV X-ray Background (XRB) into discrete sources [4, 5, 6], where the largest uncertainty is the XRB measurement itself. The field-to-field difference in the contribution to the XRB

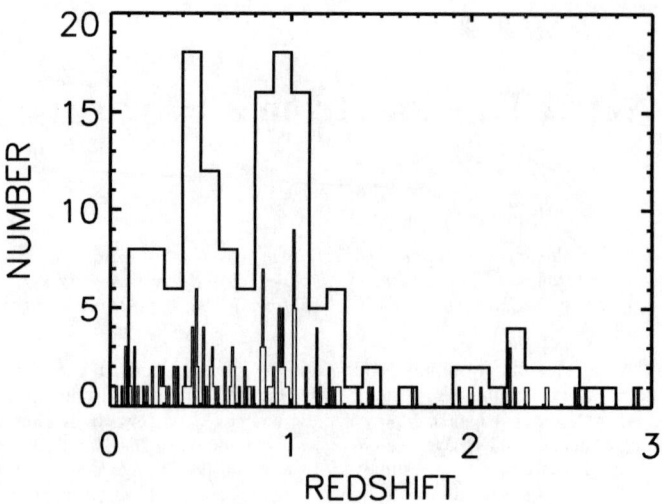

FIGURE 1. Redshift distribution in the CDF-N for two bin sizes ($\Delta z = 0.1$ and $\Delta z = 0.01$). Figure from Ref. 7.

from the X-ray sources in the CDF-N and CDF-S is 40%; this is substantially above the Poisson noise expected from the number of sources, suggesting clustering.

Ref. [7] presented an optical catalog for the 1 Ms CDF-N sample [8], including spectroscopic redshifts for 182 of the 370 point sources. In Fig. 1 we show histograms of the redshift distribution (using $\Delta z = 0.1$ and $\Delta z = 0.01$ bins). The figure suggests there may be large scale structure in the field. The two structures centered on $z = 0.843$ and $z = 1.0175$ (also detected in optical surveys[9, 10]) contain at least 10 X-ray sources within 1000 km s^{-1} of the center positions. The number of sources in these structures (a total of 24 identified sources in the $z = 0.843$ and $z = 1.0175$ structures) is sufficiently large that it could account for a part of the field-to-field variation seen in the X-ray number counts.

COSMIC EVOLUTION

With the high spatial resolution and energy sensitivity of deep *Chandra* observations, X-ray samples can be selected at both low and high redshifts in the same rest-frame *hard* energy band and their counterparts identified. Using the CDF-N sample supplemented with some shallower surveys, we constructed samples that probe to an L_x below 10^{43} erg s^{-1} in the redshift interval $z = 2 - 4$ and below 10^{42} erg s^{-1} in the interval $z = 0.1 - 1$. Both samples are highly complete in redshift identifications at high luminosities. At $L_x > 10^{44}$ erg s^{-1} in the high-redshift interval there are 22 spectroscopi-

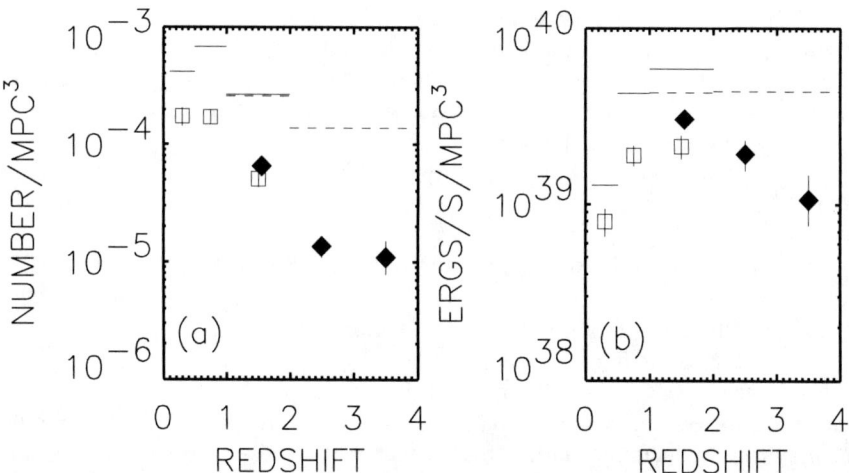

FIGURE 2. Evolution with redshift of the comoving number density (a) and the 2 − 8 keV comoving luminosity density production rate ($\dot{\lambda}_x$) of $L_x > 10^{42}$ erg s^{-1} sources (b). Open squares (solid diamonds) are measured from observed-frame 2 − 8 keV (0.5 − 2 keV). Poissonian 1σ uncertainties are based on the number of sources in each bin. Dashed (0.5 − 2 keV) and solid (2 − 8 keV) bars show upper limits found by assigning all the unidentified sources to the center of each redshift bin, so the bars are not consistent with one another. To minimize this effect at high redshifts, we use a single $z = 2 - 4$ redshift bin. At the highest redshifts a small correction (less than 20%) has been applied to extrapolate from the minimum measured luminosity to $L_x = 10^{42}$ erg s^{-1}. Note that the uncertainty in the $\dot{\lambda}_x$, which is weighted to higher luminosity sources, is smaller than the uncertainty in the number density. Figure from Ref. 11.

cally identified sources, 21 of which are broad-line AGN. At most there are a further 18 spectroscopically unidentified sources that could lie at these luminosities in this redshift interval; however, a substantial fraction of these are likely at other redshifts. At $L_x > 10^{44}$ erg s^{-1} in the low redshift interval, 85% of the identified sources are broad-line AGN, and there are only three unidentified sources that could lie at these luminosities in this redshift interval. Thus, most of the high-luminosity X-ray sources are broad-line AGN rather than obscured AGN.

In Fig. 2a we show the comoving number density evolution with redshift of $L_x > 10^{42}$ erg s^{-1} sources [11]. At these luminosities the probability that the X-ray sources are powered by anything other than AGN is extremely low, so we are mapping the evolution of intermediate luminosity AGN. In the $z = 1 - 2$ redshift bin we show the number densities from both observed-frame 2 − 8 keV (open squares) and observed-frame 0.5 − 2 keV (solid diamonds); the results are similar, although the soft band may be missing some obscured sources. The solid (2 − 8 keV) and dotted (0.5 − 2 keV) horizontal bars show the number densities obtained if all the spectroscopically unidentified sources are assigned redshifts at the center of each redshift bin. The bars are not consistent with one another because all the unidentified sources are included in all the bins.

The number density of intermediate luminosity AGN does not rise rapidly from $z \sim 0$ to $z \sim 2.5 - 3$, in contrast to that of higher luminosity AGN (e.g., [12]). In fact, the evolution with increasing redshift is flat or declining. Thus, AGN traced by X-ray luminosity show their own version of downsizing, just as star-forming galaxies do [13]. High X-ray luminosity sources, which strongly overlap with broad-line AGN (quasar) populations, peak at higher redshifts, while intermediate luminosity sources are still common now.

In Fig. 2b we show the evolution with redshift of the $2-8$ keV comoving luminosity density production rate ($\dot{\lambda}_x$) of $L_x > 10^{42}$ ergs s^{-1} sources. Integration of power-law fits to the hard X-ray luminosity functions give a similar answer: 1.9×10^{39} ergs s^{-1} Mpc^{-3} in the $z = 0.1 - 1$ redshift interval and 2.2×10^{39} ergs s^{-1} Mpc^{-3} in the $z = 2 - 4$ interval. The largest uncertainty is the redshift distribution of the unidentified sources rather than the small effects of extrapolation outside our observed luminosity range. This incompleteness uncertainty is less than a factor of 2.9 even in the high redshift interval.

We conclude that while the higher accretion mass flow rates that power the most luminous AGN peaked at higher redshift and are now much rarer, lower accretion mass flow rates are still common at the present time. These less luminous events dominate SMBH formation.

AGN FORMATION AT $Z > 5$

Most of our current information on very high redshift AGN relates to the very high luminosity tail of the luminosity function (LF). At $z > 5$ the Sloan Digital Sky Survey (SDSS) probes to an 1450 Å absolute magnitude of about -26 [14, 15], much brighter than the average luminosity of AGN at lower redshifts [16]. Calculations of the contributions of high-redshift AGN to the ionizing flux in the Universe must therefore rely on large extrapolations (e.g., [17]), and deeper observations of the LF are needed to constrain this quantity directly. Such observations can also test models of supermassive black hole formation (e.g., [18, 19, 20]).

We can efficiently search for faint, high-redshift AGN using combined ultradeep X-ray and optical imaging. Since at high redshifts observed-frame X-ray bands correspond to very high rest-frame energies, we can obtain a relatively complete sample of AGN, including sources surrounded by very high column densities of gas and dust. Some Compton-thick sources may be omitted from the sample but would not be expected to contribute much to the ionizing light. As there are only a handful of stars in X-ray samples, there is little problem in separating red stars from high-redshift AGN.

We used deep broad-band optical imaging with Suprime-Cam [21] on Subaru and follow-up optical spectroscopy with LRIS [22] on Keck to measure the $z > 5$ AGN population in the CDF-N 2 Ms exposure. Our sample probes almost two orders of magnitude fainter than the typical X-ray flux seen in the SDSS, but the area of sky covered is four orders of magnitude smaller. High-redshift sources in the X-ray sample are easily identified because the increasing line blanketing of the Lyman forest at high

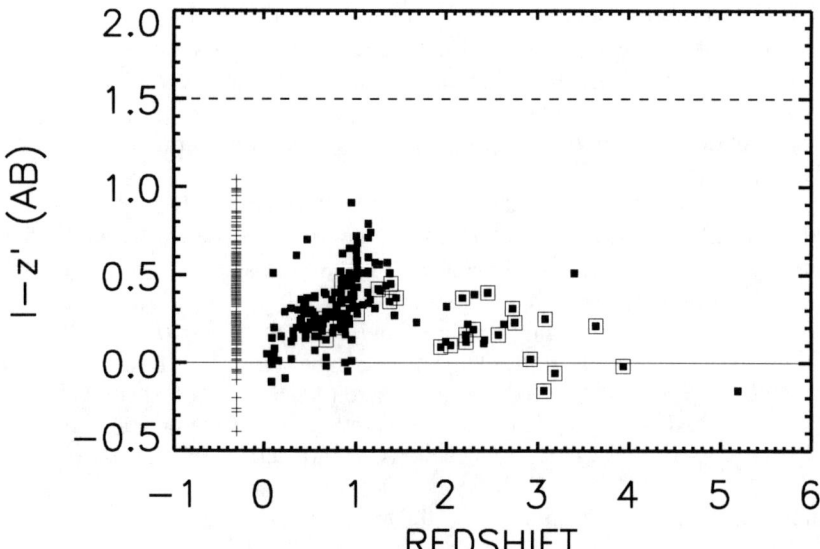

FIGURE 3. $I - z'$ in the AB system versus redshift for the 423 sources in the CDF-N with $z' < 25.2$. Sources without redshifts are shown at $z = -0.3$, and broad-line AGN are enclosed in a second, larger symbol. The dashed line ($I - z' = 1.5$) is the color of $z \sim 6$ sources in the quasar and galaxy models described in the text. None of the spectroscopically unidentified sources are red enough in $I - z'$ to be at $z > 6$.

redshifts produces extremely large breaks across redshifted Lyα (e.g., [15]). Figure 3 shows $I - z'$ versus redshift (spectroscopically unidentified sources are at $z = -0.3$) for our $z' < 25.2$ sample. We see from Fig. 3 that none of the sources lie at $z > 6$, since such sources would be redder than $I - z' > 1.5$. In Fig. 4 we show $V - I$ versus $I - z'$. The dashed line is the track with redshift of the composite SDSS quasar spectrum of [23], extrapolated with a $f_v \sim v^{-0.46}$ power-law below Lyα, modulated by the [24] forest transmissions, and convolved through the Suprime-Cam filters. The extrapolation below the Lyman continuum limit is not critical since the high incidence of Lyman limit systems at these redshifts truncates the spectrum at shorter wavelengths. The dot-dash line shows the same calculation for a $f_v \sim v^0$ galaxy with no emission lines and a cut-off at 912 Å. The deep forest absorption moves the tracks far from the galaxy populations. Apart from the one $z = 5.19$ AGN (large solid square) with $z' = 23.9$, there are no other $z > 5$ candidates.

We compute the number density of $z = 5 - 6.5$ sources in the rest-frame $2 - 8$ keV luminosity range 10^{43} to 10^{44} erg s^{-1} following the procedures described in [11]. In Fig. 5 we show the number density obtained using only the one $z > 5$ spectroscopic identification (solid diamond). We also computed an upper limit to this number density by assigning all the optically faint sources that could lie in the redshift interval redshifts at the center of the interval; the horizontal bar corresponds to those with L_x in the specified luminosity range. In Fig. 5 we compare our $z = 5 - 6.5$ number density with the lower redshift number densities from [11] for both the 10^{43} to 10^{44} erg s^{-1} (solid circles) and 10^{44} to 10^{45} erg s^{-1} ranges (open circles). Again, the symbols denote the number densities obtained from only the spectroscopically identified AGN, while the horizontal bars show the upper limits. The bars are not consistent with one another because all the unidentified sources are included in each redshift bin if they lie in the specified luminosity range.

To compare with the optically-selected samples of [25] and [15], in Fig. 6 we show the results for an $\Omega_M = 1$, $\Omega_\Lambda = 0$ cosmology with $H_o = 50$ km s^{-1} Mpc^{-1}. The small number uncertainties are large in Fig. 6, but the number densities in the 10^{43} to 10^{44} erg s^{-1} range (AGN whose X-ray luminosities would classify them as Seyferts) show a slow decrease with increasing redshift from $z = 0$ to $z = 6$. This result holds even within the systematic uncertainties shown by the the horizontal bars. In contrast, the number densities in the 10^{44} to 10^{45} erg s^{-1} range (AGN with quasar luminosities) peak at $z = 1.5 - 3$ and closely match in shape the evolution of the optically-selected samples. Figures 5 and 6 show that this conclusion does not depend on the cosmological geometry.

The number density at $z = 5 - 6.5$ for $L_x \approx 10^{43}$ to 10^{44} erg s^{-1} is about three orders of

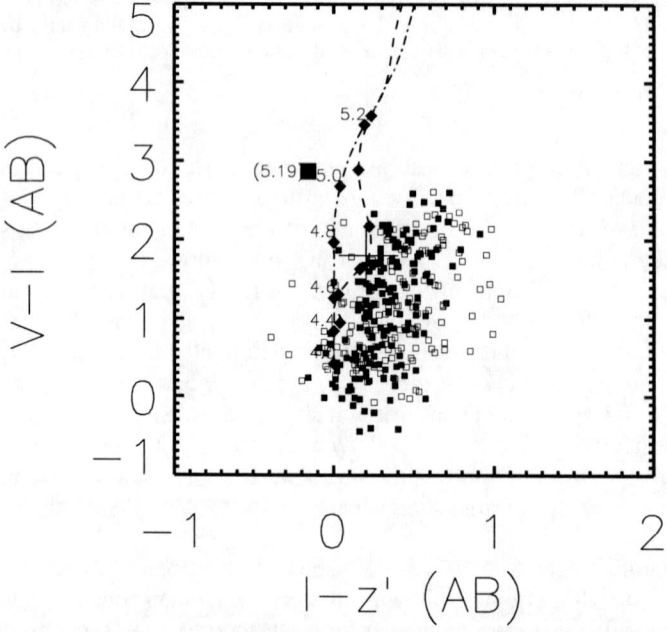

FIGURE 4. $V - I$ versus $I - z'$ for the $z' < 25.2$ sample in the CDF-N sample. Solid (open) squares denote galaxies with (without) redshift identifications. All sources are detected above the 2σ level in all three bands. One sigma uncertainties are shown for a $z' = 25.2$ source lying at $I - z' = 0.2$ and $V - I = 1.8$, typical of the location from which galaxies might scatter into regions of the color-color space and be misidentified as high-redshift objects. The one spectroscopically identified $z = 5.19$ source is shown with a large solid square. The dashed (dot-dashed) curve shows the quasar (blue galaxy) track with redshift. Redshifts are given at the positions of the diamonds on the tracks.

magnitude higher than the optically-selected SDSS sample at $L_x \approx 10^{45}$ erg s^{-1}. While this conclusion is based on the single $z > 5$ object, the fact that other high-redshift AGN have also been found in the relatively small *Chandra* fields (e.g., [26]; $z = 4.93$) and in the small (74 arcmin^2) optical field of [27] ($z = 5.5$) gives us confidence that the CDF-N field is not anomalous. If we use our spectroscopically identified point, the slope β of the LF ($\phi(L)dL = L^{-\beta}dL$) would be about 2.6, which is similar to the bright end slope at lower redshifts in both the optical [16] and X-ray [12]. If we include the unidentified objects, the maximum value of β would be about 3. A detailed LF determination for this redshift interval will require much larger area X-ray samples.

Ref. [15] investigated the AGN ionizing flux by fitting a variety of power-laws to extrapolate the bright SDSS data. A power-law with our observed slope of $\beta = 2.6$ matched to the SDSS data fails by more than an order of magnitude to ionize the intergalactic medium at these high redshifts, even extrapolating to faint magnitudes (their Fig. 10). Thus, there are too few moderate luminosity AGN at $z = 5 - 6.5$ to ionize the intergalactic medium.

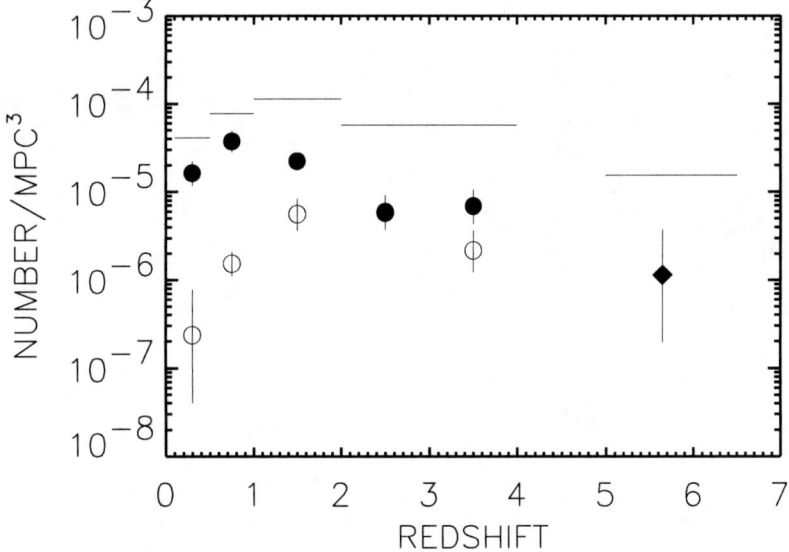

FIGURE 5. Number density of sources with rest-frame 2 – 8 keV luminosities between 10^{43} and 10^{44} erg s^{-1} (solid symbols) and between 10^{44} and 10^{45} erg s^{-1} (open symbols) versus redshift. Circles are from Ref. 11. Points below (above) $z = 2$ were determined from the observed-frame 2 – 8 keV (0.5 – 2 keV) sample. An intrinsic $\Gamma = 1.8$ was assumed, for which there is only a small differential K-correction to correct to rest-frame 2 – 8 keV. Poissonian 1σ uncertainties are based on the number of sources in each redshift interval. Horizontal bars show the maximal LF in the 10^{43} to 10^{44} erg s^{-1} range found by assigning all the sources that could lie in each redshift (and then luminosity) interval a redshift at the center of the interval.

REFERENCES

1. Barger, A. J., Cowie, L. L., Bautz, M. W., Brandt, W. N., Garmire, G. P., Hornschemeier, A. E., Ivison, R. J., and Owen, F. N., *Astronomical Journal*, **122**, 2177 (2001).
2. Alexander, D. M., et al., *Astronomical Journal*, submitted
3. Giacconi, R., et al., *Astrophysical Journal Supplement Series*, **139**, 369 (2002)
4. Campana, S., Moretti, A., Lazzati, D., and Tagliaferri, G., *Astrophysical Journal*, **560**, L19 (2001)
5. Cowie, L.L., Garmire, G.P., Bautz, M.J., Barger, A.J., Brandt, W.N., and Hornschemeier, A.E., *Astrophysical Journal*, **566**, L5 (2002)
6. Rosati, P., et al., *Astrophysical Journal*, **566**, 667 (2002)
7. Barger, A. J., Cowie, L. L., Brandt, W. N., Capak, P., Garmire, G. P., Hornschemeier, A. E., Steffen, A. T., and Wehner, E. H., *Astronomical Journal*, **124**, 1839 (2002)
8. Brandt, W. N., et al., *Astronomical Journal*, **122**, 2810 (2001)
9. Cohen, J. G., Hogg, D. W., Blandford, R., Cowie, L. L., Hu, E., Songaila, A., Shopbell, and P., Richberg, K., *Astrophysical Journal* **538**, 29 (2000)
10. Dawson, S., Stern, D., Bunker, A. J., Spinrad, H., and Dey, A., *Astronomical Journal*, **122**, 598 (2001)
11. Cowie, L. L., Barger, A. J., Bautz, M. W., Brandt, W. N., and Garmire, G. P., *Astrophysical Journal*, in press
12. Miyaji, T., Hasinger, G., and Schmidt, M., *Astronomy and Astrophysics*, **353**, 25 (2000)
13. Cowie, L. L., Songaila, A., Hu, E. M., and Cohen, J. G., *Astronomical Journal*, **112**, 839 (1996)
14. Fan, X., et al., *Astronomical Journal*, **122**, 2833 (2001)
15. Fan, X., et al., *Astronomical Journal*, **121**, 54 (2001)
16. Pei, Y.C., *Astrophysical Journal*, **438**, 623 (1995)

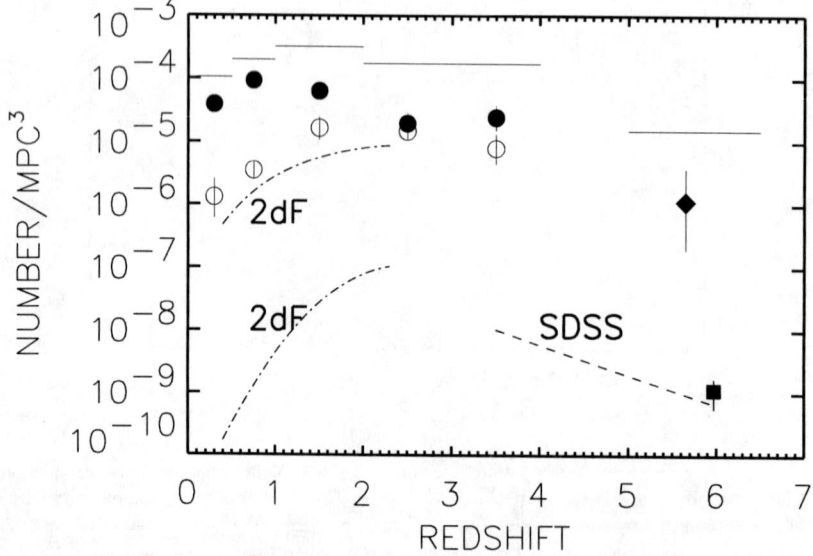

FIGURE 6. As in Fig. 5 but for an $\Omega_M = 1$, $\Omega_\Lambda = 0$ cosmology with $H_o = 50$ km s^{-1} Mpc^{-1}. Dot-dashed curves show the 2dF quasar LF of Ref. 25: upper curve is for objects with absolute 1450 Å magnitudes brighter than -23 (quasars) that roughly matches our $L_x > 10^{44}$ erg s^{-1} selection, lower curve is for objects brighter than -26.8 that matches the SDSS sensitivity to high-redshift quasars. Dashed line shows the SDSS objects brighter than -26.8 in the $z = 3.5$ to $z = 6$ range. Solid square and uncertainty is for the SDSS $z > 5.7$ quasar sample to -26.8 of Ref. 15.

17. Madau, P., Haardt, F., and Rees, M. J., *Astrophysical Journal*, **514**, 648 (1999)
18. Haiman, Z. and Loeb, A. 1998, *Astrophysical Journal*, **503**, 505
19. Haehnelt, M. G., Natarajan, P., and Rees, M. J., *Monthly Notices of the Royal Astronomical Society*, **300**, 817 (1998)
20. Haiman, Z. and Loeb, A. 1999, *Astrophysical Journal*, **521**, L9
21. Miyazaki, S., et al., *Publications of the Astronomical Society of Japan*, in press
22. Oke, J.B., et al., *Publications of the Astronomical Society of the Pacific*, **107**, 375 (1995)
23. Vanden Berk, D.E., et al., *Astronomical Journal*, **122**, 549 (2001)
24. Songaila, A. and Cowie, L. L., *Astronomical Journal*, **123**, 2183 (2002)
25. Boyle, B. J., Shanks, T., Croom, S. M., Smith, R. J., Miller, L., Loaring, N., and Heymans, C., *Monthly Notices of the Royal Astronomical Society*, **317**, 1014 (2000)
26. Silverman, J. D., et al., *Astrophysical Journal*, **569**, L1 (2002)
27. Stern, D., Spinrad, H., Eisenhardt, P, Bunker, A. J., Dawson, S., Stanford, S. A., & Elston, R., *Astrophysical Journal*, **533**, L75 (2000)

The Star Formation History of Galaxies from Infrared Observations

A. Franceschini* and G. Rodighiero*

Dipartimento di Astronomia, Vicolo Osservatorio 5, I-35122 Padova, Italy

Abstract. We exploit deep surveys performed in the IR and sub-millimeter, as well as the spectral intensity of the extragalactic IR background, to constrain galaxy evolution in an entirely complementary way compared with the usual deep optical-UV searches. These observations indicate high rates of evolution for the comoving IR luminosity density up to redshift $z \sim 1$. We also report on spectral analyses of these sources, revealing them to involve massive galaxies hosting violent starbursts. Our inferred evolutionary scheme considers phases of long-lived quiescent and enhanced star formation (SF) taking place during transient events recurrently triggered by interactions and merging. We interpret the strong observed evolution as an increase with z of the rate of galaxy interactions (*density evolution*) and of their IR luminosity due to the more abundant fuel in the past (*luminosity evolution*). The evolution of IR emissivity of galaxies from the present time to $z \sim 1$ is so strong that the constraints by the observed z-distributions and CIRB spectrum impose it to turn-over at $z > 1$: scenarios in which a relevant fraction of stellar formation occurs at very high-z are not supported by our analysis. We finally notice an interesting match between this epoch of peak SF, as revealed by IR observations, and the epoch when the galaxy mass function shows first signs of a strong negative evolution.

INTRODUCTION

There are two main approaches to detect and characterize high-redshift galaxies. The first is the well-established one looking at the integrated stellar emissions through very deep UV/optical/near-IR continuum surveys using ground-based optical telescopes or HST. By these means, large numbers of galaxies at any redshifts up to z=6 and more have been discovered during the last decade (e.g. [1, 2]). Although exploiting the most powerful instrumentation available for astronomy, the obvious limitation of this approach is that, by looking at the already formed stellar populations, it misses by definition the most active phases of star formation. Even the detection of young stars in the optical-UV is made difficult by the large amounts of absorbing dust typically present in star-forming regions of galaxies.

The alternative and entirely complementary approach to the direct detection of stellar light is to look for the emissions by diffuse media present in high-redshift galaxies: emission lines by ionized gas and molecules, hot plasma and high-energy particle emissions in X-rays and radio, or thermal continuum emissions by dust reprocessing into the IR the intense UV flux of young stars. This second channel of cosmological information is expected to provide essential data on galaxies during ancient phases in which only a minor fraction of the baryon content was already transformed into stars.

It is in any case by combining the information on already formed stellar populations from deep optical/near-IR imaging with data on diffuse media by long-wavelength surveys that we can expect to infer the most detailed and reliable picture about the formation of structure. We summarize in this paper some results of recent deep observations carried out in the mid- and far-IR with the Infrared Space Observatory to constrain the evolution of the star formation rate in galaxies over the redshift interval $0 < z < 1.3$, and mention attempts to further extend this information to higher redshifts by means of deep sub-millimetric observations from ground with the JCMT and IRAM telescopes.

We will then eventually compare the results of the far-IR observations with current tentative evidence on the evolution of the mass function of galaxies derived from deep near-IR observations.

A DEEP VIEW OF THE IR SKY: THE ACTIVE PHASE OF STAR FORMATION

A Diffuse Background of Cosmological Origin (CIRB)

The detection of the CIRB in the all-sky COBE maps brought for the first time to determine the integrated emission of distant galaxies in the form of an isotropic background signal in the far-IR and sub-mm. This discovery was made possible by the multiwavelengths maps of the whole sky between 100 and 1000 μm, including data from two independent instruments on-board of COBE, FIRAS [3, 4] and DIRBE [5]. The current information on the CIRB is summarized Fig. 1.

The integrated CIRB intensity between 100 and 1000 μm, where the present CIRB estimates are more reliable, is $\sim (30 \pm 5)\ 10^{-9}\ Watt/m^2/sr$. The avaliable constraints at shorter wavelengths (Fig. 1) imply that the total intensity of the CIRB between 7 and 1000 μm is roughly $vI(v)|_{FIR} \simeq 40\ 10^{-9}\ Watt/m^2/sr$. This flux is larger than the integrated bolometric emission by distant galaxies between 0.1 and 7 μm obtained by counting galaxies down to the faintest fluxes achiavable with HST in the Hubble Deep Fields ([9], $vI(v)|_{opt} \simeq (17 \pm 3)\ 10^{-9}\ Watt/m^2/sr$). The IR spectral domain is then the major reservoir of radiant energy by cosmic sources at any epochs.

Mid-IR Surveys by the Infrared Space Observatory (ISO)

The imaging capabilites and good sensitivity of the Infrared Space Observatory have been exploited for deep systematic explorations of the sky at mid-IR wavelengths with sensitivities sufficient to detect sources at cosmological redshifts [10, 11, 6].

Particularly in the mid-IR broad-band filter at 12-18 μm ($\lambda_{eff} = 15\ \mu$m) extragalactic surveys have been performed in the ISOCAM Guaranteed Time, over a total area of 1.5 square degrees with more than one thousand detected sources [10]. The two Hubble Deep Field areas (North and South), including the Flanking Fields for a total of $\sim$ 50 sq. arcmin, have been deeply surveyed by ISOCAM at 15 μm to a sensitivity limit of 100

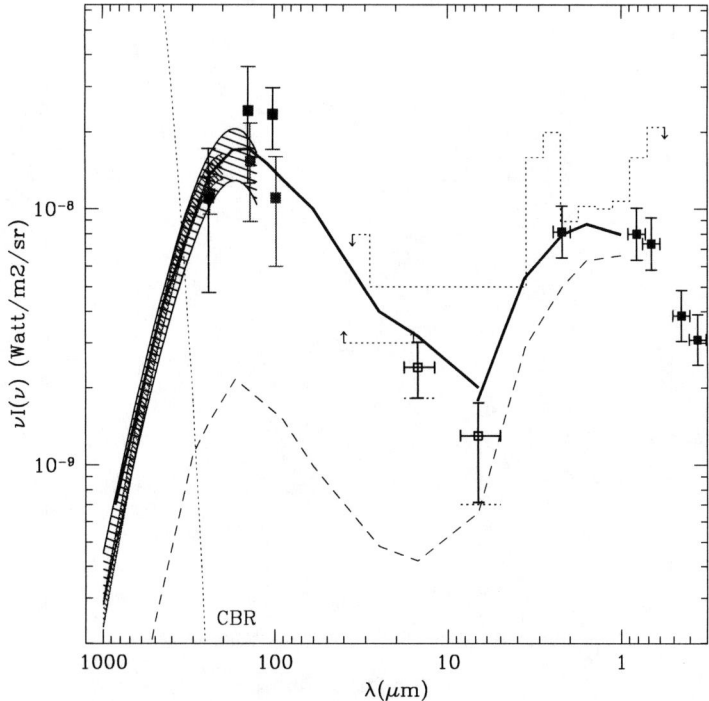

FIGURE 1. The Cosmic Infrared Background (CIRB) spectrum as measured in the all-sky COBE maps compared with estimates of the optical background based on ultradeep HST surveys (see text). The lower dashed line is the expectation based on the assumption that the IR emissivity of galaxies does not change with cosmic time. The thick line is the predicted CIRB by [6]. Between $4\mu m < \lambda < 60\mu m$, where the CIRB is dominated by the the Inter-Planetary Dust emission, indirect constraints can be inferred from measurements of the cosmic opacity for $\gamma - \gamma$ interactions ([7, 8], see the dotted histograms).

μJy. The redshift distributions show an excess number of sources between z=0.5 and z=1. At brighter fluxes, the European Large Area ISO Survey (ELAIS) observed a total of 12 square degrees at 15 μm with ISOCAM [12, 13].

The differential counts of extragalactic sources at 15 μm (stars are easily identified as bright point-like associations in the optical maps) normalized to $S^{-2.5}$ (Fig. 2) reveal a roughly euclidean slope from the brightest fluxes down to $S_{15} \sim 10$ mJy and a strong excess between $S_{15} \simeq 3$ and 0.3 mJy, where the counts increase as $dN \propto S^{-3.1} dS$. [14] estimate that of order of 50% of the CIRB intensity at 15 μm has already been resolved into sources by the deepest ISOCAM surveys.

[15] report a systematic analysis of faint mid-IR sources detected in the HDF South based on optical and near-IR spectroscopy. The baryonic masses in stars are estimated

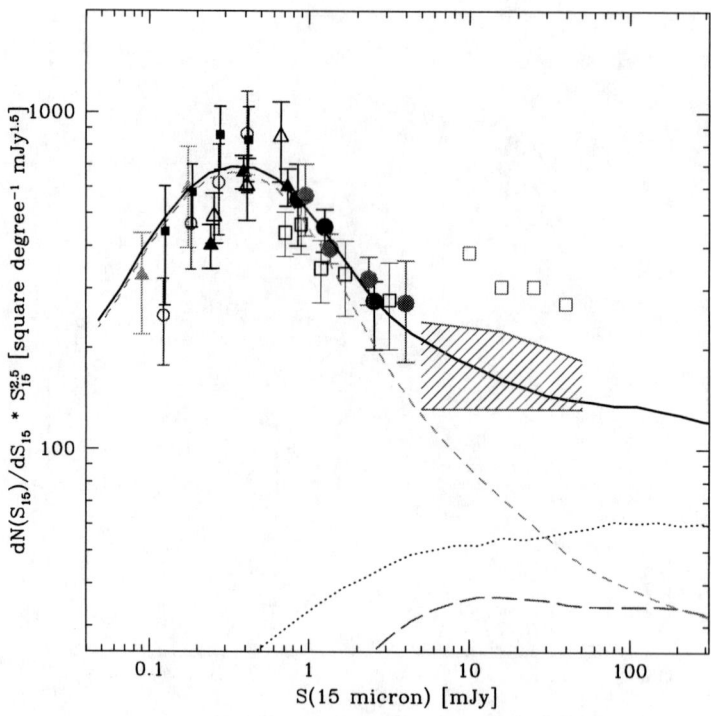

FIGURE 2. Normalized differential counts at $\lambda_{eff} = 15 \,\mu m$ (see [10]. The various lines are different populations (AGNs as long-dashed line, normal spirals as dots, and starburst galaxies as short dashes) of the model by [6].

from fits of the overall optical-IR continuum, and found to correspond to massive members of groups, as also confirmed by dynamical measures based on high-resolution IR spectroscopy [16]. The H_α line is always present and rather intense, revealing that these galaxies host strong starbursts ($SFR \sim 100 \, M_\odot/yr$). However, by comparing masses and star-formation rates we infer that only a fraction of the galactic stars can be typically formed in any single starburst event, while several of such episodes during a protracted SF history are required for the whole galactic build-up.

Far-IR Surveys with ISO: a new data-reduction tool

The Photopolarimeter on-board of ISO (ISOPHOT) included two imaging cameras with broad-band filters centered at 95 and 175 μm. The longer-wavelength camera was used by the ISOPHOT most important survey project, FIRBACK [17], to detect at 170 μm the sources of the CIRB (a sample of roughly 200 sources is reported by [18]). However, this survey is limited by extragalactic confusion due to the large ISOPHOT

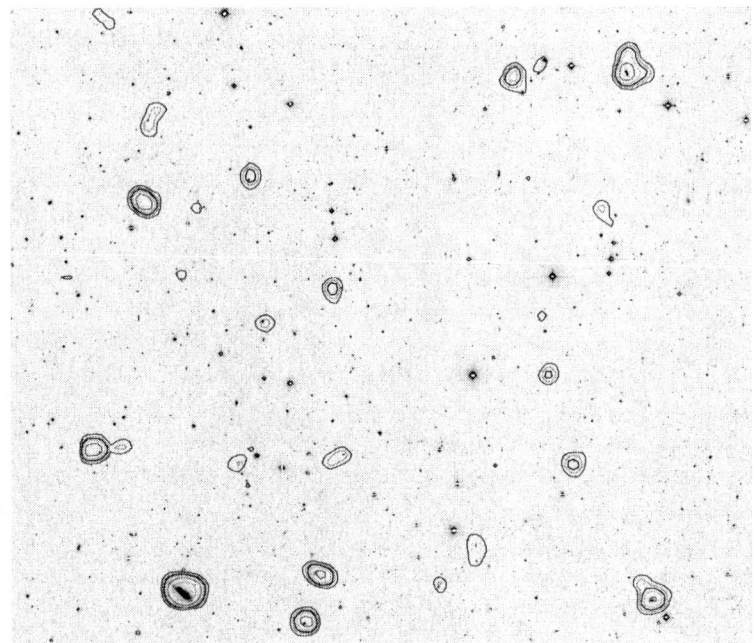

FIGURE 3. Overlay of the 95 μm contour plot based on the ISOPHOT Lockman Hole Survey with a deep r-band optical image (Rodighiero 2002, PhD Thesis, Padova University). This is likely the deepest image of the sky obtained at far-IR wavelengths.

beam (90 arcsec) to fluxes brighter than $S_{170} \geq 140$ mJy. At this limit the fraction of the CIRB intensity resolved into discrete sources was only $\sim 5\%$.

The ISOPHOT 100 μm camera was expected to deliver data with higher imaging quality and less affected by confusion. Unfortunately, its detectors turned out to suffer serious problems of hysteresis (slow response to flux variations) and variations in the baseline signal due to cosmic ray impacts (*glitches*).

For this reason, we [19] have developed a new technique for the reduction of the ISOPHOT survey data. The method is based on the assumption that the detector transients due to the charged-particle impacts follow two different timescales, a long and a short one. It consists in the analysis of the time history of each detectors to estimate the background signal level, and to identify the response variations corresponding to the *glitches* and to the real sources. Then the reconstructed time history of each detector is back-projected into the sky map, where sources are identified and registered in position and flux. As for the photometry, both the ISOPHOT internal and astronomical calibrators were used. The whole process was finally tested by means of extensive Monte Carlo simulation runs.

This reduction method turned out to be essential to derive reliable maps and source catalogues from the deep ISOPHOT 95 μm surveys, while being less important for the analysis of the 175 μm data for which the detector behaviour was less subject to artifacts.

We have exploited this method to reduce a deep raster survey performed by [21] in

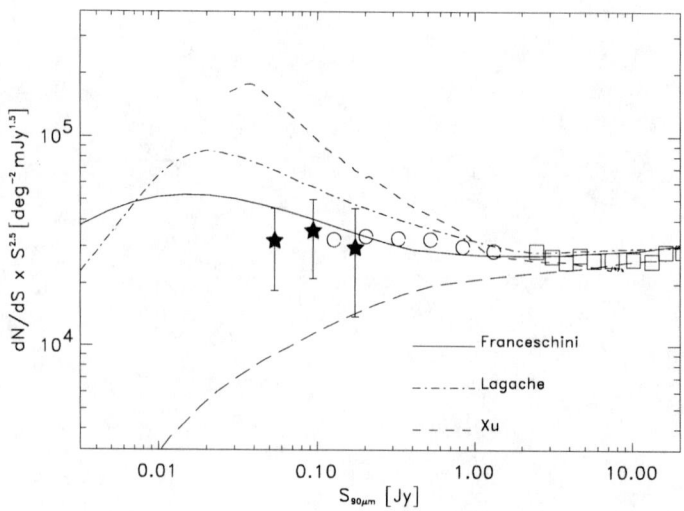

FIGURE 4. The extragalactic counts at 95 μm in differential euclidean-normalized form, based on the deep ISOPHOT survey in the Lockman Hole reduced and analysed by [19] (starred symbols). The ELAIS data come from [20] (open circles), while the open squares are based on the IRAS survey. The various lines are from published models. The long-dashed lowest line corresponds to the prediction of a model assuming the IRAS local luminosity function and no evolution.

the Lockman Hole over an area of $44' \times 44'$ minimally affected by the Galactic *cirrus* (the LHEX area). An overlay of the 95 μm contour plot with a deep optical image is shown in Fig. 3. Over this half square degree 36 sources are detected by Rodighiero et al. with S/N>3 down to a flux limit of $S_{95\mu m} \simeq 20$ mJy. This sample turns out to be 80% complete above 70 mJy, while the completeness decreases at fainter fluxes. Eighteen of the 36 sources show a radio counterpart in deep radio images covering part of the area, and are likely galaxies at moderate redshifts ($z < 0.5$), as confirmed by the 12 spectroscopic redshifts available for these counterparts. The other half of the sample, also considering the faintness of the optical counterparts, are likely luminous starbursts at higher redshifts (the stellar contamination is expected to be negligible).

[19] have derived from this sample the extragalactic source counts, shown in Fig. 4 in a differential euclidean-normalized form. The long-dashed lowest line in the figure corresponds to the prediction of a model assuming the IRAS local luminosity function without evolution: the data clearly reject it with good confidence. The continuous line is the same multi-wavelength evolution model assumed to fit the 15 μm counts in Fig. 2: the two independent datasets then provide consistent evidence for an evolutionary population of luminous starbursting galaxies at $z \sim 1$ to dominate the IR sky. The CIRB intensity is uncertain close to 100 μm due to the Zodiacal and Galactic contribution (published values ranging from 11 to 22 nanoW/m^2/sr). The Lockman Hole survey has correspondingly resolved at the 20 mJy flux limit from $\sim 9\%$ to $\sim 18\%$ of the cosmic background close to its peak emission wavelength.

Surveys at millimetric wavelengths

The sensitivities of the ISO's mid-IR and far-IR cameras have allowed to detect sources typically up to $z \sim 1.3$, while the K-corrections and cosmic dimming of the flux prevented us to cover efficiently source populations at higher z. On the other side, surveys in the sub-millimeter offer the rather unique advantage to generate volume-limited samples from flux-limited observations, due to the peculiar shape of galaxy's sub-mm spectra. Sensitive sub-mm surveys preferentially select sources at very high redshifts ($z \geq 1$), hence providing a complementary information to the ISO data.

Important discoveries have come from the implementation of arrays of bolometers (SCUBA and MAMBO) on JCMT and IRAM (e.g. [22, 23, 24]). SCUBA has allowed to resolve $\sim 20\%$ of the CIRB background at 850 μm into a population of faint distant, mostly high-z, sources. The extragalactic source counts show a dramatic departure from the Euclidean law [$dN \propto S^{-3}\,dS$ in the interval from 2 to 10 mJy], a clear signature of the strong evolution and high redshift of mm-selected sources.

The Cosmic History of star formation based on IR Data

The data on faint IR/sub-mm galaxies (some of which are reported in Figs. 2 and 4) together with the spectral intensity of the CIRB (Fig. 1) can be best reproduced [6, 25] considering the contribution of two source populations, one non evolving (with local luminosity function consistent with the IRAS 12 μm LLF), the other strongly evolving in cosmic time both in comoving number density and in luminosity as:

$$\rho(L[z],z) = \rho_0(L_0) \times (1+z)^{4.6}, \quad L(z) = L_0 \times (1+z)^2 \quad z < z_{break}$$

and $\rho(L[z],z) = const$ for $z_{break} < z < z_{max}$, with $z_{break} = 0.8$ and $z_{max} \sim 3$. A third source population (however never contributing significantly to the counts, see e.g. Fig. 2) are type-I AGNs, assumed to evolve in luminosity as $L(z) = L(0) \times (1+z)^3$ up to $z = 1.5$. The local fraction of the evolving starburst population is assumed to be ~ 10 percent of the total, roughly consistent with the observed fraction of interacting galaxies.

Deep surveys at various IR/sub-mm wavelengths can be exploited to simultaneously constrain the evolution properties and broad-band spectra of faint IR sources. Most of these observed multi-wavelength statistics can be reproduced under the assumption that the SED of the dominant IR evolving population is that of a typical starburst spectrum (e.g. M82). Best-fits to the counts based on this reference model are given in Figs. 1, 2 and 4.

The good match to the IR data found by assuming a typical starburst spectrum for the evolving sources is consistent with our finding that the faint IR-selected population is likely dominated by star-formation in galaxies rather than AGN emission, which would produce completely different IR spectra.

From luminosity-density to the comoving star-formation rate

[26] have shown that starbursts with bolometric luminosities above $10^{10} L_\odot$ produce the bulk of their energy in the far-IR. It is for this reason that the far-IR selection of high-z galaxies provides the most reliable way to estimate the galaxy's fundamental parameter, the rate of ongoing star formation (SFR), hence obtaining the comoving star-formation rate from luminosity-density.

To this end, a measure of the source flux in the rest-frame around 100 μm, where the galaxy IR SEDs are observed to peak, would be needed in principle. Of course, this quantity is available for only a small fraction of the faint IR sources, those detected by ISOPHOT at either 95 or 175 μm. The bulk of the sources was instead detected at 15 μm, due to the ISO's better spatial resolution here. There is neither much perspective of an improvement until the operation of FIRST-Herschel in 2007. An important result by [14], however, was to show that for a large variety of galaxies (from normal galaxies to luminous and ultra-luminous dusty starbursts) the mid-IR flux is extremely well correlated with the bolometric IR emission: $L_{IR} \sim 11 \cdot \nu L_\nu [15\mu m]$. They have shown that in a large sample of local objects only a very small fraction (few %) show significant departures from this correlation (a discrepant case is the highly-extinguished ULIRG Arp 220). This almost linear correlation of the bolometric far-IR and mid-IR luminosities was also confirmed by the excellent match between the SFR estimates based on the mid-IR and radio fluxes for dusty starbursts at $z \sim 1$.

Then we computed SFRs adopting the [27]'s calibration assuming a Salpeter IMF with standard stellar mass limits:

$$\frac{SFR}{M_\odot yr^{-1}} = 1.72 \cdot 10^{-10} L_{IR} [L_\odot] \qquad (1)$$

and transformed the IR-luminosity density as a function of cosmic time from our best-fit IR evolution model to the redshift-dependent comoving SFR density. This is shown as red continuous line in Fig. 5. It is interesting to note here the very fast rise of the SFR density from z=0 to z$\sim$ 1 and the flattening above. The detailed behavior of SFR at $z > 1$ is still uncertain, but a flattening seems required to match the observed z-distributions of mid-IR selected samples and the CIRB spectrum [6].

DEEP NEAR-IR GALAXY SURVEYS: A COMPLEMENTARY VIEW OF THE PASSIVE EVOLUTION PHASE

Measuring the mass function of distant galaxies and its evolution with cosmic time offers the complementary approach to recover the origin of cosmic structure compared with the direct detection of star-formation as described in the previous Section. In particular, the evolution of the massive end of the galaxy population, tracing the assembly of present-day luminous galaxies, is a critical cosmogonic observable.

Current large optical telescopes allow in principle to measure dynamical masses through high spatial and spectral resolution spectroscopy of high-z galaxies, but this requires long time integrations. A powerful alternative exploits observations of the

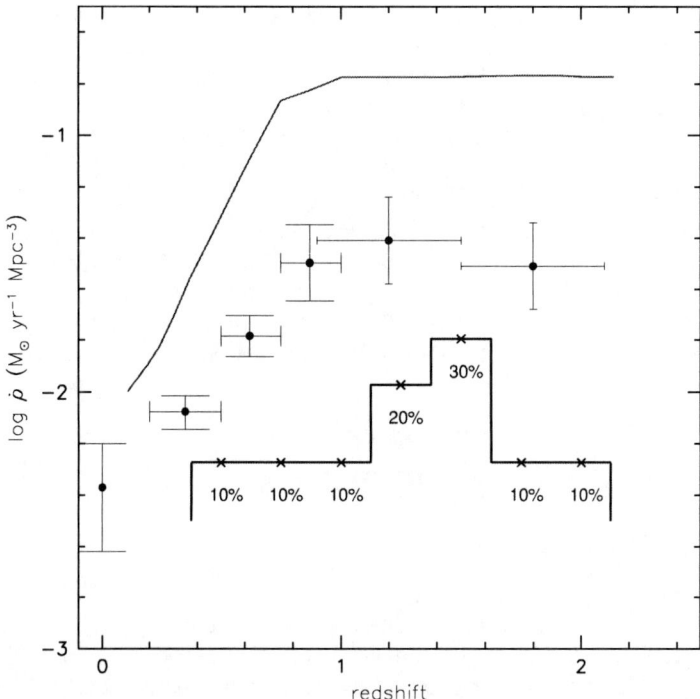

FIGURE 5. Evolution of the comoving star formation rate density for the IR-selected population based on our model of IR evolution (solid curve, see text for details). The IR evolution is compared with data coming from optical observations, transformed to our adopted $\Omega_m = 0.3$, $\Omega_\Lambda = 0.7$ cosmology. The histogram represents the percentual formation rate of field early-type galaxies (arbitrary normalization).

spectral intensity of galaxies in the near-IR and its weak dependence on the age of the contributing stars.

We have performed an extensive analysis of morphologically selected elliptical and S0 galaxies in the Hubble Deep Field (HDF) North and South and the HDFS NICMOS field, identifying 69 E/S0 galaxies with K<20.15 over an area of 11 arcmin^2. Although a moderately small number over a modest sky area, this sample benefits from the best imaging and photometric data available on high-redshift galaxies. Multi-waveband photometry allows us to estimate with good accuracy the redshifts for those galaxies lacking a spectroscopic measure. We ([28, 29]) found that massive E/S0s tend to disappear from this flux-limited sample at $z > 1.4$. This adds to the evidence that the rest-frame colours and SEDs of the numerous objects found at $0.8 < z < 1.2$ are inconsistent with a very high redshift of formation for the bulk of stars, while being more consistent with protracted (either continuous or episodic) star formation down to $z \leq 1$. A distribution of the rates of formation of E/S0 consistent with this analysis is reported in Fig. 5.

These results based on high-quality imaging on a small field can be complemented with data from colour-selected samples of red objects (EROs) on much larger sky areas. The apparent demise of E/S0s going above z~1 may be paralleled by a similarly fast decrease in the areal density of EROs when the colour limit is changed from (R-K)=5 to (R-K)=6 (corresponding to z=1 and z=1.3 respectively).

Altogether, the cosmic epochs corresponding to redshifts around 1 and up to 2 seem to correspond to a very active phase for the assembly of massive E/S0 galaxies in the field, and also probably one where a substantial fraction of their stars are formed (note however that this result rests on the assumption of a Salpeter stellar IMF, while different conclusions have been reached by [30], assuming a Scalo IMF). If our results will be confirmed, in particular by the forthcoming SIRTF and HST investigations, this will mark a very interesting match between the phase of peak SFR indicated by the far-IR observations and that inferred from the observed evolution of the mass function of galaxies based on near-IR studies.

CONCLUSIONS

Our main conclusions are the following.

1. We have made a preliminary comparison of the cosmic history of star-formation based on mid- and far-IR studies emphasizing the ongoing rate of SF and that inferred from the evolution of the mass function of galaxies based on deep near-IR surveys. We have found an interesting, although tentative, match between the two in the sense of a very active phase of SF around $z\sim 1$.
2. Our analysis of the faint IR-selected source populations do not seem to support scenarios in which a relevant fraction of stellar formation occurs at very high-z (e.g. producing the bulk of stars in spheroids at z>2). In this sense our current results agree with the view of a protracted phase for the formation of galaxies down to relatively recent cosmic epochs ($z\leq 1$) predicted by hierarchical formation models (e.g. S. White, this Conference).

ACKNOWLEDGMENTS

A.F. thanks the organizers of the Maryland October Meeting of 2002 for their kind invitation. We also benefited by many exchanges and collaborations, in particular with S. Berta, C. Cesarsky, D. Elbaz, D. Fadda, and C. Lari.

REFERENCES

1. Madau, P., Ferguson, H. C., Dickinson, M. E., Giavalisco, M., Steidel, C. C., and Fruchter, A., *MNRAS*, **283**, 1388–1404 (1996).
2. Steidel, C. C., Adelberger, K. L., Giavalisco, M., Dickinson, M., and Pettini, M., *ApJ*, **519**, 1–17 (1999).
3. Puget, J.-L., Abergel, A., Bernard, J.-P., Boulanger, F., Burton, W. B., Desert, F.-X., and Hartmann, D., *A&A*, **308**, L5 (1996).

4. Fixsen, D. J., Dwek, E., Mather, J. C., Bennett, C. L., and Shafer, R. A., *ApJ*, **508**, 123–128 (1998).
5. Hauser, M. G. e. a., *ApJ*, **508**, 25–43 (1998).
6. Franceschini, A., Aussel, H., Cesarsky, C. J., Elbaz, D., and Fadda, D., *A&A*, **378**, 1–29 (2001).
7. Stecker, F. W., de Jager, O. C., and Salamon, M. H., *ApJ*, **390**, L49–L52 (1992).
8. Stanev, T., and Franceschini, A., *ApJ*, **494**, L159 (1998).
9. Madau, P., and Pozzetti, L., *MNRAS*, **312**, L9–L15 (2000).
10. Elbaz, D. e. a., *A&A*, **351**, L37–L40 (1999).
11. Altieri, B. e. a., *A&A*, **343**, L65–L69 (1999).
12. Oliver, S. e. a., *MNRAS*, **316**, 749–767 (2000).
13. Gruppioni, C., Lari, C., Pozzi, F., Zamorani, G., Franceschini, A., Oliver, S., Rowan-Robinson, M., and Serjeant, S., *MNRAS*, **335**, 831–842 (2002).
14. Elbaz, D., Cesarsky, C. J., Chanial, P., Aussel, H., Franceschini, A., Fadda, D., and Chary, R. R., *A&A*, **384**, 848–865 (2002).
15. Franceschini, A. e. a., *A&A* (2003), in press.
16. Rigopoulou, D., Franceschini, A., Aussel, H., Genzel, R., Thatte, N., and Cesarsky, C. J., *ApJ*, **580**, 789–799 (2002).
17. Puget, J. L. e. a., *A&A*, **345**, 29–35 (1999).
18. Dole, H. e. a., *A&A*, **372**, 364–376 (2001).
19. Rodighiero, G., Lari, C., Franceschini, A., Gregnanin, A., and Fadda, D., *MNRAS* (2003), submitted.
20. Efstathiou, A. e. a., *MNRAS*, **319**, 1169–1177 (2000).
21. Kawara, K. e. a., *A&A*, **336**, L9–L12 (1998).
22. Smail, I., Ivison, R. J., and Blain, A. W., *ApJ*, **490**, L5 (1997).
23. Hughes, D. H. e. a., *Nature*, **394**, 241–247 (1998).
24. Bertoldi, F. e. a., *A&A*, **360**, 92–98 (2000).
25. Xu, C., Lonsdale, C., Shupe, D., Franceschini, A., Martin, C., and Schiminovich, D., *ApJ*, **519**, 1–17 (2003), apJ in press, astro-ph/021234.
26. Sanders, D. B., and Mirabel, I. F., *Ann. Rev. Astron. Astr.*, **34**, 749 (1996).
27. Kennicutt, R. C., *Ann. Rev. Astron. Astr.*, **36**, 189–232 (1998).
28. Franceschini, A., Silva, L., Fasano, G., Granato, L., Bressan, A., Arnouts, S., and Danese, L., *ApJ*, **506**, 600–620 (1998).
29. Rodighiero, G., Franceschini, A., and Fasano, G., *MNRAS*, **324**, 491–497 (2001).
30. Cimatti, A. e. a., *A&A*, **391**, L1–L5 (2002).

Formation and Evolution of Supermassive Black Holes in Galactic Centers: Observational Constraints

Günther Hasinger* and the CDF-S team[†]

*Max-Planck-Institut für extraterrestrische Physik
Postfach 1319, D–84541 Garching, Germany
†

Abstract.
Deep X–ray surveys have shown that the cosmic X–ray background (XRB) is largely due to the accretion onto supermassive black holes, integrated over the cosmic time. These surveys have resolved more than 80% of the 0.1-10 keV X–ray background into discrete sources. Optical spectroscopic identifications show that the sources producing the bulk of the X–ray background are a mixture of obscured (type-1) and unobscured (type-2) AGNs, as predicted by the XRB population synthesis models. A class of highly luminous type-2 AGN, so called QSO-2s, has been detected in the deepest *Chandra* and *XMM-Newton* surveys. The new *Chandra* AGN redshift distribution peaks at much lower redshifts ($z \approx 0.7$) than that based on ROSAT data, indicating that Seyfert galaxies peak at significantly lower redshifts than QSOs.

INTRODUCTION

Deep X-ray surveys indicate that the cosmic X-ray background (XRB) is largely due to accretion onto supermassive black holes, integrated over cosmic time. In the soft (0.5-2 keV) band more than 90% of the XRB flux has been resolved using 1.4 Msec observations with ROSAT [1] and recently 1-2 Msec Chandra observations [2, 3] and 100 ksec observations with XMM-Newton [4]. In the harder (2-10 keV) band a similar fraction of the background has been resolved with the above Chandra and XMM-Newton surveys, reaching source densities of about 4000 deg^{-2}.

The X-ray observations have so far been about consistent with population synthesis models based on unified AGN schemes [5, 6], which explain the hard spectrum of the X-ray background by a mixture of absorbed and unabsorbed AGN, folded with the corresponding luminosity function and its cosmological evolution. According to these models, most AGN spectra are heavily absorbed and about 80% of the light produced by accretion will be absorbed by gas and dust [7]. In particular they require a substantial contribution of high-luminosity obscured X-ray sources (type-2 QSOs), which so far have only scarcely been detected.

Optical follow-up programs with 8-10m telescopes have been completed for the ROSAT deep surveys and find predominantly AGN counterparts of the faint X–ray source population [8, 9] mainly X–ray and optically unobscured AGN (type-1 Seyferts and QSOs) and a smaller fraction of obscured AGN (type-2 Seyferts). Optical identifications for the deepest *Chandra* and *XMM-Newton* fields are now approaching a completeness of 60-80% and find a mixture of obscured and unobscured AGN with an increasing fraction of obscuration towards fainter fluxes [10, 11]. Interestingly, first examples of the long-sought class of type-2 QSO have been detected in deep *Chandra* fields [12, 13].

After having understood the basic contributions to the X-ray background, the general interest is now focussing on understanding the physical nature of these sources, the cosmological evolution of their properties, and their role in models of galaxy evolution. We know that basically every galaxy with a spheroidal component in the local universe has a supermassive black hole in its centre [14]. The luminosity function of luminous X-ray selected AGN shows strong cosmological density evolution at redshifts up to 2, which goes hand in hand with the evolution of optically selected QSO and radio quasars, as well as the cosmic star formation history [15]. At the redshift peak of optically selected QSO around z=2 the AGN space density is several hundred times higher than locally, which is in line with the assumption that most galaxies have been active in the past and that the feeding of their black holes is reflected in the X-ray background. While the comoving space density of optically and radio-selected QSO has been shown to decline significantly beyond a redshift of 2.7 [16, 17, 18], the statistical quality of X-ray selected AGN high-redshift samples still needs to be improved [15]. The new Chandra and XMM-Newton surveys are now providing strong additional constraints here.

In this review we compare the optical identification work in the two deepest Chandra fields, the Chandra Deep Field South and the Hubble Deep Field North and show preliminary results on the cosmological evolution of Seyfert galaxies.

THE DEEPEST CHANDRA FIELDS

The Chandra X-ray Observatory has performed deep X-ray surveys in a number of fields with ever increasing exposure times [19, 20, 21, 13] and has completed a 1 Msec exposure in the Chandra Deep Field South (CDF-S, [2]) and a 2 Msec exposure in the Hubble Deep Field North (HDF-N, [3]). In Figure 1 (left), we show the colour composite Chandra image of the CDF-S. This was constructed by combining images, smoothed with a Gaussian with $\sigma=1$" in three bands (0.3-1 keV, 1-3 keV, 3-7 keV), which contain approximately equal numbers of photons from detected sources. Blue sources are those undetected in the soft (0.5-2 keV) band, most likely due to intrinsic absorption from neutral hydrogen with column densities $N_H > 10^{22}$ cm^{-2}. Very soft sources appear red. A few extended low surface brightness sources are also readily visible in the image. Figure 1 (right) shows a similar image for the 970 ksec observation of the HDF-N, kindly supplied by N. Brandt [22]. The corresponding 2 Msec image will be published soon.

The CDF-S was also observed with XMM-Newton for a net exposure of 500 ksec in July 2001 and January 2002 (PI: J. Bergeron, see [23]). The EPIC cameras have a larger field-of-view than ACIS, and a number of new diffuse sources are detected just outside the Chandra image. X-ray spectroscopy of a large number of sources will ultimately be very powerful with XMM-Newton (see [24] for the Lockman Hole).

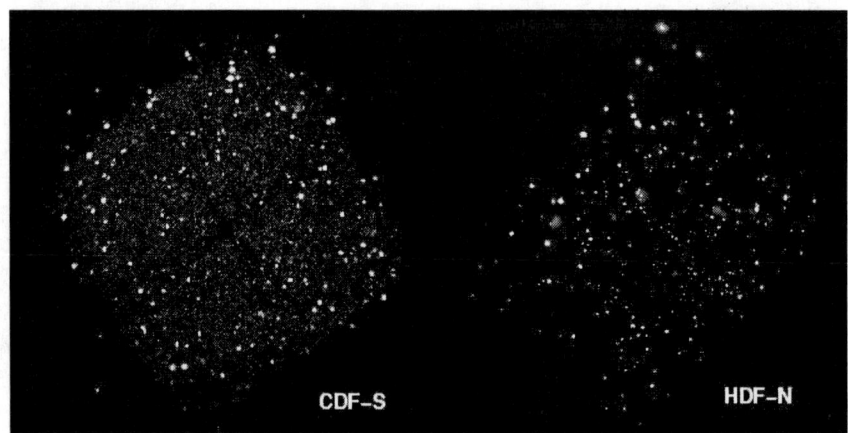

FIGURE 1. left: Composite image of the Chandra Deep Field South of 940 ks (pixel size=0.984", smoothed with a σ=1" Gaussian). The image was obtained combining three energy bands: 0.3-1 keV, 1-3 keV, 3-7 keV, from [2]. Right: Similar image for the 970 ksec Chandra observation of the Hubble Deep Field North [22].

VLT OPTICAL SPECTROSCOPY

Optical spectroscopy in the CDF-S has been carried out in 11 nights with the ESO Very Large Telescope (VLT) in the time frame April 2000 - December 2001, using deep optical imaging and low resolution multiobject spectroscopy with the FORS instruments with individual exposure times ranging from 1-5 hours. Some preliminary results including the VLT optical spectroscopy have already been presented [12, 2]. The complete optical spectroscopy will be published in [11].

Redshifts could be obtained so far for 169 of the 346 sources in the CDF-S, of which 123 are very reliable (high quality spectra with 2 or more spectral features), while the remaining optical spectra contain only a single emission line, or are of lower S/N. For objects fainter than R=24 reliable redshifts can be obtained if the spectra contain strong emission lines. For the remaining optically faint objects we have to resort to photometric redshift techniques. Nevertheless, for a subsection of the sample at off-axis angles smaller than 8 arcmin we obtain a spectroscopic completeness of about 60%. Including photometric redshifts [27, 28] this completeness increases to ≈ 80% for the CDF-S.

Figure 2 shows examples of six VLT spectra of CDF-S sources of sources which are selected from one of the two redshift spikes detected in the field at z=0.733 [23, 11, 26]. The objects are sorted according to their X-ray luminosity which is given in the figure together with the X-ray hardness ratio ($HR = (H - S)/(H + S)$), where H and S are the count rates in the 2-7 keV and 0.5-2 keV, respectively. The spectrum in the upper left corresponds to a bona fide QSO, with a broad MgII line and a blue nonthermal continuum. This object has a high X-ray luminosity and soft X-ray spectrum consistent with its optical classification. The AGN indicators (broad lines, nonthermal continuum) in the upper row decrease in strength with lower X-ray luminosity and the optical

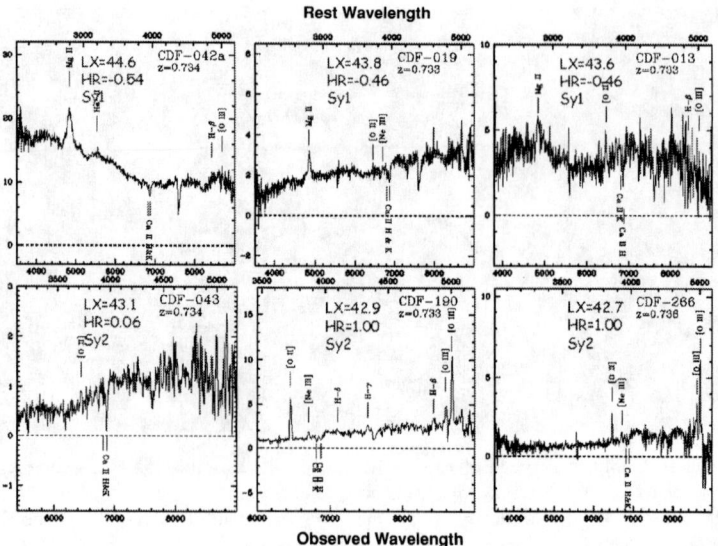

FIGURE 2. Optical spectra of six CDF-S sources selected from one of the two observed redshift spikes at z=0.733 (see [23, 26]), obtained using multiobject-spectroscopy with FORS at the VLT [11]. The objects are sorted according to their X-ray luminosity, which is given in each panel together with the X-ray hardness ratio and the X-ray/optical classification.

spectrum becomes more and more dominated by the stars in the host galaxy (see also [9]). However, the X-ray luminosity and hardness ratios still indicate that the X-ray flux is dominated by type-1 AGN emission. In the lower row the X-ray luminosity still is above that of typical starburst galaxies and in addition the X-ray spectrum gets significantly harder, with hardness ratios above 0, indicating strongly absorbed X-ray continua. These objects are classified as Seyfert-2 galaxies, although their optical spectra do not necessarily reveal any AGN features. The spectrum in the lower left shows a Seyfert-2 galaxy with significant X-ray absorption and an AGN-type luminosity. The latter spectrum is characteristic for the bulk of the detected galaxies, which show either no or very faint high excitation lines indicating the AGN nature of the object, so that we have to resort to a combination of optical and X-ray diagnostics to classify them as AGN.

X-RAY/OPTICAL CLASSIFICATION

Following [2, 23] we show in Figure 3 left the hardness ratio as a function of the luminosity in the 0.5-10 keV band for 165 sources for which we have optical spectra and rather secure classification in the CDF-S [11]. The X-ray luminosities are not corrected for internal absorption and are computed in a critical density universe with $H_0 = 50$ km s^{-1} Mpc^{-1}. Different source types are clearly segregated in this plane. Type-1 AGNs

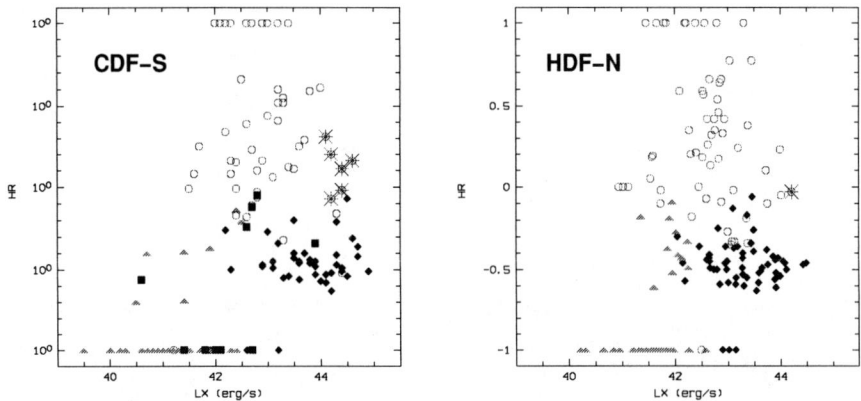

FIGURE 3. Hardness ratio versus rest frame luminosity in the total 0.5-10 keV band. Objects are coloured according to their X-ray/optical classification: filled diamonds correspond to type-1 AGN, open hexagons to type-2 AGN, triangles to galaxies and filled squares to extended X-ray sources. The large asterisks indicates type-2 QSOs. A critical density universe with $H_0 = 50$ km s^{-1} Mpc^{-1} has been adopted. Luminosities are not corrected for possible intrinsic absorption.

(black diamonds) have luminosities typically above 10^{42} erg s^{-1}, with hardness ratios in a narrow range around HR$\approx$-0.5. Type-2 AGN are skewed towards significantly higher hardness ratios (HR>0), with (absorbed) luminosities in the range 10^{41-44} erg s^{-1}. Direct spectral fits of the XMM-Newton and Chandra spectra clearly indicate that these harder spectra are due to neutral gas absorption and not due to a flatter intrinsic slope (see [24, 29]). Therefore the unabsorbed, intrinsic luminosities of type-2 AGN would fall in the same range as those of type-1's.

In Figure 3 we also indicate the type-2 QSOs (asterisks), the first one of which was discovered in the CDF-S [12]. In the meantime, more examples have been found in the CDF-S and elsewhere [13]. It is interesting to note that no high-luminosity, very hard sources exist in this diagram. This is a selection effect of the pencil beam surveys: due to the small solid angle, the rare high luminosity sources are only sampled at high redshifts, where the absorption cut-off of type-2 AGN is redshifted to softer X-ray energies. Indeed, the type-2 QSOs in this sample are the objects at $L_X > 10^{44}$ erg s^{-1} and HR>-0.2. The type-1 QSO in this region of the diagram is a BAL QSO with significant intrinsic absorption.

About 10% of the objects have optical spectra of normal galaxies (marked with triangles), luminosities below 10^{42} erg s^{-1} and very soft X-ray spectra (several with HR=-1), as expected in the case of starbursts or thermal halos. Those at $L_X < 10^{41.5}$ erg s^{-1} and HR<-0.7 are at particularly low redshifts. However, a separate subset has harder spectra (HR>-0.5), and luminosities $> 10^{41}$ erg s^{-1}. In these galaxies the X-ray emission is likely due to a mixture of low level AGN activity and a population of low mass X-ray binaries (see [25]). Therefore the deep Chandra and XMM-Newton surveys detect for the first time the population of normal starburst galaxies out to intermediate redshifts [19, 21, 30]. These galaxies might become an important means to study the star

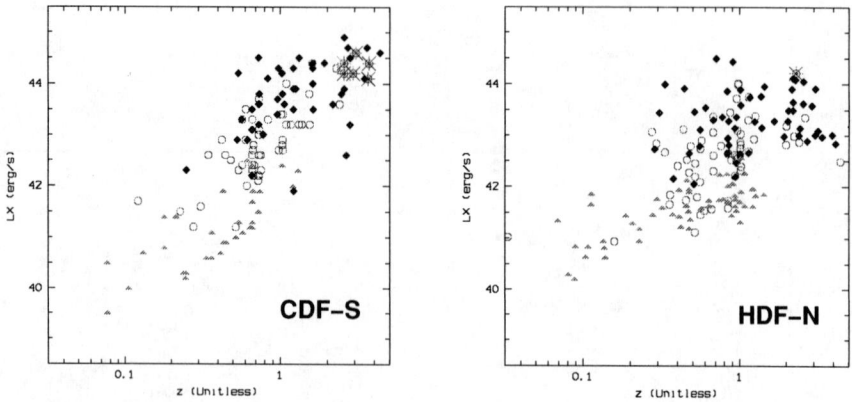

FIGURE 4. X-ray luminosity versus redshift magnitude for the CDF-S-sources (left) and the HDF-N sources (right). Symbols are the same as in Figure 3.

formation history in the universe completely independently from optical/UV, sub-mm or radio observations.

Figure 3 (right) shows the same diagram for the spectroscopic identifications in the HDF-N [10]. While the authors give only purely optical classification information for the X-ray counterparts (basically "galaxy" or "broad line object"), we have applied the above X-ray/optical classification scheme also to their catalogue. The corresponding diagram shows basically the same features: the X-rays show that type-1 (mainly broadline) AGN cluster around HR$\approx$-0.5 and break the degeneracy between type-2 Seyferts and normal galaxies.

There is, however, one significant difference between the two identification samples: while the CDF-S identification catalogue [11] includes 6 type-2 QSOs (see figure 3 (left)), which all are characterised by strong and narrow UV emission lines (Lyman-α, CIV etc.) with almost absent continuum, the corresponding HDF-N catalogue [10] lists only one possible type-2 QSO. A closer look to Figure 4 and the optical magnitudes of the CDF-S type-2 QSO shows that these objects are predominantly detected at redshifts z>2 and optical magnitudes R> 24. One reason for the relative absence of this population in the HDF-N could be that a smaller number of identifications at R> 24 exist in this survey compared to the CDF-S. However, a true field to field variation (cosmic variance) cannot be ruled out. At least in the CDF-S there is no significant variation of the ratio of type-1 to type-2 objects over the redshift range z=0.5-4.

THE REDSHIFT DISTRIBUTION

The current spectroscopic/photometric completeness of the CDF-S and HDF-N identifications allows to compare the observed redshift distribution with predictions from X–ray background population synthesis models [6], which, due to the saturation of the QSO

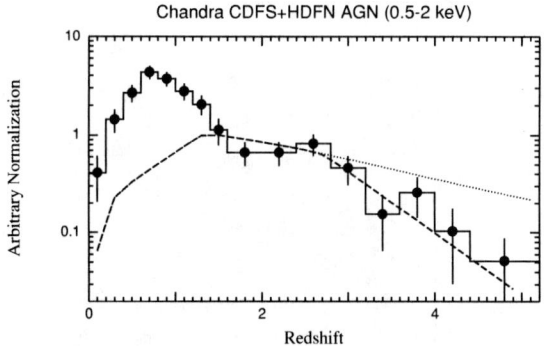

FIGURE 5. Redshift distribution of 243 AGN selected in the 0.5-2 keV band from the inner 10 arcmin radius of the Chandra CDF-S and HDF-N survey samples (solid circles and histogram), compared to model predictions from population synthesis models [6]. The dashed line shows the prediction for a model, where the comoving space density of high-redshift QSO follows the decline above z=2.7 observed in optical samples [16, 18]). The dotted line shows a prediction with a constant space density for $z > 1.5$. The two model curves have been normalized to their peak at z=1, while the observed distribution has been normalized to roughly fit the models in the redshift range 1.5–2.5

evolution predict a maximum at redshifts around z=1.5. Figure 5 shows two predictions of the redshift distribution from the Gilli et al. model for a flux limit of 2.3×10^{-16} erg cm^{-2} s^{-1} in the 0.5-2 keV band with different assumptions for the high-redshift evolution of the QSO space density. The two models have been normalized at the peak of the distribution.

The actually observed redshift distribution of AGN selected from the HDF-N and CDF-S Chandra deep survey samples at off-axis angles below 10 arcmin and in the 0.5-2 keV band has been arbitrarily normalized to roughly fit the population synthesis models in the redshift range 1.5 - 2.5 and shown in Fig. 5 as histogram and data points. In the redshift range below 1.5 it is radically different from the prediction, with a peak at a redshift at z≈0.7. This low redshift peak is dominated by Seyfert galaxies with X-ray luminosities in the range $L_X = 10^{42-44}$ erg/s. Since the peak in the observed redshift distribution is expected at the redshift, where the strong positive evolution of AGN terminates, we can conclude that the evolution of Seyfert galaxies is significantly different from that of QSOs, with their evolution saturating around a redshift of 0.7, compared to the much earlier evolution of QSOs which saturates at z≈1.5. The statistics of the two samples is now sufficient to rule out the constant space density model at redshifts above 3, clearly indicating a decline of the X-ray selected QSO population at high redshift consistent with the optical findings. However, the statistical errors and the likely spectroscopic incompleteness still preclude a more accurate determination.

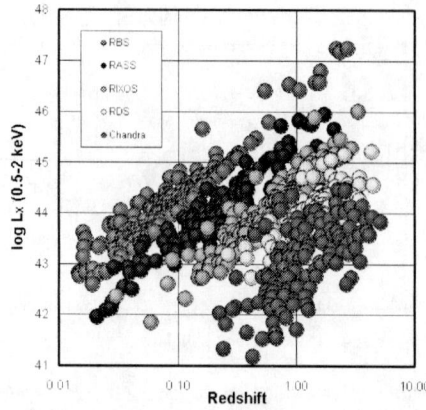

FIGURE 6. Luminosity as a function of redshift for different survey samples of type-1 AGN selected in the 0.5-2 keV band. The ROSAT surveys correspond to the same samples utilized in [15]: i.e., (from top-left to bottom-right) ROSAT Bright Survey (RBS; [32]), ROSAT All-Sky-Survey northern selected areas (RASS; [33]), ROSAT International X-ray Optical Survey (RIXOS; [34]), ROSAT Deep Surveys (NEP, [35]; Marano field, [36]; UKDS, [37]; Lockman Hole, [8, 9]); Chandra objects from the CDF-S and HDF-N ([11, 10]).

THE ROSAT/CHANDRA LUMINOSITY FUNCTION

To investigate in more detail, where the difference between the Chandra objects and the predictions, which are based on the ROSAT data originate from, we have calculated preliminary luminosity functions. For the first time the Chandra deep survey data were merged with the whole body of previously identified ROSAT AGN samples, used in [15] to compute the AGN luminosity function. To make the analysis as complete and homogeneous as possible, we have selected only the type-1 AGN in all samples and treated only the detections and X-ray fluxes in the 0.5-2 keV band. Figure 6 shows the X-ray "Hubble-Diagram" for the different samples, covering an unprecedented six orders of magnitude in flux limit and seven orders of magnitude in survey solid angle between the ROSAT Bright Survey and the Chandra Deep Survey. This diagram shows, that the new Chandra sources are predominantly Seyfert galaxies at a median luminosity of $\approx 10^{43}$ erg s^{-1} and a median redshift around 0.7.

A preliminary luminosity function was calculated using the V/V_a method and is shown in two redshift shells (z=0.015-0.2 and z=1.6-2.3) in Figure 7 (left). The shape of the two luminosity function is significantly different, so that the cosmological evolution can be described neither by pure luminosity nor pure density evolution. The surprising result is, however, that the high-redshift luminosity function is almost horizontal at luminosities below $\approx 10^{44}$ erg s^{-1} and approaches the local space density in the Seyfert range. The strong positive density evolution, well known from previous AGN samples in the optical, radio and X-ray range, therefore only holds for relatively luminous AGN (i.e. QSOs), while the lower luminosity AGN (Seyfert galaxies) show much less or even negative density evolution.

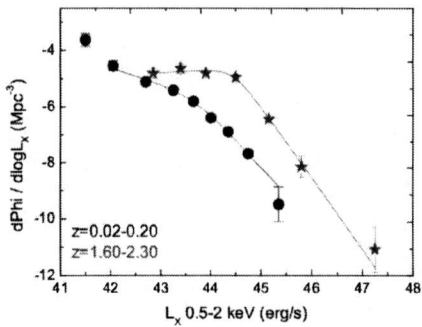
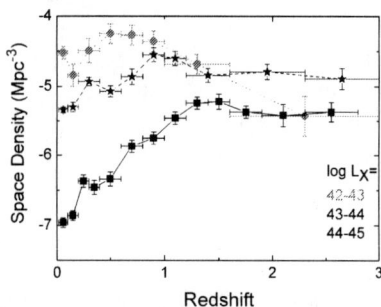

FIGURE 7. Left: Luminosity function of type-1 AGN selected in the 0.5-2 keV band in two redshift shells: z=0.02-0.2 [41] and z=1.6-2.3 (this work). Right: Space density as a function of redshift for different luminosity classes.

To illustrate the different evolutionary behaviour for different luminosity classes in more detail we calculated space densities as a function of redshift for different luminosity classes: $log(L_X)$ =42-43, 43-44 and 44-45, shown in Figure 7 (right). While the evolution of the highest luminosity class (44-45), the QSOs, follows very well the known strong positive evolution with an increase of almost two orders of magnitude in space density, saturating at z≈ 1.5, the evolution of lower luminosity classes is weaker and saturates at significantly later redshifts. The highest space density is achieved for the Seyferts of luminosity class 42-43 at redshifts around 0.7 at a space density about a factor of 10 higher than that of QSOs. Beyond z=0.7 there is a significant decline of the Seyfert space density. This is the reason, why the Chandra deep surveys are dominated by this type of object in this redshift shell and not, as originally expected by Seyfert galaxies at higher redshifts.

These, still preliminary, new results paint a dramatically different evolutionary picture for low-luminosity AGN compared to the high-luminosity QSOs. While the rare, high-luminosity objects can form and feed very efficiently rather early in the universe, the bulk of the AGN has to wait much longer to grow. This could indicate two modes of accretion and black hole growth with different accretion efficiency, as e.g. proposed in [38]. The late evolution of the low-luminosity Seyfert population is very similar to that which is required to fit the Mid-infrared source counts and background (see e.g. [39]), however, contrary to what has been assumed by Franceschini et al., this evolution applies to all low-luminosity AGN (type-1 and type-2).

These results, however, have still to be taken with a grain of salt. First, the spectroscopic incompleteness in the Chandra samples is still substantial and before a formal publication of all spectroscopic and photometric redshifts is available, it is too early to draw final conclusions. Also, the derivation of the luminosity function and its evolution still needs to be confirmed by more modern methods (see e.g. [40, 41]).

ACKNOWLEDGMENTS

I thank the Chandra Deep Field South team for the good cooperation and the permission to use some data in advance of publication. I am grateful to Maarten Schmidt and Takamitsu Miyaji for the collaboration on the preliminary Chandra/ROSAT luminosity function.

REFERENCES

1. Hasinger, G., Burg, R., Giacconi, R., et al., 1998, A&A 329, 482
2. Rosati P., Tozzi P., Giacconi R., et al., 2002, ApJ 566, 667
3. Brandt W.N., Alexander D.M., Bauer, F.E., Hornschemeier A.E., 2002, astro-ph/0202311
4. Hasinger, G., Altieri, B., Arnaud, M., et al., 2001, A&A 365, 45
5. Comastri, A.; Setti, G.; Zamorani, G.; Hasinger, G., 1995, A&A 296, 1
6. Gilli, R., Salvati, M., Hasinger, G., 2001, A&A 366, 407
7. Fabian A.C., Barcons X., Almaini O., Iwasawa K., 1998, MNRAS 297, L11
8. Schmidt, M., Hasinger, G., Gunn, J.E., et al., 1998, A&A 329, 495
9. Lehmann, I., Hasinger, G., Schmidt, M., et al., 2001, A&A 371, 833
10. Barger, A. J., Cowie, L. L., Brandt, W.N. et al., 2002, AJ 124, 1839
11. Szokoly, G., Hasinger G., Rosati, P. et al., 2003 (in prep.)
12. Norman C., Hasinger G., Giacconi R., et al. 2002, ApJ 571, 218
13. Stern D., Moran E.C., Coil A.L., et al., 2002, ApJ 568, 71
14. Gebhardt K., Bender R., Bower G., et al., 2000, ApJ 539, 13
15. Miyaji, T., Hasinger, G., Schmidt, M., 2000, A&A 353, 25
16. Schmidt, M., Schneider, D.P. & Gunn J.E., 1995, AJ 114, 36
17. Shaver P.A. et al., 1996, Nature 384, 439
18. Fan X., Strauss M., Richards G., et al., 2001, AJ 121, 31
19. Mushotzky, R.F., Cowie L.L., Barger, A.J., Arnaud, K.A., 2000, Nature 404, 459
20. Hornschemeier, A.E., Brandt, W.N., Garmire, G.P., et al., 2000, ApJ 541
21. Giacconi, R., Rosati P., Tozzi P., et al., 2001, ApJ 551, 624
22. Brandt W.N., Alexander D.M., Hornschemeier A.E., et al., 2001, AJ 122, 2810
23. Hasinger, G., Bergeron, J., Mainieri, V., et al., 2002, ESO Messenger 108, 11
24. Mainieri V., Bergeron J., Rosati P., et al., 2002, A&A 393, 425
25. Barger, A. J., Cowie, L. L., Mushotzky, R. F., Richards, E. A., 2001, AJ 121, 662
26. Gilli, R., Cimatti, A., Daddi, E., et al., 2003, ApJ submitted
27. Wolf, C., Dye, S., Kleinheinrich, M., et al., 2001, A&A 377, 442
28. Mainieri V., et al., 2002, priv. comm.
29. Bauer F.E., Vignali C., Alexander D.M., et al., 2003, AN in press (astro-ph/0210310
30. Lehmann I., Hasinger G., Murray S.S, Schmidt M., 2002, astro-ph/0109172
31. Cimatti, A., Daddi E., Mignoli M., et al., 2002, A&A 381, L.68
32. Schwope A., Hasinger G., Lehmann I., et al., 2000, AN 321, 1
33. Appenzeller, I., Thiering, I., Zickgraf, F.-J., et al., 1998, ApJS 117, 319
34. Mason, K.O., Carrera, F.J., Hasinger G., et al., 2000, MNRAS 311, 456
35. Bower, R.G., Hasinger, G., Castander, F.J., et al., MNRAS 281, 59
36. Zamorani, G., Mignoli, M., Hasinger G., et al., 1999, A&A 346, 731
37. McHardy, I.M., Jones L.R., Merrifield M.R., et al., 1998, MNRAS 295, 641
38. Duschl, W.J., Strittmatter, P.A., in Active Galactic Nuclei: from Central Engine to Host Galaxy Abstract Book, meeting held in Meudon, France, July 23-27, 2002, Eds.: S. Collin, F. Combes and I. Shlosman. To be published in ASP (Astronomical Society of the Pacific), Conference Series, p. 76.
39. Franceschini, A.; Braito, V.; Fadda, D., 2002, MNRAS 335, L51
40. Schmidt M., Hasinger G., Miyaji, T. in: Symposium in honour of Joachim Trümpert's 65th birthday, MPE report 272, 213
41. Miyaji, T., Hasinger, G., Schmidt, M., A&A 369, 49 (2001)

Formation of Supermassive Black Holes

Selig Kainer* and William K. Rose[†]

*BKG Research, 3622 Ordway Street, NW, Washington, DC 20066
[†]Department of Astronomy, University of Maryland, College Park, MD 20742

Abstract. Three different proposed mechanisms for supermassive black hole (SMBH) formation have been presented. Firstly, it has been suggested that SMBHs form in galactic nuclei by coherent collapse. Secondly, it has been suggested that SMBHs form by stellar collapse and then grow in mass. A third scenario is that massive black holes have a primordial origin and then grow in mass by accretion. This latter scenario is consistent with some hybrid versions of inflationary cosmology. In this paper we discuss the above three scenarios for SMBH formation with emphasis on the third possibility.

INTRODUCTION

Massive black holes are believed to be present at the centers of elliptical galaxies and spiral galaxies with bulges. In a recent publication Gebhart et al. 2000 (and references within) have shown that there is a correlation between central black hole mass and velocity dispersion. A relationship between black hole mass and the bulge luminosities of their host galaxies also exists (Richstone et al. 1998). It follows that black hole masses are approximately proportional to the host-galaxy spheroid mass. In addition, the mass equivalent energy density of quasar background light is roughly consistent with the estimated total mass density in black holes. The number density of bright quasars in the universe peaked at $z \simeq 2 - 3.5$. This epoch is somewhat before the epoch of maximum star formation (i.e. $z \simeq 1 - 2$). The above discussion indicates that massive black holes in relatively nearby galaxies are primarily fossil black holes from the quasar era and that they form either before or concomitant with the oldest and densest regions of galaxies.

Three different scenarios for the formation of massive black holes have been advanced. The first of these scenarios asserts that massive black holes form in galactic nuclei by coherent collapse probably during the epoch of spheroid formation. A second scenario maintains that black holes form by stellar collapse and then grow in mass by merger with other compact objects and/or mass accretion. A third scenario asserts that massive black holes have a primordial origin and then grow in mass by accretion. This latter point of view is consistent with some hybrid versions of inflationary cosmology (Garcia-Bellido, Linde and Wands 1996, Yokoyama 1997).

RESULTS

The number of quasars in the universe increases rapidly with redshift until z = 2, it increases more gradually until z = 3 - 3.5 and then decreases sharply after z = 4. Relatively faint quasars increase in number much more gradually between z = .5 and z = 2 than those of high luminosity. It is widely believed that most quasars represent an early phase in the formation of bright galaxies and that phase lasts for about 10^8 years (Haehnelt and Rees 1993). Apparently there is little time delay between the formation of bright galaxies, which probably occur in the potential wells of dark matter halos, and the appearance of quasars. In the hierarchical model for structure formation small structures appear first whereas larger structures are formed by mergers.

It is possible that star formation may be inefficient in primordial gas clouds and consequently gas collapses toward the center of the protogalaxy without much fragmentation into stars. The collapsing gas cloud has obtained angular momentum by tidal torques. If a thin, self- gravitating disk is formed then it will be unstable to non-axisymmetric gravitational instabilities because it violates the Toomre criterion. Angular momentum loss may be rapid (Lin and Pringle 1987). Supernova ejecta may also create viscosity and angular momentum loss by producing random bulk motions of gas clouds (Silk and Rees 1998). Supermassive objects that are self- gravitating do not have stable, equilibrium states and therefore may collapse into supermassive black holes. Dynamical friction can cause supermassive black holes to migrate to the centers of protogalaxies even if they are not formed at galactic centers.

The bimodial spatial distribution of quasars can be explained as follows (Haehnelt and Rees 1993). At high and intermediate redshifts the formation of supermassive black holes occurs along with the oldest subgalactic structures and before star formation has peaked the number and luminosities of quasars then decrease because of fuel depletion. Spiral galaxies have much lower mass spheroids and central black holes than giant elliptical galaxies. At low redshifts there is sufficient gas in spiral galaxies for mass accretion to occur intermittently at the Eddington rate and therefore the AGN phenomenon is primarily due to Seyfert galaxies whereas giant elliptical galaxies, which contain more massive black holes and are usually gas depleted, only accrete mass at the Eddington rate when a merger occurs or a dwarf galaxy is cannibalized. The latter comment explains why luminous quasars are not numerous at low redshifts.

A second scenario for the formation of massive black holes asserts that initially a much lower mass black hole is formed by stellar collapse and subsequently its mass increases substantially. Black holes formed in a stellar cluster are more massive than an average star in the cluster and consequently will migrate to its core where they tend to exchange into binaries that may merge by gravitational radiation. Although three body interactions tighten a binary system each tightening causes recoil and it has been shown (e.g. Sigurdson and Hernquist 1993) that before a binary becomes sufficiently hard to merge rapidly by gravitational radiation it will be expelled from the cluster unless its mass exceeds about 50 $M_\odot$ (Miller and Hamilton 2002). If a > 50 $M_\odot$ black hole is produced in a cluster then it can merge with lighter black holes and remain in the cluster. The cluster may be at the galactic center or may later merge with its host galaxy and subsequently reach the galactic center.

Gephart et al. have argued that $\sim 10^3$ $M_\odot$ black holes may be present at the centers of some globular clusters. Moreover, the discovery of discrete sources in starburst galaxies with X-ray luminosities up to 10^{41} ergs^{-1} can be interpreted as due to black holes with masses $> 10^3$ M$\odot$ if luminosities are sub-Eddington (Kaart et al. 2001). Dynamical friction would cause M $> 10^5$ - 10^6 M$\odot$ objects to migrate to the center of starburst galaxies. Therefore, the non-nuclear locations of these X-ray sources implies that their upper mass limit is $\simeq 10^5$ - 10^6 M$\odot$.

Early versions of inflationary cosmology (Guth 1981, Linde 1982, Aldrich and Steinhardt 1982) introduce a single scalar field and do not lead to primordial black hole production except possibly for mini black holes, which would evaporate quickly on evolutionary timescales. More recently Garcie-Bellide et al. 1996 have examined a hybrid version of inflation with two scalar fields. Density fluctuations of order unity can lead to the formation of primordial black holes when such fluctuations enter the horizon. In the above version of hybrid inflation and some other versions with multiple scalar fields (Yokoyama 1997) primordial black holes of a particular mass spectrum can be formed with a number density that is consistent with cosmological constraints on total black hole mass and number. In other words topological catastrophes can be eliminated and black holes with masses of $\sim 10^6$ M$\odot$ that become the central black holes of quasars and other AGNs can be primordial.

One of the most puzzling results from the Chandra X-ray telescope is the discovery in starburst galaxies of unresolved X-ray sources whose luminosities appear to be more than ten and perhaps hundreds of times greater than the Eddington luminosity of a neutron star. A luminous, off-centered X-ray source in the nuclear region of M82 is an interesting example of such a source (Kaaret et al. 2001). Its luminosity varied between 4.5×10^{40} and 1.6×10^{41} erg s^{-1} on timescales that ranged between 10^4s and a month (Matsumoto et al. 2001). We can conclude that this X-ray source is $< 10^5$ M$\odot$ if its age is $\sim 10^{10}$ years because dynamical friction will cause it to spiral into the galactic center in the latter timescale. A 500 - 10^5 M$\odot$ accreting black hole is probably the most plausible interpretation of these observations. As discussed above this source could be an intermediate mass black hole that was formed at the center of a cluster and then escaped from the cluster when it merged with the galaxy. However, a primordial black hole is another possibility. Observations at high redshifts may make it possible to distinguish between these two possibilities.

Attempts to explain the luminosities and number of quasars as a function of redshift have been interpreted as providing evidence that some accretion flows are advective (Begelman 1978). In an advective accretion disk the viscous timescale is shorter than the radiation timescale and therefore radiation is advected into the black hole rather than escaping. This implies that the mass accretion rate can substantially exceed the rate necessary to generate the Eddington luminosity. The existence of primordial black holes of $\sim 10^6$ M$\odot$, however, provides an alternate explanation for interpreting the space density of quasars (Kauffmann and Haehnelt 2000) and consequently additional support for the point of view that massive black holes are primordial.

REFERENCES

1. Aldrich, A. and Steinhardt, P. J. 1982, Phys. Rev. Lett., 48, 1220.

2. Begelman, M. 1978, MNRAS, 184, 53.
3. Garcia-Bellido, J., Linde, A., and Wands, D. 1996, Phys. Rev. D, 23, 347.
4. Gephart, K. et al. 2000, Ap. J., 539, L13.
5. Guth, A. H. 1981, Phys. Rev. D, 23, 347.
6. Haehnelt, M. G. and Rees, M. J. 1993, MNRAS, 263, 168.
7. Kaart, Y. et al. 2001, MNRAS, 321, L29.
8. Kauffmann, G. and Haehnelt, M. 2000, MNRAS, 311, 576.
9. Lin, D. N. C. and Pringle, J. E. 1987, MNRAS, 225, 607.
10. Linde, A. D. 1982, Phys. Lett, 108B, 389.
11. Matsumoto, H. et al. 2001, Ap. J., 547, L25.
12. Miller, M. C. and Hamilton, D. P. 2002, MNRAS, 330, 232.
13. Richstone, D. et al. 1998, Nature, 395, A14.
14. Sigurdson, S. and Hernquist, L. 1993, Nature, 364, 423.
15. Silk, J. and Rees, M. J. 1998, A and A, 331, L1.
16. Yokoyama, J. 1997, A and A, 318, 673.

Discovery of Galaxies in the z=1.5-2.5 "Bright Ages"

James W. Colbert*, Matthew Malkan* and Michael Rich*

*UCLA Dept. of Physics & Astronomy, University of California, Los Angeles, CA 90095, USA

Abstract.
We present preliminary results from a multi-year optical and infrared photometry and spectroscopy project, with the goal of examining galaxy evolution through the relatively unexplored z=1.5-2.5 "Bright Ages". This substantial percentage of cosmic time may well be when the peak of the universe's star formation occurred, but we know little about it. We have completed an extensive imaging program, including over 30 nights on the Lick 3m and several HST WFPC2 fields. We emphasize deep IR photometry in order to measure the second strongest spectral feature in galaxies–the Balmer break, at rest wavelength 4000 to 3650 Å. We are now following up all interesting candidates using the Keck LRIS instrument, including its new blue sensitive CCD. We present spectra from three star-forming galaxies with redshifts of z=2.1, 2.4, and 2.6, one AGN at z=2.4, and an ERO with redshift z=1.0.

THE BRIGHT AGES

Present results indicate that luminosity density and global star formation rates increase with redshift out past z=1 and are falling or have leveled off when measured around z∼3 [1, 2, 3]. This may make this intermediate redshift range the epoch during which the familiar components of modern galaxies assembled, a possible "Bright Ages". However, despite this epoch being the most energetic period in a typical galaxy's history, very few of these galaxies have ever been spectroscopically observed. Difficulties with both candidate selection and redshift confirmation have led to a significant gap in the number of spectroscopically confirmed galaxies between z's of 1.5 and 2.5. For example, only 8 have been spectroscopically confirmed in the Hubble Deep Field after extensive efforts, and no Bright Ages galaxy candidates in HDF-S have confirming spectra yet. This puts us in the odd position of knowing more about the most distant galaxies than those closer. This redshift range also represents a significant percentage of the universe's evolution (∼3 Gyr), as opposed to the more distant and better studied z=3-4 range, which covers about 0.5 Gyr.

Hercules Deep Field. We have observed ten fields with both the Lick 3m Prime Focus and its infrared UCLA Gemini cameras, obtaining both optical (BVRI) and infrared (JHK) images, but the initial data presented here comes from a single field. Assembling archival data from GO programs 5308, 5985, and 7459 (PIs=Windhorst and Keel), we have produced an HST deep field which is the third deepest HST field imaged to date, centered on the z=2.39 radio galaxy 53w002. We find that the summed

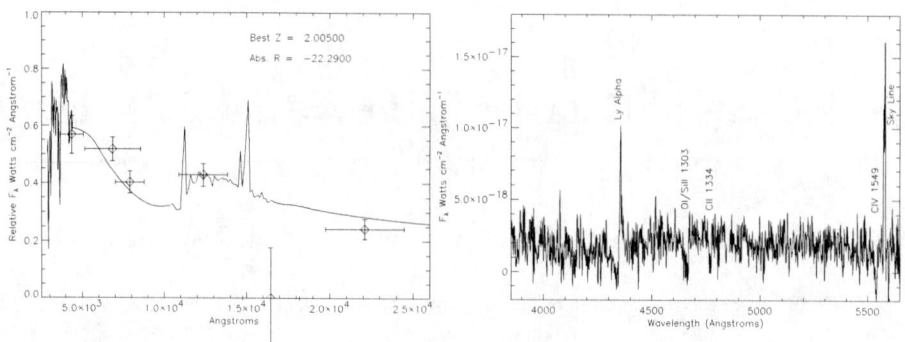

FIGURE 1. Left: Photometric Redshift fit for Bright Ages galaxy, "B85". The best fit is a star forming galaxy template with a dust extinction of Av=0.4 at a redshift of z=2.01. Right: Spectrum for z=2.58 galaxy "B85", identified from Lyman α, CIV 1549, CII 1334, & OI/Si II 1303. While not in perfect agreement with its photo-z, it was correctly identified as a "Bright Ages Galaxy".

images in F450W, F606W and F814W reach within 0.5-1 mag of the depth of the Hubble Deep Field North. In all, the 53w002 field has over 60 hours of integration with WFPC2 and NICMOS, with 20 nights of ground-based JHK infrared images from the 2.4m MDM/TIFKAM telescope (to K=21.0), producing a full set of multi-wavelength photometry. Using this database, we have obtained photometric redshifts for all galaxies with both infrared and HST detections.

PHOTOMETRIC REDSHIFTS FROM IR PHOTOMETRY

To select our z=1.5-2.5 galaxy candidates, we estimate the photometric redshifts of all IR detected objects by fitting the measured galaxy spectral energy distributions (SEDs) to redshifted galaxy templates [4, 5]. The key spectral feature for these fits is the 3600/4000 Angstrom Balmer break, which moves out of the optical and into the IR at redshifts above z~1.5 and through the infrared J-band for the redshifts of z=2-3. This creates an infrared analog to the U-band break method, or basically a J-band break. This also requires good photometry in the K filter, so that at least one point in the spectral energy distribution is cleanly on the red side of the Balmer break. Figure 1 shows an example fit of a Bright Ages galaxy, with a photometric redshift found by using the photometric redshift code *hyperz* [6].

LRIS-B. Thanks to the development of the LRIS-B multi-object spectrograph with its new blue-sensitive CCD, the Keck telescope is now optimized to take spectra below 4000 Angstroms. This is blue enough to make useful spectroscopic measurements for objects at the key z=1.5-2.5 redshifts. To demonstrate the power of this new instrument, we present the confirming spectroscopic redshift for object "B85" next to its photometric redshift fit in Figure 1.

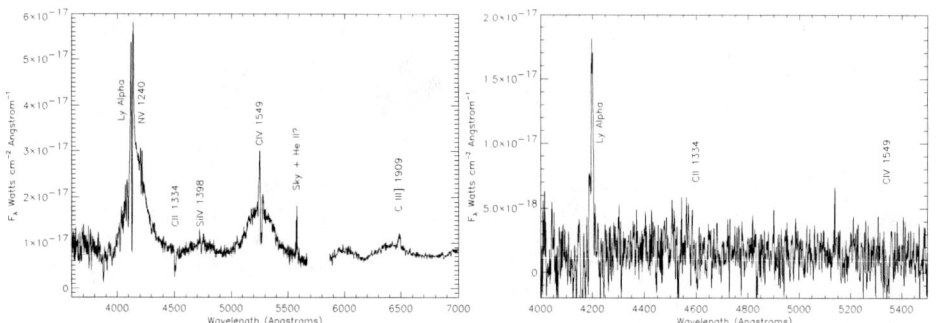

FIGURE 2. Left: Spectrum for "object 19" F410m candidate. It is a confirmed AGN with a redshift of z=2.45. Right: Spectrum for "object 11" F410m candidate with a confirmed redshift of z=2.45.

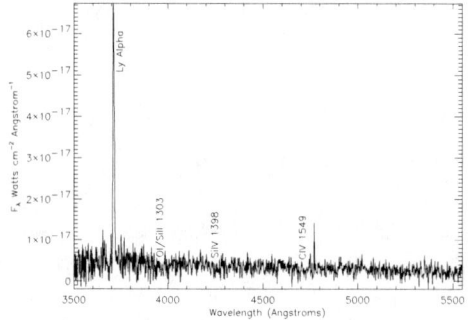

FIGURE 3. Spectrum for "object 12" F410m candidate. We did not detect an emission line near 4100Å, but instead made the serendipitous discovery of a Lyman α emission line at 3710Å, giving a redshift of z=2.05.

PROTO-GALACTIC CLUMPS

Deep narrow-band F410m imaging in the Hercules Deep Field identified 18 candidate emission-line galaxies which would all have z=2.4, if they have a Lyman α emission line in that filter [7]. However, of our 11 spectra of those candidates, only 2 objects show any line emission (Ly α or otherwise) at that wavelength, both of which we present in Figure 2.. For another candidate, shown in Figure 3, we instead serendipitously discovered Ly α emission at 3710 Å.

EXTREMELY RED OBJECTS

Another advantage of deep IR photometry is the discovery of Extremely Red Objects (EROs), defined in this case as having F606w-K > 6. We have found six EROs in two deep F606w WFPC2 fields covering just over 10 square arcmin of sky. This gives

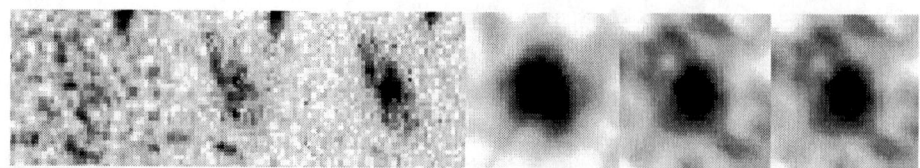

FIGURE 4. Mosaic of reduced multicolor images of a 4" x 4" box around a galaxy in the 53W002 field with photometric redshift z=1.12. Wavelength increases from B at the left to K on the right; the object can qualify as an "Extremely Red Object" (ERO), with V-K>6. The smooth round bulge becomes progressively more dominant at long wavelengths, while at short wavelengths the galaxy appears patchy with irregular arms.

a density of EROs down to K=20.5 in the Hercules Deep Field of 0.5 per square arcmin. Four of the six have very compact morphology through all bands. The fifth is slightly elongated, but the last and brightest one appears patchy and disk-like at shorter wavelengths, but with a bulge that becomes progressively more dominant with longer wavelength. In Figure 4 we show its 6-color image. This ERO has a spectroscopic redshift placing it z=1.03.

SUMMARY

- We used photometric redshifts, including key near-infrared photometry, to select high-redshift targets. We then took spectroscopy of these candidates with the blue sensitive Keck LRIS-B spectrograph. We presented four LRIS-B spectra from the z=1.5=2.5 "Bright Ages", of which only one is an AGN. This is already a significant addition to spectroscopically confirmed galaxies from these redshifts.
- Of the 11 F410m emission-line candidate galaxies examined, only 2 had emission lines in the predicted wavelength region.
- We found 6 EROs, defined as F606w-K>6, for an ERO density of 0.5 per square arcmin (K<20.5). One has a measurable line which we identify as [OII] 3727, placing the ERO at a z=1.03.

REFERENCES

1. Lilly, S.J., Le Fevre, O., Hammer, F., & Crampton, D. 1996, ApJL, 460, L1
2. Lowenthal, J.D. et al. 1996, ApJ, 481, 673
3. Madau, P., Pozzetti, L., & Dickinson, M. 1998, ApJ, 498, 106
4. Koo, D.C. 1985, AJ, 90, 418
5. Fernandez-Soto, A., Lanzetta, K.M., & Yahil, A. 1999, ApJ, 513, 34
6. Bolzonella, M., Miralles, J.-M., Pelló, R. 2000, A&A, 363, 476
7. Pascarelle, S.M., Windhorst, R.A., Keel, W.C., & Odewahn, S.C. 1996, Nature, 383, 45

Multi-wavelength Luminosity Functions of Galaxies

Jonathan P. Gardner

NASA's GSFC, Code 681, Greenbelt MD 20771

Abstract. Multivariate or multi-wavelength luminosity functions will reveal the interplay between star formation, chemical evolution, and absorption and re-emission of dust within evolving galaxy populations. By using principal component analysis to reduce the dimensionality of the problem, I optimally extract the relevant photometric information from large galaxy catalogs. As a demonstration of the technique, I derive the multi-wavelength luminosity function for the galaxies in the released SDSS catalog, and compare the results with those obtained by traditional methods. This technique will be applicable to catalogs of galaxies from datasets obtained by 2MASS, and the SIRTF and GALEX missions.

INTRODUCTION

The luminosity function (LF) of a population of galaxies is fundamental to understanding the emission properties within the population. Ideally, one would construct the multivariate distribution of the galaxies over a range of properties, including luminosity, color, morphological type and surface brightness. In the past, however, the relatively small samples of galaxies with measured redshifts has limited LFs to one or two properties, typically luminosity in a single filter. With the advent of large, wide-area galaxy surveys such as the Sloan Digital Sky Survey (SDSS), with photometry and measured redshifts for tens of thousands of galaxies, it is now possible to explore the distribution of the galaxies as a function of several variables.

A luminosity function is essentially a histogram, and therefore must either be parameterized, or binned. The Schechter parameterization [1] is successful at fitting most LFs measured in a single filter; however there is no equivalent parameterization for multivariate LFs, and so they must be binned. If the variables are strongly correlated, however, appropriate binning is not straight-forward. Principal Component Analysis (PCA) determines an orthogonal set of axes in a multi-dimensional space which maximize the variance for each axis, and is ideally suited for finding the appropriate dimensionality of the distribution.

DETERMINING LUMINOSITY FUNCTIONS

Several methods have been developed to determine the LF in a single filter. The maximum likelihood method [2] fits a parameterized model, such as a Schechter function, avoids binning, and is independent of clustering (under the assumption that the LF is

TABLE 1. Single filter luminosity functions

Limit	Redshift	N_{gal}	M*	α	M*	α	M*	α
			Limited in Each Band		All Data		Blanton et al.	
u<18.4	z<0.1	5125	-18.35	-1.28	-18.84	-1.17	-18.34	-1.35
g<17.65	z<0.16	12612	-19.95	-1.21	-20.04	-0.99	-20.04	-1.26
r<17.77	z<0.2	32782	-20.90	-1.16	-20.90	-1.16	-20.83	-1.20
i<16.90	z<0.2	19828	-21.25	-1.25	-21.33	-1.22	-21.26	-1.25
z<16.50	z<0.2	16499	-21.48	-1.21	-21.60	-1.25	-21.55	-1.24

independent of the density.) The step-wise maximum likelihood avoids the assumption of a particular parameterization, but bins the data [3]. In either of these methods, it is necessary to know the minimum luminosity, L_{min}, that a galaxy could have, and still be detected in the sample. For a flux-limited galaxy catalog selected in the filter of interest, L_{min} is easily calculated from the limiting flux. If the galaxies have been selected in a different filter, one can either discard the galaxies fainter than the limit at which the catalog is complete (see, e.g., the discussion of the catalog from [4] in [5]), or one can determine L_{min} by adding the color of the galaxy to the limiting magnitude in the selection filter [6]. The latter method is subject to bias if the color and luminosity are correlated.

In Table 1, I use the maximum likelihood method to determine the single-filter Schechter luminosity function parameters M^* and α for a sample of 32782 galaxies from the SDSS early release dataset [7]. I used both methods, first limiting the sample in each band, and using all the data. For comparison, I show the results of [8], who used an earlier subset of the data to determine the LF, limiting the sample in each band.

To determine the bivariate or multi-variate luminosity function it is possible to make a straight-forward modification of the step-wise maximum likelihood method. If there is significant correlation between the variables, however, then it is difficult to determine the optimal binning.

PRINCIPAL COMPONENT ANALYSIS

PCA reduces the dimensionality of a dataset by identifying "principal components", a projection onto axes which maximize the variance. Components, or dimensions, which do not carry much weight can be discarded, and the system can be transformed back onto the original axes. The principal components can be determined through singular value decomposition, $A = U \bullet diag(w) \bullet V^T$, where A is the matrix of absolute magnitudes. The eigenvectors V are the principal component axes, and the eigenvalues w are the relative weights of those components.

In Figure 1 I plot the principal component eigenvectors of the SDSS absolute magnitudes. The first principal component is dominated by luminosity, comprising 90% of the information. The second component is a measure of overall color, red or blue, although it is dominated by the $u - g$ color. The third component is the shape of the spectrum, "hump" vs. "bowl", as discussed by [9]. The fourth and fifth components are dominated

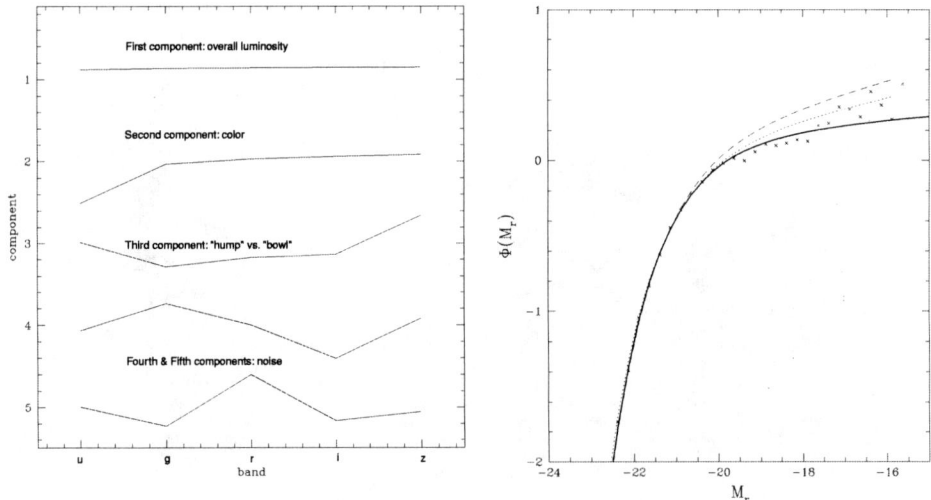

FIGURE 1. The left panel shows principal component analysis of the SDSS absolute magnitudes, as discussed in the text. The right panel shows the "luminosity function" of the first principal component, transformed back to absolute r magnitude (solid line). For comparison, I plot the Blanton r-band LF (dashed line) and the LF determined through the traditional method (dotted line). At bright magnitudes the LFs are nearly identical, but they diverge at faint magnitudes, where the color-magnitude relation plays a larger role.

by the noise.

By inverting the singular value decomposition equation, it is possible to discard the components that are not needed, and transform the data back onto the original axes. On the right side of Figure 1, I have measured the luminosity function of the first principal component, and plot a comparison to the r-band LF measured by [8], as well as my determination from Table 1. At bright magnitudes, the LFs are nearly identical, but they diverge at faint magnitudes, where the color-magnitude relation plays a larger role. The color-magnitude relation dominates the second principal component.

In Figure 2 I show the joint luminosity-color distribution for the SDSS galaxies (left panel), and the distribution of the first two principal components transformed back to the original axes (right panel.) In both panels, the color-luminosity relation is clearly evident. In the original data the E/S0 ridgeline is visible at $g - r \simeq 0.7$ (Blanton et al. 2001), however, this is not seen in the reconstruction. Although I have converted the second principal component back to $g - r$ rest frame color, this color component is dominated by $u - g$, as can be seen in the left panel of Figure 1. The color-luminosity relation is strong in this color, but the red galaxies do not have much effect. The E/S0 ridge is a subtlety that requires 3 or more principal components.

This is work in progress, and I have not yet included a proper treatment of errors. The photometric errors in the SDSS dataset are small in comparison to the intrinsic distribution, and to the bin size. However, the u-band photometry of the reddest galaxies have the highest noise, and this might have a significant effect on the second principal

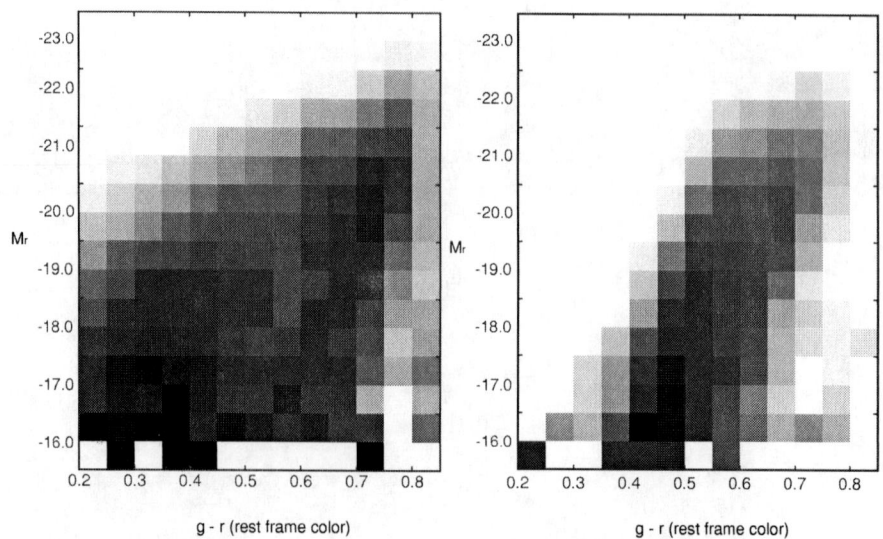

FIGURE 2. The left panel shows the joint luminosity-color distribution $\Phi(M_r, g-r)$ for the SDSS galaxies, plotted on a log scale. The right panel shows distribution of the 1st and 2nd principal components, converted back to M_r and $g-r$ color.

component. I have also not yet included normalization constraints, or ϕ^*. Finally, in order to apply this technique to other datasets such as the 2MASS survey, I will need to take into account the multiple selection functions, and different magnitude limits in different bands. Adding the SIRTF and GALEX data to this analysis will enable a comparison of the ultraviolet light that escapes the galaxies with the light that is absorbed and re-emitted at longer wavelengths.

REFERENCES

1. Schechter, P. 1976, ApJ, 203, 297
2. Sandage, A., Tamann, G. A., & Yahil, A. 1979, ApJ, 232, 352
3. Efstathiou, G., Ellis, R. S., & Peterson, B. A. 1988, MNRAS, 232, 431
4. Mobasher, B., Ellis, R. S., & Sharples, R. M. 1986, MNRAS, 223, 11
5. Gardner, J. P., Cowie, L. L., & Wainscoat, R. J. 1993, ApJ, 415, L9
6. Mobasher, B., Sharples, R. M., & Ellis, R. S. 1993, MNRAS, 263, 560
7. Stoughton, C., et al. 2002, AJ, 123, 485
8. Blanton, M. R., et al. 2001, AJ, 121, 2358
9. Koo, D. C. 1985, AJ, 90, 418

Galaxy Clusters to $z \leq 1$ from the Oxford Dartmouth Thirty Degree Survey

Molly Hammell*, Gary Wegner*, Leonidas Moustakas[†], Paul Allen**, Gavin Dalton** and Edward Olding**

Department of Physics & Astronomy, Dartmouth College, Hanover, NH 03755
[†]*Space Telescope Science Institute, Baltimore, MD 21218*
**Astrophysics, University of Oxford, Keble Road, OXFORD, OX1 3RH, UK.*

Abstract.
The properties of galaxy clusters in the local universe have been fairly well determined in the past few decades, and wide field surveys in the near infrared are converging on a statistically significant sample of high redshift clusters. These catalogs may soon allow discrimination between the competing models of galaxy formation and evolution [1]. The Oxford-Dartmouth Thirty Degree Survey (ODT) will span four widely separated 3° x 3° fields, to B < 26 in UBVRi'Z with an extension in the near-infrared to K < 19. With more than half of the survey completed, this deep, wide-area, multi-color dataset has yielded a large sample of K-selected clusters to probe the formation and evolution history of galaxies in dense environments. An exploration of cluster color-magnitude slopes and intercepts [2], luminosity functions [3], and morphological distributions [4, 5] should constrain the relative dominance of star formation rates and merger events on cluster galaxy evolution. Here, we present our cluster-finding method and preliminary results.

INTRODUCTION

Clusters of galaxies can be powerful probes of cosmologically interesting questions. In addition to their use in tracing the large-scale distribution of matter, clusters provide large samples of galaxies at the same redshift and with similar (recent) environments. Before questions about the evolutionary history of cluster galaxies can be answered, however, one needs (at the very least), a large sample of relatively unbiased clusters. To that end, we present a simple algorithm for selecting intermediate redshift clusters in a multi-color survey. This introduction provides a brief summary of ODT survey parameters. More information on the ODT is given in Wegner et al., in these proceedings. Section 2 describes the cluster-finding algorithm. Section 3 gives a brief summary of the results.

The optical images were obtained with the Oxford Wide Field Camera on the 2.5 m Isaac Newton Telescope at La Palma. Approximately 15 square degrees of the sky has been observed to 5σ detection limits of $U = 26.0$, $B = 26.0$, $V = 25.5$, $R = 25.25$, $i' = 24.5$, and $Z = 23.0$. Meanwhile, approximately 3.3 square degrees has been completed in the near infrared to a 3σ limit of $K = 18.75$ using the TIFKAM instrument on the 1.3m McGraw-Hill telescope at MDM Observatory. These images were reduced using a standard package of *IRAF* scripts. The *SExtractor* package was used for star/galaxy separation and as part of the general reduction, astrometric and photometric calibration scripts. Starlink's *ASTROM* package formed the basis of the astrometric calculations. The publicly available *hyperz* package[7] was used to determine photometric redshifts. A detailed description of image observation and reduction for the survey will be found in several papers now in preparation[8, 9, 10].

THE CLUSTER FINDING ALGORITHM

Cluster identification algorithms begin with a set of definitions for cluster candidacy. At a minimum, a cluster can be defined as a region of high density, but such a basic definition would include coincidental projections in two dimensional surveys. To maximize the number of clusters identified while minimizing contamination from projections, we use the well known cluster color-magnitude (CM) relation (motivated by the algorithm of Gladders & Yee [11]). While the CM relation aided in selecting candidates, the core of our algorithm simply selects small regions of relatively high density, then merges "deblended" cluster candidates. This is done in the 4 simple steps outlined below.

Step One: a catalog of objects with low contamination from distant stars or local galaxies is compiled by rejecting all objects with:

- a *SExtractor* calculated stellarity, $CLASS\_STAR > 0.8$
- an (aperture magnitude computed) $R - K < 2$
- an (apparent isophotal) $K < 14.0$

This reduced catalog is split into two components. All objects with an apparent K magnitude of less than 17.5 are identified as tracer candidates—those most likely to represent cluster core ellipticals (if not intrinsically bright field objects). The rest are put into a general list of possible members.

In *Step Two*, the algorithm loops through the list of tracer and general objects to identify cluster candidates. Any tracer objects separated by less than 60 arcseconds (a generous angular size of 0.25 Mpc at z=0.3 and 0.5 Mpc at z=1) are merged into a list of brightest cluster objects. Any objects from the general list within 120 arcseconds of a tracer position are then included as members. The tracer objects needed a tighter radial constraint than the general objects to avoid blending close individual clusters. In case this leaves larger clusters in several detected groups, the algorithm loops back through the new list to merge any objects whose borders overlap by more than 60 arcseconds.

Step Three computes the slope of the CM ridge using an iterative, linear least-squares fitting/clipping routine. The first fit is computed using only those cluster members within 1 (biweighted scale of) deviation from the (biweight estimated) average color (in $V - R$ and $R - I$).[1] Since the slope of the color-magnitude relation is typically very small and very tight (slopes of ≈ -0.05 and scatter of less than 0.1 [11]), this step generally rules out only a few objects—those most likely to be foreground/background galaxies or unresolved stars. The iterative part of this step simply recalculates the CM slope using

---

[1] The biweight estimator was chosen over the mean statistic since it has been shown to more reliably identify trends in small data samples with a few points of large deviation[12]

any members within 3 (biweighted) deviations from the previous fit until the scatter is less than 0.25 rms (or until routine runs out of members to fit).

At this point, the candidate cuts are introduced. *Step Four* simply identifies those candidates whose density is at least 1.5σ above the mean density (not a pure mean – averaged over the number of objects within 120 arcseconds of each "tracer" group), whose CM slope is between -.2 and +0.1 units of color per magnitude, and whose scatter about this CM slope is less than 0.25 rms. This results in a surface density of ≈ 20 objects per square degree whose images confirm their cluster candidacy, in line with the surface density of clusters and rich groups found by other methods [13, 14].

RESULTS

Old, massive, evolved clusters are the most likely cluster type to have enough bright core members visible at intermediate redshifts for the red sequence to stand out in a noisy CM Diagram (CMD). Accordingly, this algorithm is best at finding such systems. Figure 1 shows three examples of objects selected by the method described above. The slopes and scatter in these CMDs are neither the best nor worst examples of objects returned. Small, 1x1 arcminute, R-band pictures of the cluster cores are shown above each corresponding set of CMDs.

An additional bias was introduced by using "tracer" objects with K magnitudes between 14.0 and 17.5. Assuming a maximum brightest cluster absolute K magnitude near -26, and a maximum distance modulus of ≈ 44, a naive calculation (without SED convolution with our filters) translates these magnitudes to clusters in the redshift range, $z \approx 0.3 - 1.2$. Results from photometric redshift estimates put most selected clusters in the redshift range, $z \approx 0.2 - 1.0$. This does not significantly weaken the sample since lower redshift clusters have most likely been detected by previous surveys and, higher redshift clusters are more likely to show up in a selection of extremely red objects.

Lastly, these results reflect analysis of the 3.3 square degrees currently observed in the K-band. A forthcoming analysis [9] will take advantage of the full 15 square degree region covered by optical observations. An analysis of how these clusters aid in probing galaxy evolution will also be discussed.

ACKNOWLEDGMENTS

This research was supported in part by the NSF, NASA, and the UK PPARC.

REFERENCES

1. Kauffmann, G., and Charlot, S., *Monthly Notices of the Royal Astronomical Society* **197**, L23–L28 (1998).
2. Bower, R. G., Kodama, T., and Terlevich, A. I., *Monthly Notices of the Royal Astronomical Society*, **299**, 1193–1208 (1998).

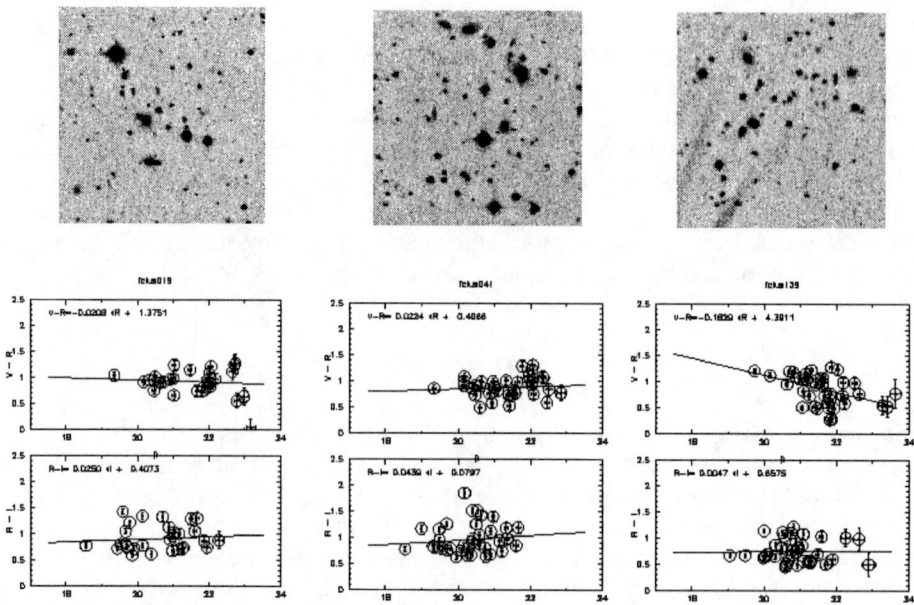

FIGURE 1. Three clusters found by the ODT method described herein. Shown at the top are R-band images taken with the 2.5m INT. Each picture is approximately 60 arcseconds on a side, cropped to show the brightest members (and for size). Below each image are the corresponding color-magnitude diagrams ($V - R$ vs. R, top, and $R - I$ vs. I, bottom) used to identify the cluster. The scatter and slopes of these diagrams are typical of the clusters selected by the algorithm.

3. de Propris, R., Stanford, S. A., Eisenhardt, P. R., Dickinson, M., and Elston, R., *The Astronomical Journal*, **118**, 719–729 (1999).
4. Postman, M., and Gellar, M. J., *The Astrophysical Journal*, **281**, 95–99 (1984).
5. Whitmore, B. C., Gilmore, D. M., and Jones, C., *The Astrophysical Journal*, **407**, 489–509 (1993).
6. Bertin, E., and Arnouts, S., *Astronomy and Astrophysics Supplement*, **117**, 393–404 (1996).
7. Bolzonella, M., Miralles, J. M., and Pello, R., *Astronomy and Astrophysics*, **363**, 476–492 (2000). See also URL http:webast.ast.obs-mip.fr/hyperz/.
8. Allen, P. D., Olding, E. J., Dalton, G. B., Moustakas, L. A., Wegner, G. A., and Hammell, M. C., in preparation (2003).
9. Hammell, M. C., Wegner, G. A., Moustakas, L. A., Allen, P. D., Olding, E. J., and Dalton, G. B., in preparation (2003).
10. Olding, E. J., Allen, P. D., Dalton, G. B., Moustakas, L. A., Booth, J., Wegner, G. A., and Hammell, M. C., in preparation (2003).
11. Gladders, M., and Yee, H.K.C., *The Astronomical Journal*, **120**, 2148–2162 (2000).
12. Beers, T. C., Flynn, K., and Gebhart, K., *The Astronomical Journal*, **100**, 32–46 (1990).
13. Olsen, L. F., Scodeggio, M., da Costa, L., Benoist, C., Bertin, E., Deul, E., Erben, T., Guarnieri, M. D., Hook, R., Nonino, M., Prandoni, I., Slijkhuis, R., Wicenec, A., Wichmann, R., *Astronomy and Astrophysics*, **345**, 681–690, (1999).
14. Postman, M., Lauer, T. R., Oegerle, W., and Donahue, M., *The Astrophysical Journal*, **579**, 93–126 (2002).

Rest Frame Optical Spectra of Lyman Break Galaxies: Other Lensing Arcs around MS1512-cB58

Matthew A. Malkan[*], Harry I. Teplitz[†] and Ian S. McLean[*]

[*]*UCLA Astronomy, LA CA 90095-1562*
[†]*SIRTF Science Center, Pasadena CA 91125*

Abstract. We have obtained near-infrared spectra of two images of the galaxy at z=2.72 which is gravitationally lensed by the foreground cluster MS1512+36. The brighter arc, cB58, is an image of only the nucleus and the southern half of the background galaxy, while the fainter image, A2, encompasses the entire background galaxy. Thus the gravitational lensing provides spatial resolution on a smaller scale than is routinely available by other methods.

Our observations indicate no evidence for any systematic rotational velocity gradient across the face of this galaxy. The nucleus and outer regions of the galaxy do not differ in their gas reddening or excitation level, based on the identical Hα/5007 ratios. cB58 (which is more dominated by the nucleus) has relatively stronger continuum emission, perhaps because of a higher ratio of old to young stars, compared to the outer parts of the galaxy.

A second emission line source, denoted as K1, at a slightly lower redshift was serendipitously detected in the slit. It appears to be the gravitationally lensed image of another background galaxy in the same group as cB58.

A SECOND IMAGE OF THE BRIGHTEST KNOWN LBG

Most of the detailed spectral information on $z > 2$ galaxies has come from observation of the rest-frame ultraviolet redshifted into the optical passband. The Lyman Break Galaxies (LBGs; [1]) are strongly starbursting galaxies, and in principle may be the tracers of the global star formation history of the universe, if the effects of dust extinction on the UV continuum can be quantified. However, the average attenuation from dust in these galaxies is uncertain to almost an order of magnitude ([2], [3]): ([4]) suggest a factor of >6, while [5] argue no correction is needed.

The need for measurements in the optical rest frame is clear, but even with the advent of IR spectrographs on large telescopes, only about two dozen $z > 2$ starbursts have been spectroscopically observed in the rest frame optical (e.g. [2], [6], [7]).

The z=2.72 galaxy MS1512-cB58 ([8]) is the apparently brightest Lyman Break Galaxy known, due to lensing magnification (of 50X) by a foreground galaxy cluster at z=0.37. Lensing reconstruction ([9]) shows that cB58 is a magnification of only half of the source galaxy, and parts of the nucleus are not included. The brightest counter arc to cB58, "A2", is a magnification of a the entire source. Thus spectroscopy of the two arcs provides measurement of different spatial regions in the source galaxy. This kind of high spatial resolution is not routinely available to intrinsically faint, unlensed, LBGs even with AO.

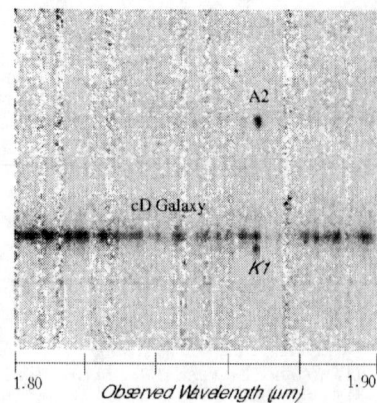

FIGURE 1. The two dimensional, rectified, atmospheric absorption corrected spectrum of A2 and other objects in the slit. The strong line in the A2 spectrum is redshifted [OIII]5007. The object labeled K1 was not previously known. Note that the apparent bump at a similar wavelength in the continuum of the cD galaxy is not an emission line, but merely a local maximum in the atmospheric transmission which, unlike the new redshifted 5007 line, goes away when divided by an A star spectrum.

Keck-II Infrared Spectroscopy. On 4 April 2002, we used NIRSPEC ([10]) to obtain a spectrum of A2, in the H and K bands (1.5-2.5 μm), with R$\sim$ 1500 (FWHM$\sim$ 4 pixels = 16 Angstroms). In 20 and 30 minutes of observation in H and K respectively, we clearly detect the OIII doublet (see figures 1 and 2) and Hα. However, Hβ falls partially on a night sky line.

COMPARISON OF THE CB-58 ARC AND ITS COUNTER-ARC A2

Redshift: The redshift calculated from the emission lines in cB58 is 2.7290 ± 0.0007, in contrast to the $z_{abs} = 2.7233 \pm 0014$ reported from the rest-frame UV interstellar absorption lines ([8]). This systematic shift is typical LBGs ([2],[5]). The velocity of A2 is almost identical (within the noise of the A2 spectrum, $\Delta V \leq 100$ km/sec.) indicating a narrow range of radial velocities across the entire object.

Line Widths: Emission lines allow us to measure the velocity dispersion. Using the lensing reconstruction, we obtain a virial mass $M_{vir} = 1.0 \times 10^{10} M_\odot$ for cB58, in good agreement with the masses of LBGs at $z \sim 3$ ([5]). None of the emission lines in A2 are resolved in velocity (due to the use of a wider slit), consistent with the small FWHM measured for cB58 ($\Delta v \sim 175$ km/sec)

Extinction and Chemical Abundance: In cB58 we measure Hα:Hβ=3.23 implying E(B-V)$\simeq$ 0.27 using the [11] reddening law (and very little change using the LMC law). Its derived one-third solar metallicity is consistent with the $Z/Z_\odot \sim 0.25$ found by [12] and 0.4 by [13]. Since the [OIII]5007/Hα ratio in both cb58 and A2 is the same (0.85 for cB58 and 0.89 +/- 0.06 for A2), there is no evidence for any large-scale spatial variation of extinction or metallicity across this galaxy.

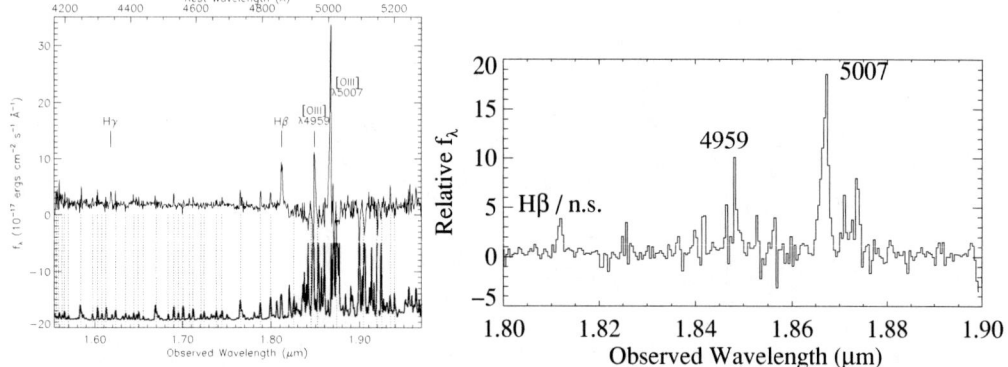

FIGURE 2. left: The H-band spectrum of the main arc, cB58. The 1σ error bars are plotted underneath, showing the sharp increase longward of 1.8 μm, where the atmospheric transmission cuts off. right: The extracted H-band spectrum of the counter-arc image A2.

Star Formation Rate: The most dramatic difference between our spectra of the cB58 arc and the A2 image is that the latter, which includes more of nucleus of the lensed galaxy, shows twice as large equivalent widths of Hα and OIII. The local star formation is proceeding at different rates in different parts of this object.

SERENDIPITOUS DISCOVERY OF A NEW ARCLET AT Z=2.69

In addition to A2, we serendipitously detected a new emission line source in the field. Figure 3 shows the location of the NIRSPEC slit we used to measure the A2 candidate, superposed on the F555W HST image of the field. On the other side of the cD galaxy, which we used for the alignment and setup of the slit, we found another high-z line-emitting galaxy; this object, "K1", was independently confirmed by the detection of both Hα and OIII5007 in nodded pairs of spectra. The two dimensional spectrum shows the clearly detected 5007 line in the counter arc (A2) and K1. [OIII]4959 is detected, but Hβ falls on an OH line and is not convincingly detected above the noise in this short (20 minutes) integration.

Comparison of K1 to cB58/A2

K1 is blue-shifted from cB58 and A2 by ∼ 150 km/sec. The Hα/[OIII] ratio is much lower in K1, 0.25 ± 0.05, which could be explained by some combination of lower reddening and/or lower metal abundance. K1 is likely to be a lensed image of a different

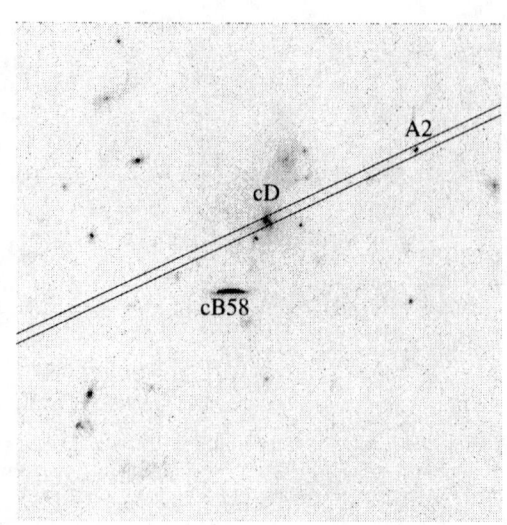

FIGURE 3. The NIRSPEC slit, superposed on the F555W WPFC2 image of the field. The image is trimmed to 32"x32". North is down.

galaxy in the same group, rather than another counter image of cB58.

ACKNOWLEDGMENTS

We thank Stella Seitz for useful discussions.

REFERENCES

1. Steidel, C.C., Giavalisco, M., Pettini, M., Dickinson, M., & Adelberger, K.L., 1996, ApJ Letters 462, L17
2. Pettini, M., Kellogg, M., Steidel, C.C., Dickinson, M., Adelberger, K.L., & Giavalisco, M., 1998, ApJ 508, 539; P98
3. Trager, S.C., Faber, S.M., Dressler, A., & Oemler, A., 1997, ApJ 485, 92
4. Meurer, G.R., Heckman, T.M., & Calzetti, D. 1999, ApJ, 521, 64
5. Pettini, M. et al. 2001, ApJ, 554, 981
6. Teplitz, H., et al. 2000, ApJL, 533, 65
7. Pettini, M. et al. 2002, ApJ, 569, 742
8. Yee, H.K.C., Ellingson, E., Bechtold, R.G., Carlberg,R.G., Cuillandre, J.-C., 1996, AJ 111, 1783
9. Seitz, S., Saglia, R., Bender, R., Hopp, U., Belloni, P., Ziegler, B., 1998, MNRAS 298, 945
10. McLean, I. S., et al. 1998, SPIE, vol. 3354, 566
11. Calzetti, D.A., Kinney, A.L., & Storchi-Bergmann, T., 1994, ApJ, 429, 482
12. Pettini, M. et al. 2000, ApJ, 528, 96
13. Frayer, D.T., et al. 1997, AJ, 113, 562

The Clustering of Massive Galaxies at $z \sim 1$

R. A. Overzier*, H. J. A. Röttgering*, R. J. Wilman* and R. B. Rengelink*

Leiden Observatory, University of Leiden, P.O. BOX 9513, 2300 RA, Leiden, The Netherlands

Abstract. We use the angular two-point correlation function to estimate the spatial correlation length of radio sources taken from the large-area 1.4 GHz NVSS radio survey. At the median survey redshift of $z \sim 1$, r_0 is found to be increasing with flux density. This is consistent with a scenario in which powerful (i.e. FRII) radio galaxies probe significantly more massive spatial structures than less powerful radio galaxies. The large spatial correlation length that we derive for FRIIs is remarkably close to that of extremely red objects (EROs). This implies that powerful radio galaxies and EROs trace equally massive structures at $z \sim 1$. Moreover, because powerful radio galaxies and EROs are both associated with luminous early-type galaxies we propose that they could be the same objects seen at different evolutionary stages. The correlation length of massive, luminous galaxies at $z \sim 1$ is comparable to that of bright ellipticals locally, suggesting that r_0 (comoving) of these massive galaxies has changed little from $z \sim 1$ to $z \approx 0$. This is in excellent agreement with current ΛCDM hierarchical model predictions.

THE CLUSTERING OF RADIO SOURCES

Studying the clustering of galaxies as a function of redshift can shed light on some fundamental questions in cosmology concerning galaxy formation and evolution. For example, which of the galaxies observed at high redshift are the progenitors of local galaxies, and which galaxies contain the remnant black holes that once powered high redshift active galactic nuclei (AGN)? Radio surveys can make an important contribution to this study: in contrast to magnitude-limited surveys that mainly probe the local universe, they easily provide statistical samples of objects situated at cosmological distances, making them ideal for studying a population of galaxies in the epoch of galaxy formation at $z \sim 1 - 2$.

Measurements of the angular two-point correlation function have shown that radio galaxies are biased tracers of the galaxy population as a whole, with inferred spatial correlation lengths ranging from $r_0 \sim 5 - 15$ h^{-1} Mpc at $z \sim 1$ [e.g. 1, 2, 3, 4, 5]. Rengelink [4] and Rengelink and Röttgering [6] first pointed out that the broad range in r_0 measured can be explained by a scenario in which powerful (i.e. FRII-type) radio galaxies have a larger r_0 than less powerful radio galaxies, consistent with the mounting evidence that powerful high-redshift radio galaxies are the progenitors of local cD-galaxies residing in some of the most massive environments, and are hence strongly clustered.

The large-area, medium-deep 1.4 GHz NVSS survey [7] is the first survey that allows us to conclusively study this effect *within a single survey*, whereas the hypothesis of Rengelink and Röttgering [6] was based on the comparison of surveys of different frequencies and resolution. In addition, in our analysis we have carefully dealt with double

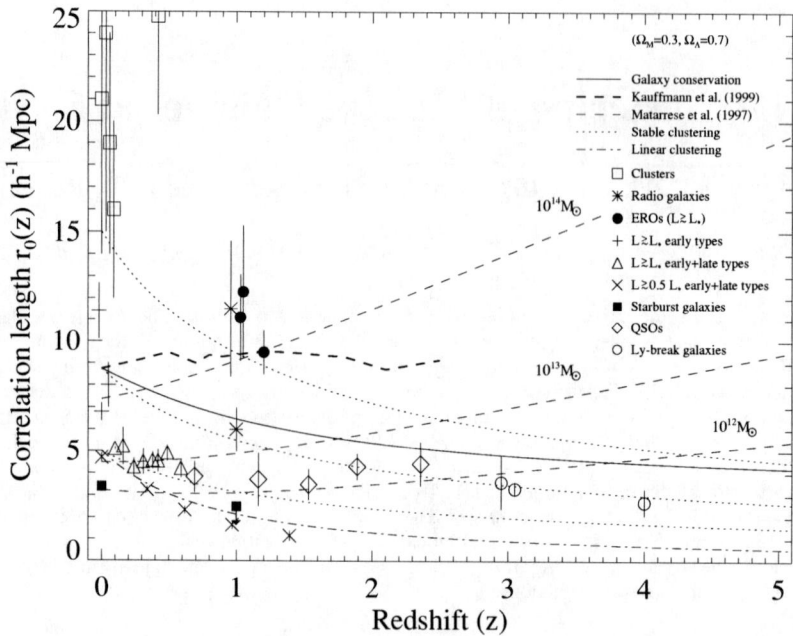

FIGURE 1. The redshift evolution of galaxy clustering. Lines represent the following models: stable clustering (dotted lines), linear clustering (dot-dashed lines), galaxy conservation model normalized to local ellipticals (solid line, see [11]), hierarchical model for clustering evolution of early-type galaxies (thick dashed line, from [12]), clustering evolution as a function of minimum dark matter halo mass (dashed lines, from [13]). See [14] for a detailed description and for references to data taken from literature.

component radio sources that have proven to be an important spurious contribution to the clustering signal on small angular scales (see also [8, 9]). Following standard procedures involving the inversion of the angular correlation function using the cosmological Limber equation [10], we indeed find $r_0 \sim 6$ for relatively low power radio galaxies, and $r_0 \sim 12$ for powerful (FRII) radio galaxies (see the forthcoming paper by Overzier et al. for details).

THE EVOLUTION OF GALAXY CLUSTERING

Fig. 1 shows the observed clustering as a function of redshift and galaxy-type compiled from literature and our own measurements for NVSS radio galaxies. Our measurements suggest that bright FRII radio galaxies and EROs [15] trace equally biased structures at $z \sim 1$, and both are associated with luminous early-type galaxies. Based on their clustering, their relative number densities and assuming that the typical AGN lifetime is relatively short compared to cosmological timescales, we conclude that powerful radio galaxies and EROs may very well be identical systems seen at different stages of their evolution, in support of a similar conclusion by Willott et al. [16] who found ERO-like host galaxies for a sub-sample of 7C radio galaxies at $1 < z < 2.5$.

While linear or stable evolutionary models generally best fit measurements from samples of ordinary, optically-selected galaxies (e.g. [17, 18, 19]), the clustering evolution of bright early type galaxies as inferred from the observed clustering of local $L \geq L_*$ ellipticals and $z \sim 1$ FRII radio galaxies and EROs is best described by a model in which r_0 (comoving) is relatively constant with redshift. Such a description of clustering can be provided if the assumption that all galaxies are conserved over time is relaxed, and is the basis of the hierarchical models for structure formation. These models predict that for certain types of objects bias can grow stronger with redshift than the growth of perturbations, giving the effect that r_0 is constant or even increases with redshift (e.g. [20, 13, 12, 21]). The consistency of the observed clustering with such models ties in closely with the popular idea that the most massive galaxies form during major merging events that trigger a burst of AGN activity, as evidenced by the dormant black holes that remain in the centers of present-day massive galaxies.

REFERENCES

1. Kooiman, B. L., Burns, J. O., and Klypin, A. A., *ApJ*, **448**, 500 (1995).
2. Loan, A. J., Wall, J. V., and Lahav, O., *MNRAS*, **286**, 994–1002 (1997).
3. Cress, C. M., Helfand, D. J., Becker, R. H., Gregg, M. D., and White, R. L., *ApJ*, **473**, 7 (1996).
4. Rengelink, R. B., *Ph.D. Thesis, Rijksuniversiteit Leiden, Leiden, The Netherlands* (1998).
5. Magliocchetti, M., Maddox, S. J., Lahav, O., and Wall, J. V., *MNRAS*, **300**, 257–268 (1998).
6. Rengelink, R. B., and Röttgering, H. J. A., "Clustering evolution in the radio surveys WENSS and GB6," in *The Most Distant Radio Galaxies*, 1999, p. 399.
7. Condon, J. J., Cotton, W. D., Greisen, E. W., Yin, Q. F., Perley, R. A., Taylor, G. B., and Broderick, J. J., *AJ*, **115**, 1693–1716 (1998).
8. Blake, C., and Wall, J., *MNRAS, accepted for publication (astro-ph/0111328)* (2002).
9. Wilman, R. J., Röttgering, H. J. A., Overzier, R. A., and Jarvis, M. J., *MNRAS, accepted for publication (astro-ph/0210679)* (2002).
10. Peebles, P. J. E., *The large-scale structure of the universe*, Princeton University Press, 1980.
11. Fry, J. N., *ApJ*, **461**, L65 (1996).
12. Kauffmann, G., Colberg, J. M., Diaferio, A., and White, S. D. M., *MNRAS*, **303**, 188–206 (1999).
13. Matarrese, S., Coles, P., Lucchin, F., and Moscardini, L., *MNRAS*, **286**, 115–132 (1997).
14. Overzier, R. A., Röttgering, H. J. A., Rengelink, R. B., and Wilman, R. J., *A&A, submitted* (2003).
15. Daddi, E., Broadhurst, T., Zamorani, G., Cimatti, A., Röttgering, H., and Renzini, A., *A&A*, **376**, 825–836 (2001).
16. Willott, C. J., Rawlings, S., and Blundell, K. M., *MNRAS*, **324**, 1–17 (2001).
17. Carlberg, R. G., Cowie, L. L., Songaila, A., and Hu, E. M., *ApJ*, **484**, 538 (1997).
18. Carlberg, R. G., Yee, H. K. C., Morris, S. L., Lin, H., Hall, P. B., Patton, D., Sawicki, M., and Shepherd, C. W., *ApJ*, **542**, 57–67 (2000).
19. McCracken, H. J., Le Fèvre, O., Brodwin, M., Foucaud, S., Lilly, S. J., Crampton, D., and Mellier, Y., *A&A*, **376**, 756–774 (2001).
20. Mo, H. J., and White, S. D. M., *MNRAS*, **282**, 347–361 (1996).
21. Moustakas, L. A., and Somerville, R. S., *ApJ*, **577**, 1–10 (2002).

Faint AGN and the Ionizing Background

Michael Schirber* and James S. Bullock[†]

*The Ohio State University
[†]Harvard-Smithsonian Center for Astrophysics

Abstract.
We determine the evolution of the faint, high-redshift, optical luminosity function of Active Galactic Nuclei (AGN) implied by several observationally-motivated models of the ionizing background from $3 < z < 5$. Our results depend crucially on whether we use the total ionizing rate measured by the proximity effect technique or the lower determination favored by the flux decrement distribution of Lyα forest lines. In addition, there is an unknown contribution to the ionizing background from stars, which we parameterize by the escape fraction of ionizing photons, f_{esc} (recent estimates have it at 16% for $z \sim 3$). Increasing the contribution from stars further limits the number of AGN necessary for a given ionizing background. By comparing our expectations to faint AGN searches in the HDF and high-z galaxy fields, we find that typically-quoted proximity effect estimates of the background imply an over-abundance of AGN compared to the faint counts (even with $f_{esc} = 1$). Even adopting the lower bound on proximity effect measurements, the stellar escape fraction must be high: $f_{esc} \gtrsim 0.2$. Conversely, the lower flux-decrement-derived background requires a smaller number of ionizing sources, and faint AGN counts are consistent with this estimate only if there is a limited stellar contribution, $f_{esc} \lesssim 0.05$. Our full development and results can be found in [1].

INTRODUCTION

At low redshift, the AGN luminosity function (LF) has been well-observed and is fit with a double power law (see e.g. [2]):

$$\phi(L,z) = \frac{\phi_*/L_*}{(L/L_*)^{\gamma_f} + (L/L_*)^{\gamma_b}} \quad (1)$$

At high redshift ($z \gtrsim 2.5$) only the bright end of the LF has been characterized (see e.g. [3]). As can be seen in Figure 1, where we represent the current state of observations, even if we assume that the shape of the faint end of the LF can be described by Eq. 1, there is no constraint on where the break occurs, i.e. the values of ϕ_* and L_*.

SCHEME

One possible way to constrain the faint end of the LF is to use background measurements in wavelength bands that AGN are believed to be dominant sources. The hydrogen ionizing background is one such band, and, fortunately, its redshift evolution can be measured from Lyα forest observations. Unfortunately, there are two *discrepant* mea-

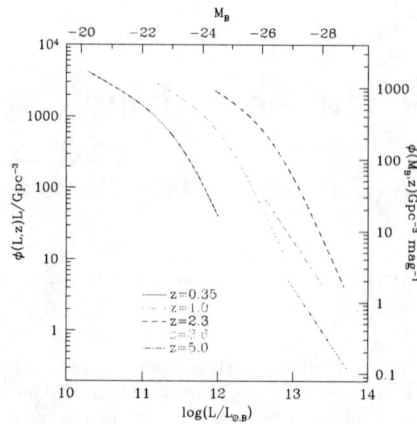

FIGURE 1. Fits to the observed LF for several redshifts. The three lowest redshift measurements come from [2] for $16.50 < M_B < 20.85$. The two highest come from [3] for $18 < i^* < 20$.

surements: the proximity effect (see [4]) and the flux decrement analysis (see [5]). We have employed both measurements to generate separate LF models.

But even if these measurements agreed, the UV light from starbursting galaxies would presumably imply a significant contribution to the ionizing background from starlight, as opposed to AGN. The fraction of ionizing radiation that is able to escape starforming regions (f_{esc}) is highly uncertain. Recent estimates at high redshift have come up with $f_{esc} \simeq 0.16$ [6], but in our analysis we have allowed this to be a free parameter.

So to reiterate, our AGN LF models are constructed to reproduce the ionizing background (either from the proximity effect or flux decrement), with some addition of starlight, parmaterized by the escape fraction, f_{esc}. We plot three of our models in Fig. 2, and compare to [7].

To evaluate different LF models we have compared them to the number of QSOs seen in faint surveys. In the HDF, there were no $z > 3.5$ QSOs seen down to $V = 27$ [8]. There were, however, a handful of $z \sim 3$ AGN found in a recent Keck survey out to $R = 25.5$ [9]. In Fig. 3, we plot the predicted number counts vs. the stellar escape fraction. To understand this plot, recall that the more ionizing photons from stars, the less number of AGN are necessary to produce the measured ionizing background. The different points in the figure are for three separate ionizing background observations.

CONCLUSIONS

From Fig. 3, it appears that the best fit value of the proximity effect is incompatible with the Keck survey results, even allowing for a very large stellar contribution. If we take the lower limit on the proximity effect, then still a large escape fraction ($f_{esc} \gg 0.16$) is needed. Conversely, adopting the flux decrement value of the ionizing background would require f_{esc} to be smaller than the presumed value ($f_{esc} < 0.16$). These limits

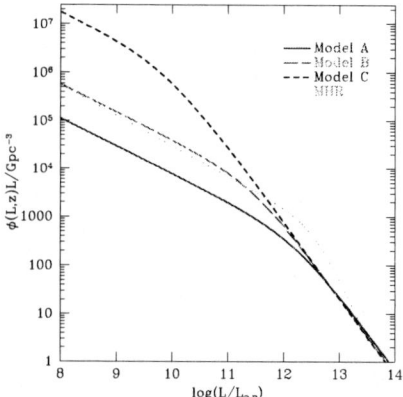

FIGURE 2. Our Luminosity Functions at $z = 3.5$. For model C, the ionizing background is from the proximity effect, whereas models A and B it is from the flux decrement analysis. In models B and C, there is no contribution from stars, but in model A we have $f_{esc} = 0.16$. Notice that all three models agree at the bright-end (where there are measurements) and then diverge at the faint end due to the location of the break. For comparison, we also plot the LF from [7].

are interesting in light of the fact that measurements of helium [10] and metal ratios 1998AJ....115.2184S in the IGM seem to point to a high stellar contribution for $z \gtrsim 3$.

Given the current paradigm of black hole accretion, we are able to limit the time averaged efficiency ($\varepsilon \equiv L/\dot{M}$) from the relic black hole density. We find $\varepsilon \lesssim 0.05$, even for the proximity effect models with the highest luminosity density. From X-ray surveys [12], the efficiency has been measured to be greater than 10%. Since we are dealing with optical/UV light, we might explain this discrepancy by assuming that $\gtrsim 50 - 70\%$ of AGN are obscured.

Finally, assuming near Eddington luminosities, we can limit the typical AGN lifetime by postulating that the AGN number density cannot exceed the relic black hole number density. We find $t_{agn} \gtrsim 10^7$ yr for our favored models.

ACKNOWLEDGMENTS

We would like to acknowledge helpful discussions with Tom Abel, Alberto Conti, Piero Madau, Smita Mathur, Pat McDonald, Jordi Miralda-Escudé, Pat Osmer, Rick Pogge, Marianne Vestergaard, Terry Walker and David Weinberg. We were supported by U.S. DOE Contract No.DE-FG02-91ER40690.

REFERENCES

1. Schirber, M., Bullock, J. S. 2002, astro-ph/0207200, ApJ accepted

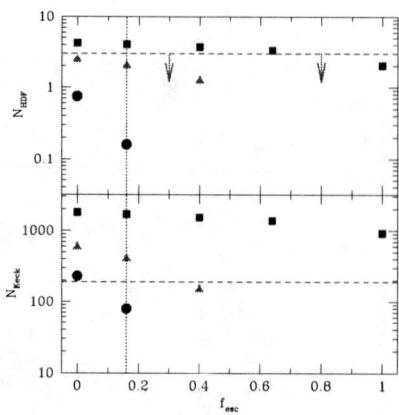

FIGURE 3. The expected number counts in the HDF (*top*) and Keck (*bottom*) surveys as a function of the escape fraction. The upper bound from [8] and the observed number from [9] are plotted as dashed horizontal lines. The squares are for an ionizing background from the best fit of the proximity effect, whereas the triangles are for the lower bound on the proximity effect [4]. The circles are from the flux decrement measurement [5]. Note that we do not plot points at high escape fraction for the flux decrement (and for the lower bound on the proximity effect, as well), since the resulting LF breaks would contradict observations from [3].

2. Boyle, B. J., Shanks, T., Croom, S. M., Smith, R. J., Miller, L., Loaring, N., & Heymans, C. 2000, MNRAS, 317, 1014
3. Fan, X. et al. 2001, AJ, 121, 54
4. Scott, J., Bechtold, J., Dobrzycki, A., & Kulkarni, V. P. 2000, ApJS, 130, 67
5. McDonald, P. & Miralda-Escudé, J. 2001, ApJ, 549, L11
6. Steidel, C. C., Pettini, M., & Adelberger, K. L. 2001, ApJ, 546, 665
7. Madau, P., Haardt, F., & Rees, M. J. 1999, ApJ, 514, 648
8. Conti, A., Kennefick, J. D., Martini, P., & Osmer, P. S. 1999, AJ, 117, 645
9. Steidel, C. C., Hunt, M. P., Shapley, A. E., Adelberger, K. L., Pettini, M., Dickinson, M., & Giavalisco, M. 2002, ApJ, 576, 653
10. Heap, S. R., Williger, G. M., Smette, A., Hubeny, I., Sahu, M. S., Jenkins, E. B., Tripp, T. M., & Winkler, J. N. 2000, ApJ, 534, 69
11. Songaila, A. 1998, AJ, 115, 2184
12. Elvis, M., Risaliti, G., & Zamorani, G. 2002, ApJ, 565, L75

On the Nature of Lyman-α Emitters

Junxian Wang*, Sangeeta Malhotra†, James Rhoads†, Michael Brown**,
Timothy Heckman* and Colin Norman*

Johns Hopkins University, Baltimore, MD 21218
†*Space Telescope Science Institute, Baltimore, MD 21218*
**Kitt Peak National Observatory*

Abstract. More than half of the approximately 360 Lyman-α emitters found by the LALA survey show rest equivalent widths (EWs) larger than 200 Å, which is the maximum EW expected for normal stellar population. The high EWs can be reproduced by stellar populations with high proportions of young, massive stars, or by type 2 AGNs, which are supposed to be detectable in X-ray by deep *Chandra* images. An 180 ks *Chandra* ACIS exposure was obtained to study the X-ray properties of these emitters. However, none of the 49 sources imaged was significantly detected (over 2σ level) in X-ray. These sources cannot be detected in the stacked X-ray image either, which has an effective *Chandra* exposure time of 6.5 million seconds. The 3σ upper limit of the average X-ray flux is 7.9×10^{-17} erg.cm^{-2}.s^{-1} in the 0.5 – 10.0 keV band, which is much fainter than the type 2 quasars discovered by Chandra so far. Optical spectra also show no AGN signatures.

INTRODUCTION

More than three decades ago Partridge & Peebles (1967)[1] predicted that galaxies undergoing their first throes of star formation should be strong emitters in the Lyα line: these galaxies are expected to contain many hot, young, massive stars, which ionize interstellar gas, and the interstellar medium that gave birth to the stars can then convert 2/3 of the ionizing continuum into Lyman-α photons as neutral hydrogen atoms are ionized and recombine. Searches for these high redshift Lyman-α emitters have last 30 years, and recently, they were finally discovered (see [2] and references therein).

The Large Area Lyman Alpha (LALA) Survey was designed to search and build a large sample of the Lyman-α emitters at high redshifts through narrowband imaging. The survey comprises two primary fields, located in Boötes (at 14:25:57 +35:32 J2000.0) and in Cetus (at 02:05:20 -04:55 J2000.0). Each field is 36 × 36 arcminutes in size. Along with five broadband filters: B_W, V, R, I, and z', eight narrowband filters were designed to search Lyman-α emitters at z $\sim$ 4.5, 5.7, and 6.5 respectively[2, 3]. The limiting Kron-Cousins system magnitudes of the broadband filters are $B_W < 26.77$, $R < 25.24, I < 24.96, V < 25.88$ and $z' < 24.53$ (where the z' limit is on the AB system).

It's interesting to notice that more than half of the Lyman-α emitters detected in LALA field show rest equivalent widths (of Lyα line) larger than 200 Å[4], while normal stellar populations are expected to produce peak equivalent widths 100Å$<$ EW$_{max} <$ 200Å[5]. Stellar populations with high proportions of young, massive stars [4] can explain the larger equivalent widths. The high EWs can also be reproduced by active galactic nuclei, most likely type 2 AGNs, since the emitters seem not to have broad

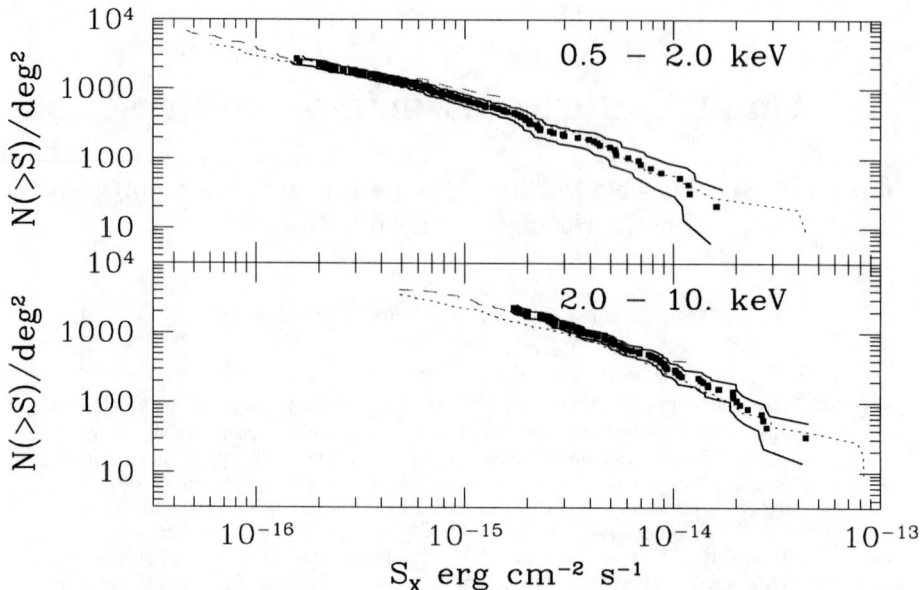

FIGURE 1. LogN-LogS in the soft (0.5 – 2.0 keV) and hard band (2.0 – 10.0 keV) bands from 180ks *Chandra* observations. Data are plotted as filled squares with solid lines enclosing 1σ errors. Number counts from two 1Ms *Chandara* deep surveys are also plotted: dashed line from CDFN [9], and dotted line from CDFS [10].

emission lines which would show up in more than one narrowband filters. Deep X-ray survey can help us distinguish AGNs from normal star forming galaxies. Type 2 quasars, which are predicted to comprise as much as 90% of high redshift quasar population [6], were recently discovered by deep X-ray surveys [7, 8]. It's thus highly valuable to check if some of the Lyman-α emitters are type 2 quasars.

CHANDRA IMAGING OBSERVATION

In order to study the X-ray properties of the Lyman-α emitters, *Chandra* imaging observations were proposed on both LALA fields. 180 ks exposure on the Boötes field, composed of two individual *Chandra* observations, was obtained using the Advanced CCD Imaging Spectrometer (ACIS) in 2002. Three X-ray images were extracted from the combined event file: a soft image (0.5 – 2.0 keV), a hard image (2.0 – 7.0 keV) and a total image (0.5 – 7.0 keV). WAVDETECT was run on each image with a probability threshold of 1×10^{-7} (correspondent to 0.5 false sources expected per image), and scales of 1,2,4,8,16 pixels. Totally 168 X-ray sources were detected, with $\sim$ 80% having optical counterparts detected in one or several of the five optical broadband images (B_W,V,R,I, and z'). The LogN-LogS derived was present in Figure 1, comparing with those from CDFS and CDFN.

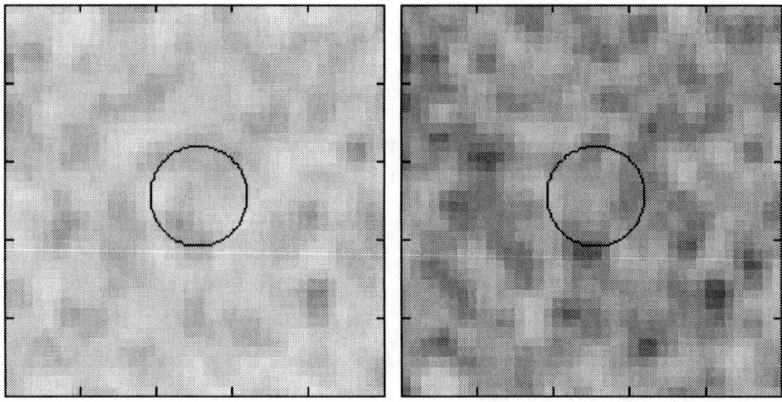

FIGURE 2. Stacked *Chandra* images of 49 Lyman-α emitters. Left: soft band (0.5 – 2.0 keV); Right: total band (0.5 – 7.0 keV). The effective exposure time of the stacked images is 6.5 Ms. The images are 40 × 40 pixels in size, and the circles are centered on the stacking position having a radius of 5 pixels. 1 pixel = 0.492".

NON-DETECTION IN X-RAY AND OPTICAL SPECTRA

None of the 49 Lyman-α emitters covered by the *Chandra* exposure was detected by our detection program. And after performing X-ray photometry we found none of the sources has X-ray net counts with significant level $> 2\sigma$ in either band. We conclude that none of the 49 Lyman-α emitters was individually detected by the 180ks *Chandra* ACIS exposure. For the Lyman-α emitter closest to the axis of the X-ray exposure, the 3σ upper limits for X-ray fluxes are 1.7×10^{-16} erg.cm^{-2}.s^{-1} for 0.5 – 2.0 keV band, 4.7×10^{-16} erg.cm^{-2}.s^{-1} for 0.5 – 10.0 keV band (assuming a powerlaw spectrum with photon index of 1.4). Other sources have higher upper limits due to the lower effective areas and larger PSF sizes.

We also stacked the X-ray imaging data at each position of Lyman-α emitters. The stacked images have an effective exposure time of 6.5 Ms (75 days). However, no X-ray emission can be detected at the stacking position in either band (soft, hard and total bands). We present the stacked images for soft and total bands in Figure 3. Simply summing up the source counts and expected background counts for 49 sources, we got 55/54.2 source/expected-background counts in soft band, and 164/163.1 in total band. It's obvious that we did not detect the Lyman-α emitters in X-ray even after stacking them. The 3σ upper limits of summed soft and total band net counts are 26 and 42, correspondent to the average X-ray flux upper limits for each of the 49 Lyman-α emitters: 1.8×10^{-17} erg.cm^{-2}.s^{-1} in 0.5 – 2.0 keV band, and 7.9×10^{-17} erg.cm^{-2}.s^{-1} in 0.5 – 10.0 keV band.

We present here the optical spectrum of one of the brightest Lyman-α emitters which clearly shows no AGN feature (only strong Lyα line is seen, not CIV and HeII line). Optical spectra available now for other sources are similar.

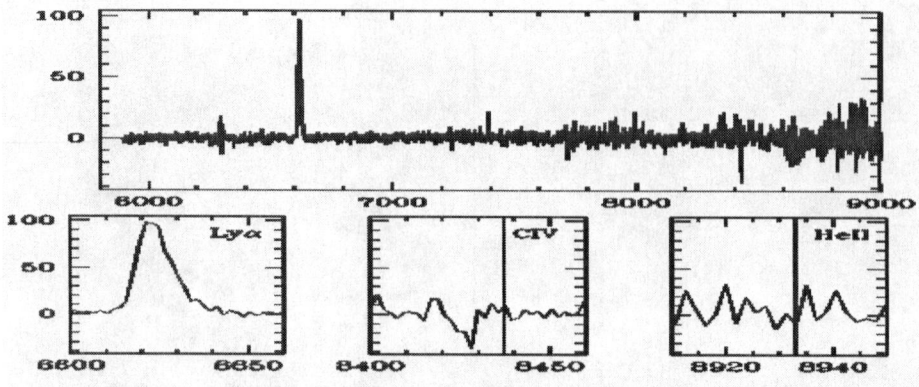

FIGURE 3. Optical spectrum for one of the brightest Lyman-α emitters obtained from Keck

CONCLUSION

Assuming we have a type 2 AGN with 2.0 – 10.0 keV luminosity (rest frame, absorption corrected) of 10^{44} erg.s^{-1} (H0=50, q0=0.5), and absorption of 10^{24} cm^{-2}, we expect 6.6 photons in 0.5 – 7.0 keV band from 180 ks *Chandra* observation. Most of the 49 Lyman-α emitters have net counts lower than this value. Furthermore, the stacking results gave a 3σ upper limit (average) of 42 photons for 6.5 Ms effective exposure time, which can convert to a luminosity of 1.5×10^{43} erg.s^{-1}, much lower than those of the previous discovered type 2 quasar (the type 2 quasar at z=3.7 detected in *Chandra* Deep Field South has a rest frame 2.0 – 10.0 keV luminosity of 6.3×10^{44} erg.s^{-1} [7], and the one at z=3.288 discovered by Stern et al. has that of 3.3×10^{44} erg.s^{-1} [8]).

We conclude that none of the covered Lyman-α emitters covered was detected by 180ks *Chandra* ACIS exposure. They are not detected either from stacked X-ray images which have 6.5 Ms effective *Chandra* exposure time. Together with optical spectra obtained, we believe that most of the Lyman-α emitters are star forming galaxies but not AGNs.

REFERENCES

1. Partridge, R.B. and Peebles, P.J.E., *ApJ*, **147**, 868 (1967).
2. Rhoads, J.E., Malhotra, S. and Dey, A., *ApJ*, **545**, L85 (2000).
3. Rhoads, J.E. and Malhotra, S., *ApJ*, **563**, L5 (2001).
4. Malhotra, S. and Rhoads, J.E., *ApJ*, **565**, L71 (2002).
5. Charlot, S. and Fall, S.M., *ApJ*, **415**, 580 (1993)
6. Gilli, R., Salvati, M. and Hasinger, G., *A&A*, **366**, 407 (2001)
7. Norman, C. et al., *ApJ*, **571**, 218 (2002)
8. Stern, D. et al., *ApJ*, **568**, 71 (2002)
9. Brandt, W.N. et al., *AJ*, **122**, 2810 (2001)
10. Rosati, P. et al., *ApJ*, **566**, 667 (2002)

Some Results from the Oxford-Dartmouth Thirty Degree Survey

G. Wegner*, M. Hammell*, G. B. Dalton[†], P. D. Allen[†], E. J. Olding[†] and L. Moustakas**

*Department of Physics & Astronomy, 6127 Wilder Laboratory, Dartmouth College, Hanover, NH 03755
[†]Department of Physics, Astrophysics, Nuclear and Astrophysics, Keble Road, Oxford, OX1 3RH, UK
**Space Telescope Science Institute, 3700 San Martin Drive, Baltimore, MD 21218

Abstract. The Oxford-Dartmouth Thirty Degree Survey (ODTS) is a deep-wide multicolor imaging survey, designed to probe structure and study objects with redshifts $z > 1$. When completed, the ODTS will cover over 30 square degrees in $UBVRI'ZK$. Here we describe some of the results on extremely red objects (EROs) and blue dropout galaxies.

1. INTRODUCTION

The observations are already well advanced in several colors and in three separate 10 square degree regions near:

00:18 24 +34:52, 09:09:45 +40:50, and 16:39:30 +45:24.

An additional field centered at 13:30 +02:30 is being added near the celestial equator. The Wide Field Camera on the 2.5-m Isaac Newton Telescope has been used to observe the visible bands and K band observations are being obtained with the MDM Observatory 1.3-m and 2.4-m telescopes. Our 5 sigma limiting isophotal magnitudes on the Vega scale are: $U = 26, B = 25.75, V = 25.5, R = 25.25, I' = 24.5, Z = 23$ with a 3 sigma limit for $K = 18.75$. For image detection, and constructing an object catalog, the *SExtractor* software [2] was employed. Stars and galaxies are separated using the star-galaxy classifier which uses a neural net to classify each object with a "stellarity" index running from 1 for stars to 0 for galaxies. Photometric redshifts are also being measured using programs like *hyperz* [3]. We have detected large numbers of extremely red galaxies, EROs, with $(R - K) > 6$, as well as B-dropout galaxies and high z quasar candidates. A significant fraction of the EROs occur in groups and appear to be tracers of clusters at $z \sim 1 - 2$. Here we briefly present our latest results on number counts, angular correlation functions, and distributions of some of these objects. The galaxy clusters work is described elsewhere in these proceedings [8].

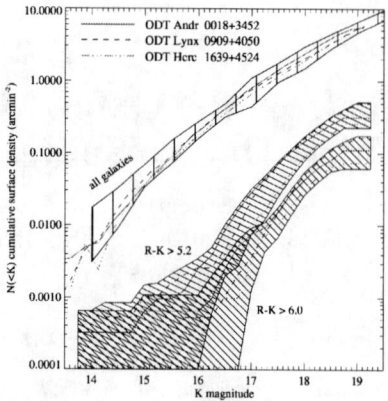

FIGURE 1. The cumulative surface density of galaxies in the K-band, for each of the three main ODTS fields [11]. From top-to-bottom are depicted the counts for galaxies with no color-cut (with the range of counts from the literature overplotted); with $R - K > 5.2$; and with $R - K > 6.0$. The color-selected galaxy counts are based on both R- and K-detections, and are therefore a lower limit. It is estimated that the stellar contamination in the Andromeda data is >15% at K >17; contamination is much less in Lynx and Hercules.

2. RESULTS ON THE EROS

The extremely red objects or EROs are sources with unusually red colors, *viz.* $(R - K) > 5$, well beyond the expected colors of normal field galaxies. It is now generally recognized that extragalactic EROs are produced in two ways: (1) From old elliptical galaxies with a strong 4000 Åbreak at redshifts $1 \leq z \leq 2$ and (2) dusty galaxies with a combination of starburst and AGN activity also occurring in this redshift range. Other kinds of objects may also contribute including low surface brightness objects [14] nearby L and T dwarfs, e.g. [10].

Understanding the extragalactic EROs and how they divide between the two classes above is an important clue to the riddle of galaxy formation. If the EROs were dominated by ellipticals, the monolithic collapse model [6] would be favored, while the hierarchical scenarios require higher numbers of dusty starburst galaxies [9], [15].

The situation is complicated however. Several investigators find that the EROs are primarily passively evolving ellipticals (cf. [12]) while [7] conclude that ERO number counts are inconsistent with this. Such disagreement is likely to be caused by strong spatial clustering [5].

Figure 1 shows cumulative surface density counts for the three main ODTS fields based on the the K-band data reported in [11]. The top set of curves shows the counts in each of the three fields for galaxies of all colors compared with literature values. The lower curves compare the color selected counts in the same three fields for $(R-K) > 5.2$ and 6.0. Stellar contamination can contribute to the fainter counts and is being studied further. These data again agree with literature values, but here there is clear evidence for cosmic variance in the ERO surface densities between fields, even for the large areas

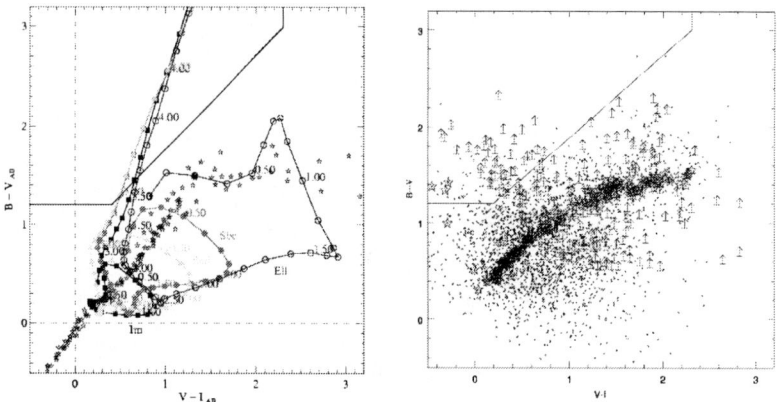

FIGURE 2. Comparison of the colors of over 5000 objects in a small region of the ODTS from Allen [1]. (Left) Colors of the empirical Coleman, Wu, & Weedman templates [4] in the $(B-V)-(V-I)$ plane for redshifts up to $z = 5$. The black stars are stellar colors from [13]. This predicts that objects with $z > 4$ lie above the black line. (Right) Observed colors. Stellar objects are shown as stars. Galaxies detected in $BVRI$ are black circles. Objects with no B detection are shown as arrows computed on the assumption that $B = 26$ and the line corresponding to the limit set in the left diagram was used to select the LBG sample in the 0^h Andromeda field.

and sample sizes of the ODTS data. The Hercules field is particularly 'deficient,' even though the all-galaxy counts are identical for all three fields. This is also reflected in the ERO correlation functions for each field which also indicate strong clustering properties.

3. LYMAN BREAK GALAXIES

We are currently searching for different types of objects using two color diagrams with our existing multicolor data in small regions of the sky, which to date yield approximately 20,000 Lyman Break Galaxies (LBGs). Figure 2, indicates the region in the $(V-I), (B-V)$ plot used to find LBGs [1] where plots of model galaxies using the ODTS filter transmissions and data from [4] are compared with meaured colors. In the 0^h field, about 1000 objects have been found to date.

Figure 3 from Allen [1] shows the measured angular correlation function for 327 $z \sim 4$ LBG candidates with $23 < I' < 24.5$ from a single ODT pointing measuring $17' \times 17'$. The correlation function is best fitted by a power law $w(\theta) = A_W \theta^{-0.8}$ with $A_W = 2.2$. This confirms previous measurements of the strong clustering of Lyman Break Galaxies in the literature. Using a redshift distribution based on photometric redshifts, deprojecting this correlation function yields $r_0 = 5.6 \text{Mpc} h^{-1}$. While slightly higher than previous measurements of r_0, this is not unexpected since the ODTS is slightly shallower than other surveys that have measured LBG clustering.

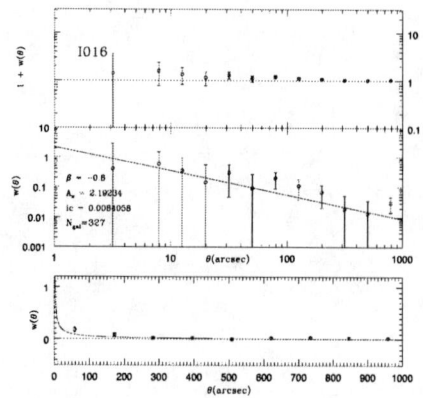

FIGURE 3. The angular correlation function for 327 $z \sim 4$ LBG candidates with $23.0 < i'(AB) < 24.5$ from a single ODT pointing covering 17×17 arcmin^2. The top two panels show logarithmic plots of $w(\theta)$ and $1 + w(\theta)$ while a linear plot is shown in the bottom panel [1].

ACKNOWLEDGMENTS

This research has been partially supported by the NSF, NASA, and PPARC.

REFERENCES

1. Allen, P. D. 2003, D.Phil. Thesis, Oxford.
2. Bertin, E. & Arnouts, S. 1996, *A&AS*, **117**, 393.
3. Bolzonella, M. et al. 2000, *A&A*, **363**, 476.
4. Coleman, G. D., Wu, C.-C., & Weedman, D. W. 1980, *ApJS*, **43**, 393.
5. Daddi, E. et al. 2002, *A&A*, **384**, L1.
6. Eggen, O. J., Lynden-Bell, D., & Sandage, A. 1962, *ApJ*, **136**, 748.
7. Firth, A. E. et al. 2002, *MNRAS*, **332**, 617.
8. Hammell, M., Wegner, G., Moustakas, L., Allen, P., Dalton, G., & Olding, E. 2003, These proceedings.
9. Kauffmann, G. & Charlot, S. 1998, *MNRAS*, **297**, L23.
10. Kirkpatrick, J. D. et al. 1999, *ApJ*, **519**, 802.
11. Olding, E. J. 2002, D. Phil. Thesis, Oxford Univ.
12. Peebles, P. J. E. 2002, astro-ph/0201015.
13. Pickles, A. J. 1998, *PASP*, **110**, 863.
14. Smith, G. P. et al 2002, *MNRAS*, **330**, 1.
15. White, S. D. M. & Frenk, C. S. 1991, *ApJ*, **379**, 52.

LOW-REDSHIFT STRUCTURE

Implications of 2d FGRS Results on Cosmic Structure

J.A. Peacock*

Institute for Astronomy, University of Edinburgh,
Royal Observatory, Edinburgh EH9 3HJ, UK

Abstract. The 2dF Galaxy Redshift Survey is the first to observe more than 100,000 redshifts, making possible precise measurements of many aspects of galaxy clustering. The spatial distribution of galaxies can be studied as a function of galaxy spectral type, and also of broad-band colour. Redshift-space distortions are detected with a high degree of significance, confirming the detailed Kaiser distortion from large-scale infall velocities, and measuring the distortion parameter $\beta \equiv \Omega_m^{0.6}/b = 0.49 \pm 0.09$. The power spectrum is measured to $\lesssim 10\%$ accuracy for $k > 0.02 \, h \, \text{Mpc}^{-1}$, and is well fitted by a CDM model with $\Omega_m h = 0.18 \pm 0.02$ and a baryon fraction of 0.17 ± 0.06. A joint analysis with CMB data requires $\Omega_m = 0.31 \pm 0.05$ and $h = 0.67 \pm 0.04$, assuming scalar fluctuations. The fluctuation amplitude from the CMB is $\sigma_8 = 0.76 \pm 0.04$, assuming reionization at $z \lesssim 10$, so that the general level of galaxy clustering is approximately unbiased, in agreement with an internal bispectrum analysis. Luminosity dependence of clustering is however detected at high significance, and is well described by a relative bias of $b/b^* = 0.85 + 0.15(L/L^*)$. This is consistent with the observation that L^* in rich clusters is brighter than the global value by 0.28 ± 0.08 mag.

1. AIMS AND DESIGN OF THE 2DFGRS

The 2dF Galaxy Redshift Survey (2dFGRS) was designed to study the following key aspects of the large-scale structure in the galaxy distribution:

(1) To measure the galaxy power spectrum $P(k)$ on scales up to a few hundred Mpc, bridging the gap between the scales of nonlinear structure and measurements from the the cosmic microwave background (CMB).

(2) To measure the redshift-space distortion of the large-scale clustering that results from the peculiar velocity field produced by the mass distribution.

(3) To measure higher-order clustering statistics in order to understand biased galaxy formation, and to test whether the galaxy distribution on large scales is a Gaussian random field.

---

* On behalf of the 2dF Galaxy Redshift Survey team: Matthew Colless (ANU), Ivan Baldry (JHU), Carlton Baugh (Durham), Joss Bland-Hawthorn (AAO), Terry Bridges (AAO), Russell Cannon (AAO), Shaun Cole (Durham), Chris Collins (LJMU), Warrick Couch (UNSW), Gavin Dalton (Oxford), Roberto De Propris (UNSW), Simon Driver (St Andrews), George Efstathiou (IoA), Richard Ellis (Caltech), Carlos Frenk (Durham), Karl Glazebrook (JHU), Carole Jackson (ANU), Ofer Lahav (IoA), Ian Lewis (AAO), Stuart Lumsden (Leeds), Steve Maddox (Nottingham), Darren Madgwick (IoA), Peder Norberg (Durham), Will Percival (ROE), Bruce Peterson (ANU), Will Sutherland (ROE), Keith Taylor (Caltech).

The survey is designed around the 2dF multi-fibre spectrograph on the Anglo-Australian Telescope, which is capable of observing up to 400 objects simultaneously over a 2 degree diameter field of view. For details of the instrument and its performance see http://www.aao.gov.au/2df/, and also Lewis et al. (2002).

The source catalogue for the survey is a revised and extended version of the APM galaxy catalogue (Maddox et al. 1990a,b,c); this includes over 5 million galaxies down to $b_J = 20.5$ in both north and south Galactic hemispheres over a region of almost $10^4 \deg^2$ (bounded approximately by declination $\delta \leq +3°$ and Galactic latitude $b \gtrsim 20°$). This catalogue is based on Automated Plate Measuring machine (APM) scans of 390 plates from the UK Schmidt Telescope (UKST) Southern Sky Survey. The b_J magnitude system for the Southern Sky Survey is defined by the response of Kodak IIIaJ emulsion in combination with a GG395 filter, and is related to the Johnson–Cousins system by $b_J = B - 0.304(B-V)$, where the colour term is estimated from comparison with the SDSS Early Data Release (Stoughton et al. 2002) The photometry of the catalogue is calibrated with numerous CCD sequences and has a precision of approximately 0.15 mag for galaxies with $b_J = 17$–19.5. The star-galaxy separation is as described in Maddox et al. (1990b), supplemented by visual validation of each galaxy image.

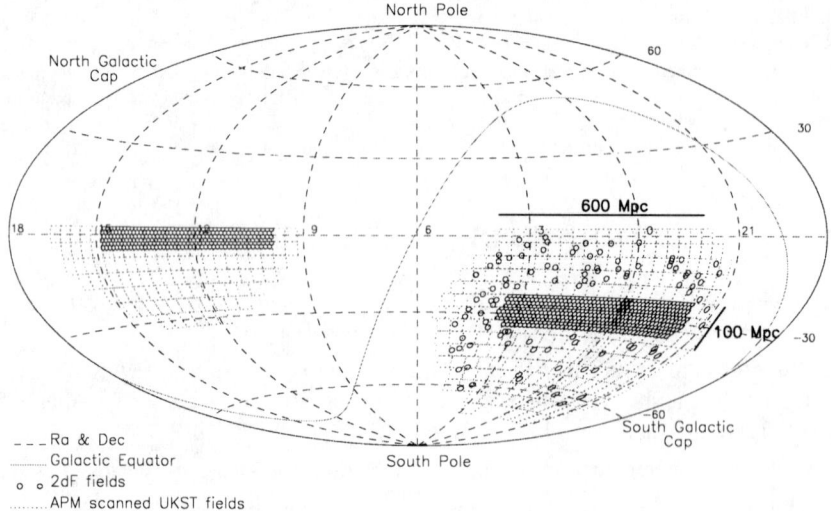

FIGURE 1. The 2dFGRS fields (small circles) superimposed on the APM catalogue area (dotted outlines of Sky Survey plates). There are approximately 140,000 galaxies in the 75° × 15° southern strip centred on the SGP, 70,000 galaxies in the 75° × 7.5° equatorial strip, and 40,000 galaxies in the 100 randomly-distributed 2dF fields covering the whole area of the APM catalogue in the south.

The survey geometry is shown in Figure 1, and consists of two contiguous declination strips, plus 100 random 2-degree fields. One strip is in the southern Galactic hemisphere and covers approximately 75°×15° centred close to the SGP at $(\alpha,\delta)=(01^h,-30°)$; the other strip is in the northern Galactic hemisphere and covers 75° × 7.5° centred at $(\alpha,\delta)=(12.5^h,+0°)$. The 100 random fields are spread uniformly over the 7000 $\deg^2$ region of the APM catalogue in the southern Galactic hemisphere. At the median redshift of the survey ($\bar{z} = 0.11$), $100 h^{-1}$ Mpc subtends about 20 degrees, so the two strips are

$375\,h^{-1}$ Mpc long and have widths of $75\,h^{-1}$ Mpc (south) and $37.5\,h^{-1}$ Mpc (north).

The sample is limited to be brighter than an extinction-corrected magnitude of $b_J = 19.45$ (using the extinction maps of Schlegel et al. 1998). This limit gives a good match between the density on the sky of galaxies and 2dF fibres. Due to clustering, however, the number in a given field varies considerably. To make efficient use of 2dF, we employ an adaptive tiling algorithm to cover the survey area with the minimum number of 2dF fields. With this algorithm we are able to achieve a 93% sampling rate with on average fewer than 5% wasted fibres per field. Over the whole area of the survey there are in excess of 250,000 galaxies.

2. SURVEY STATUS

After an extensive period of commissioning of the 2dF instrument, 2dFGRS observing began in earnest in May 1997, and terminated in April 2002. In total, observations were made of 899 fields, yielding redshifts and identifications for 232,529 galaxies, 13976 stars and 172 QSOs, at an overall completeness of 93%. The galaxy redshifts are assigned a quality flag from 1 to 5, where the probability of error is highest at low Q. Most analyses are restricted to $Q \geq 3$ galaxies, of which there are currently 221,496. An interim data release took place in July 2001, consisting of approximately 100,000 galaxies (see Colless et al. 2001 for details). A public release of the full photometric and spectroscopic database is scheduled for July 2003.

The Colless et al. (2001) paper details the practical steps that are necessary in order to work with a survey of this sort. The 2dFGRS does not consist of a simple region sampled with 100% efficiency, and it is therefore necessary to use a number of masks in order to interpret the data. Two of these concern the input catalogue: the boundaries of this catalogue, including 'drilled' regions around bright stars where galaxies could not be detected; also, revisions to the photometric calibration mean that in practice the survey depth varies slightly with position on the sky. A futher mask arises from the way in which the sky is tessellated into 2dF tiles: near the survey edges and near internal holes, a lack of overlaps mean that the sampling fraction falls to about 50%. Finally, the spectroscopic success rate of each spectroscopic observation fluctuated according to the observing conditions. The median redshift yield was approximately 95%, but with a tail towards poorer data. The terminal stages of 2dFGRS observing were in fact devoted to re-observing these fields of low completeness; nevertheless, approximately 10% of fields have completeness lower than 80%. This variable sampling makes quantification of the large scale structure more difficult, particularly for any analysis requiring relatively uniform contiguous areas. However, the effective survey 'mask' can be measured precisely enough that it can be allowed for in analyses of the galaxy distribution.

3. GALAXY SPECTRA AND COLOURS

Beyond the basic data of positions, magnitudes and redshifts, it is important on physical grounds to be able to divide the 2dFGRS database into different categories of galaxies. This has been done in two different ways. Spectral classification of 2dFGRS galaxies was performed by Folkes et al. (1999) and Madgwick et al. (2002). Principal component analysis was used to split galaxies into a superposition of a small number of templates. Not all of these are robust, owing to uncalibrated spectral distortions in the 2dF instrument, but it was possible to derive a robust classification parameter (termed η) from the templates, which effectively measures the emission-line strength (closely related to the star-formation rate). Galaxies were divided into four spectral classes; their mean spectra and separate luminosity functions are shown in Figure 2.

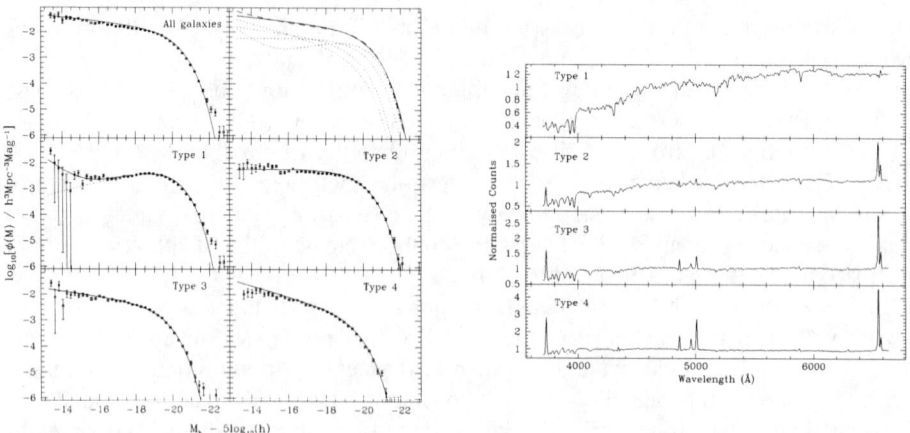

FIGURE 2. The type-dependent galaxy luminosity function according to Madgwick et al. (2002). Principal component analysis was used to split galaxies into a superposition of a small number of templates, and a categorization made based on the decomposition. Type 1 galaxies are generally E/S0, while later types range from Sa to Irr.

This classification method has the drawback that it cannot be used beyond $z = 0.15$, where Hα is lost from the spectra. Also, the fibres do not cover the whole galaxy (although Madgwick et al. 2002 show that aperture corrections are not large in practice). More recently, we have been able to obtain total broad-band colours for the 2dFGRS galaxies, using the data from the SuperCOSMOS sky surveys (Hambly et al. 2001). These yield B_J from the same UK Schmidt Plates as used in the original APM survey, but with improved linearity and smaller random errors (0.09 mag rms relative to the SDSS EDR data). The R_F plates are of similar quality, so that we are able to divide galaxies by colour, with an rms in photographic $B - R$ of about 0.13 mag. The systematic calibration uncertainties are negligible by comparison, and are at the level of 0.04 mag. rms in each band. Figure 3 shows that the colour information divides 'passive' galaxies with little active star formation cleanly from the remainder, uniformly over the whole redshift range of the 2dFGRS.

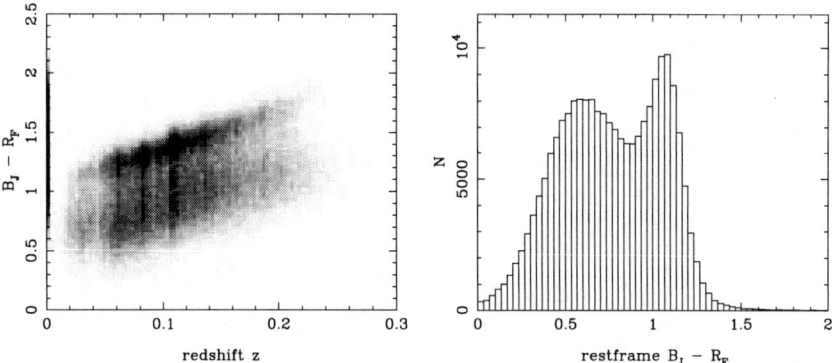

FIGURE 3. Photographic $B-R$ colour versus redshift for the 2dFGRS. The separation between 'passive' (red) and 'active' (blue) galaxies is very clear. Empirically, $B-R-2.8z$ defines a 'restframe' colour whose distribution is independent of redshift, and very clearly bimodal. This is strongly reminiscent of the distribution of the spectral type, η, and we assume that a division at $(B-R)_0 = 0.85$ achieves a separation of 'class 1' galaxies from classes 2–4, as was done using spectra by Madgwick et al. (2002).

As an immediate application, we can display the spatial distribution of 2dFGRS galaxies divided according to colour (Figure 4). The most striking aspect of this image is how closely both sets of galaxies follow the same structure. The passive subset display a more skeletal appearance, as expected owing to morphological segregation of ellipticals. A red-selected survey such as SDSS will appear more similar to the passive subset of the 2dFGRS, with relatively low sampling of the more active spectral type 2–4.

4. REDSHIFT-SPACE CORRELATIONS

The simplest statistic for studying clustering in the galaxy distribution is the the two-point correlation function, $\xi(\sigma,\pi)$. This measures the excess probability over random of finding a pair of galaxies with a separation in the plane of the sky σ and a line-of-sight separation π. Because the radial separation in redshift space includes the peculiar velocity as well as the spatial separation, $\xi(\sigma,\pi)$ will be anisotropic. On small scales the correlation function is extended in the radial direction due to the large peculiar velocities in non-linear structures such as groups and clusters – this is the well-known 'Finger-of-God' effect. On large scales it is compressed in the radial direction due to the coherent infall of galaxies onto mass concentrations – the Kaiser effect (Kaiser 1987).

To estimate $\xi(\sigma,\pi)$ we compare the observed count of galaxy pairs with the count estimated from a random distribution following the same selection function both on the sky and in redshift as the observed galaxies. We apply optimal weighting to minimise the uncertainties due to cosmic variance and Poisson noise. The redshift-space correlation function for the 2dFGRS computed in this way is shown in Figure 5. The correlation-function results display very clearly two signatures of redshift-space distortions. The 'fingers of God' from small-scale random velocities are very clear, as indeed has been the case from the first redshift surveys (e.g. Davis & Peebles 1983). However, this is

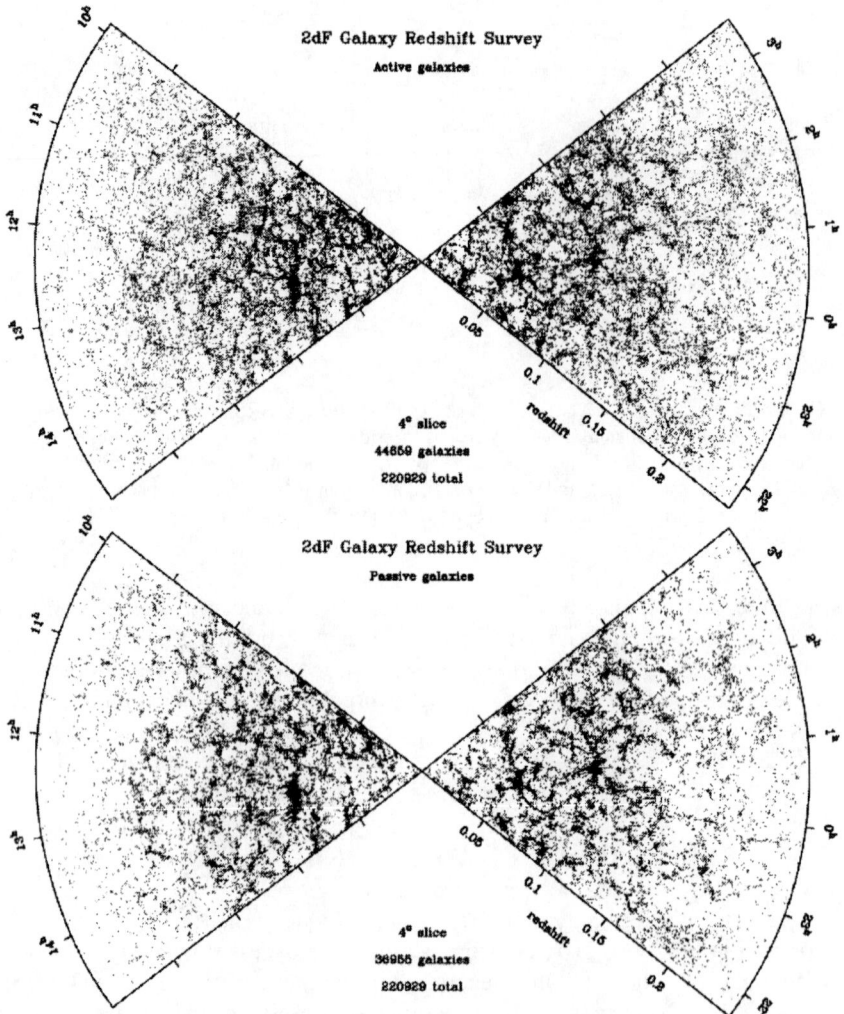

FIGURE 4. The distribution of galaxies in part of the 2dFGRS, drawn from a total of 221,496 galaxies: slices 4° thick, centred at declination −2.5° in the NGP and −27.5° in the SGP. The survey is divided at a rest-frame colour of photographic $B - R = 0.85$, into galaxies with and without active star formation. The This image reveals a wealth of detail, including linear supercluster features, often nearly perpendicular to the line of sight. It appears that these transverse features have been enhanced by infall velocities.

the first time that the large-scale flattening from coherent infall has been seen in detail. An initial analysis of this effect was performed in Peacock et al. (2001), and the final database was analysed by Hawkins et al. (2002).

The degree of large-scale flattening is determined by the total mass density parameter, Ω_m, and the biasing of the galaxy distribution. On large scales, it should be correct to assume a linear bias model, with correlation functions $\xi_g(r) = b^2 \xi(r)$, so that the

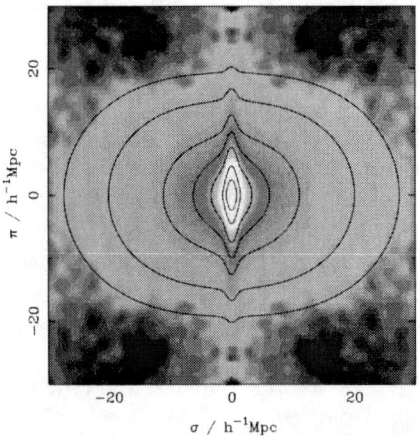

FIGURE 5. The galaxy correlation function $\xi(\sigma,\pi)$ as a function of transverse (σ) and radial (π) pair separation is shown as a greyscale image. It was computed in $0.2\,h^{-1}$ Mpc boxes and then smoothed with a Gaussian having an rms of $0.5\,h^{-1}$ Mpc. The contours are for a model with $\beta = 0.4$ and $\sigma_p = 400\,\mathrm{km\,s^{-1}}$, and are plotted at $\xi = 10, 5, 2, 1, 0.5, 0.2$ and 0.1.

redshift-space distortion on large scales depends on the combination $\beta \equiv \Omega_m^{0.6}/b$. This is modified by the Finger-of-God effect, which is significant even at large scales and dominant at small scales. The effect can be modelled by introducing a parameter σ_p, which represents the rms pairwise velocity dispersion of the galaxies in collapsed structures, σ_p (see e.g. Ballinger et al. 1996). Considering both these effects, and marginalising over σ_p, the best estimate of β and its 68% confidence interval according to Hawkins et al. (2002) is

$$\beta = 0.49 \pm 0.09 \qquad (1)$$

The quoted error is slightly larger than in Peacock et al. (2001): mainly, this reflects the decision of Hawkins et al. to concentrate on the better sampled volume at $z < 0.2$, although a more detailed comparison with mock data also indicates that the previous errors were too small by a factor of about 1.2.

Our measurement of $\Omega^{0.6}/b$ can only be used to determine Ω if the bias is known. We discuss below two methods by which the bias parameter may be inferred, which in fact favour a low degree of bias. Nevertheless, there are other uncertainties in converting a measurement of β to a figure for Ω. The 2dFGRS has a median redshift of 0.11; with weighting, the mean redshift in Hawkins et al. is 0.15, and our measurement should be interpreted as β at that epoch. The optimal weighting also means that our mean luminosity is high: it is approximately 1.4 times the characteristic luminosity, L^*, of the overall galaxy population (Folkes et al. 1999; Madgwick et al. 2002). This means that we need to quantify the luminosity dependence of clustering.

5. REAL-SPACE CLUSTERING AND ITS DEPENDENCE ON LUMINOSITY

The dependence of galaxy clustering on luminosity is an effect that was controversial for a number of years. Using the APM-Stromlo redshift survey, Loveday et al. (1995) claimed that there was no trend of clustering amplitude with luminosity, except possibly at the very lowest luminosities. In contradiction, the SSRS study of Benoist et al. (1996) suggested that the strength of galaxy clustering increased monotonically with luminosity, with a particularly marked effect for galaxies above L^*. The latter result was arguably more plausible, based on what we know of luminosity functions and morphological segregation. It has been clear for many years that elliptical galaxies display a higher correlation amplitude than spirals (Davis & Geller 1976). Since ellipticals are also more luminous on average, as shown above, some trend with luminosity is to be expected, but the challenge is to detect it.

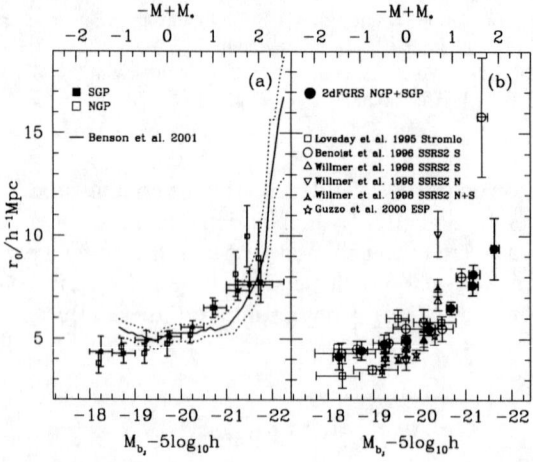

FIGURE 6. (a) The correlation length in real space as a function of absolute magnitude. The solid line shows the predictions of the semi-analytic model of Benson et al. (2001), computed in a series of overlapping bins, each 0.5 magnitudes wide. The dotted curves show an estimate of the errors on this prediction, including the relevant sample variance for the survey volume. (b) The real space correlation length estimated combining the NGP and SGP (filled circles). The open symbols show a selection of recent data from other studies.

The difficulty with measuring the dependence of $\xi(r)$ on luminosity is that cosmic variance can mask the signal of interest. It is therefore important to analyse volume-limited samples in which galaxies of different luminosities are compared in the same volume of space. This comparison was undertaken by Norberg et al. (2001), who measured real-space correlation functions via the projection $\Xi(\sigma) = \int \xi(\sigma,\pi)\,d\pi$, demonstrating that it was possible to obtain consistent results in both NGP and SGP. A very clear detection of luminosity-dependent clustering was achieved, as shown in Figure 6. The results can be described by a linear dependence of effective bias parameter on luminosity:

$$b/b^* = 0.85 + 0.15\,(L/L^*), \qquad (2)$$

and the scale-length of the real-space correlation function for L^* galaxies is approximately $r_0 = 4.8\,h^{-1}$ Mpc. This trend is in qualitative agreement with the results of Benoist et al. (1996), but in fact these workers gave a stronger dependence on luminosity than is indicated by the 2dFGRS. Finally, with spectral classifications, it is possible to measure the dependence of clustering both on luminosity and on spectral type, to see to what extent morphological segregation is responsible for this result. Norberg et al. (2002) show that, in fact, the principal effect seems to be with luminosity: $\xi(r)$ increases with L for all spectral types. This is reasonable from a theoretical point of view, in which the principal cause of different clustering amplitudes is the mass of halo that hosts a galaxy (e.g. Cole & Kaiser 1989; Mo & White 1996; Kauffman, Nusser & Steinmetz 1997).

Finally, these results would lead us to infer that the LF must change in strongly clumped regions, shifting to higher luminosities. Such an effect has been sought for many years, but always yielded null results. However, De Propris et al. (2002) have shown that L^* in rich clusters does obey a shift with respect to the global value, being brighter by 0.28 ± 0.08 mag.

6. THE 2DF GRS POWER SPECTRUM

Perhaps the key aim of the 2dFGRS was to perform an accurate measurement of the 3D clustering power spectrum, in order to improve on the APM result, which was deduced by deprojection of angular clustering (Baugh & Efstathiou 1993, 1994). The results of this direct estimation of the 3D power spectrum are shown in Figure 7. This power-spectrum estimate uses the FFT-based approach of Feldman, Kaiser & Peacock (1994), and needs to be interpreted with care. Firstly, it is a raw redshift-space estimate, so that the power beyond $k \simeq 0.2\,h\,\text{Mpc}^{-1}$ is severely damped by fingers of God. On large scales, the power is enhanced, both by the Kaiser effect and by the luminosity-dependent clustering discussed above. Finally, the FKP estimator yields the true power convolved with the window function. This modifies the power significantly on large scales (roughly a 20% correction). We have made an approximate correction for this in Figure 7. The precision of the power measurement appears to be encouragingly high, and the next task is to perform a detailed fit of physical power spectra, taking full account of the window effects. We summarize here results from this analysis (Percival et al. 2001).

The fundamental assumption is that, on large scales, linear biasing applies, so that the nonlinear galaxy power spectrum in redshift space has a shape identical to that ow linear theory in real space. We believe that this assumption is valid for $k < 0.15\,h\,\text{Mpc}^{-1}$; the detailed justification comes from analyzing realistic mock data derived from N-body simulations (Cole et al 1998). The model free parameters are thus the primordial spectral index, n, the Hubble parameter, h, the total matter density, Ω_m, and the baryon fraction, Ω_b/Ω_m. Note that the vacuum energy does not enter. Initially, we show results assuming $n = 1$; this assumption is relaxed later.

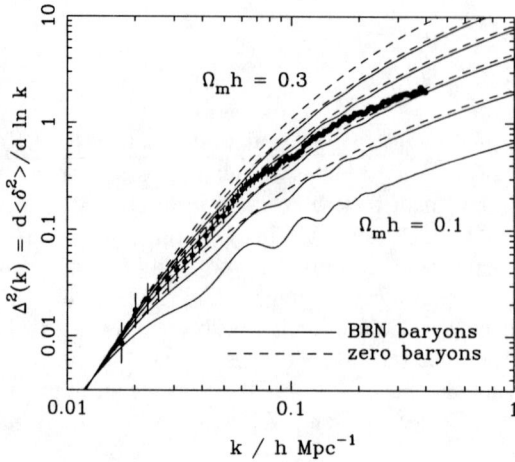

FIGURE 7. The 2dFGRS redshift-space dimensionless power spectrum, $\Delta^2(k)$, estimated according to the FKP procedure. The solid points with error bars show the power estimate. The window function correlates the results at different k values, and also distorts the large-scale shape of the power spectrum An approximate correction for the latter effect has been applied. The solid and dashed lines show various CDM models, all assuming $n = 1$. For the case with non-negligible baryon content, a big-bang nucleosynthesis value of $\Omega_b h^2 = 0.02$ is assumed, together with $h = 0.7$. A good fit is clearly obtained for $\Omega_m h \simeq 0.2$. Note that the observed power at large k will be boosted by nonlinear effects, but damped by small-scale random peculiar velocities. It appears that these two effects very nearly cancel, but model fitting is generally performed only at $k < 0.15\,h\,\mathrm{Mpc}^{-1}$ in order to avoid these complications.

In order to compare the 2dFGRS power spectrum to members of the CDM family of theoretical models, it is essential to have a proper understanding of the full covariance matrix of the data: the convolving effect of the window function causes the power at adjacent k values to be correlated. This covariance matrix was estimated by applying the survey window to a library of Gaussian realisations of linear density fields. Similar results were obtained using a covariance matrix estimated from a set of mock catalogues. It is now possible to explore the space of CDM models, and likelihood contours in Ω_b/Ω_m versus $\Omega_m h$ are shown in Figure 8. At each point in this surface we have marginalized by integrating the likelihood surface over the two free parameters, h and the power spectrum amplitude. Assuming a uniform prior for h over a factor of 2 is arguably over-cautious, and we have therefore added a Gaussian prior $h = 0.7 \pm 10\%$. This corresponds to multiplying by the likelihood from external constraints such as the HST key project (Freedman et al. 2001); this has only a minor effect on the results.

Figure 8 shows that there is a degeneracy between $\Omega_m h$ and the baryonic fraction Ω_b/Ω_m. However, there are two local maxima in the likelihood, one with $\Omega_m h \simeq 0.2$ and $\sim 20\%$ baryons, plus a secondary solution $\Omega_m h \simeq 0.6$ and $\sim 40\%$ baryons. The high-density model can be rejected through a variety of arguments, and the preferred solution is

$$\Omega_m h = 0.20 \pm 0.03; \qquad \Omega_b/\Omega_m = 0.15 \pm 0.07. \tag{3}$$

The 2dFGRS data are compared to the best-fit linear power spectra convolved with

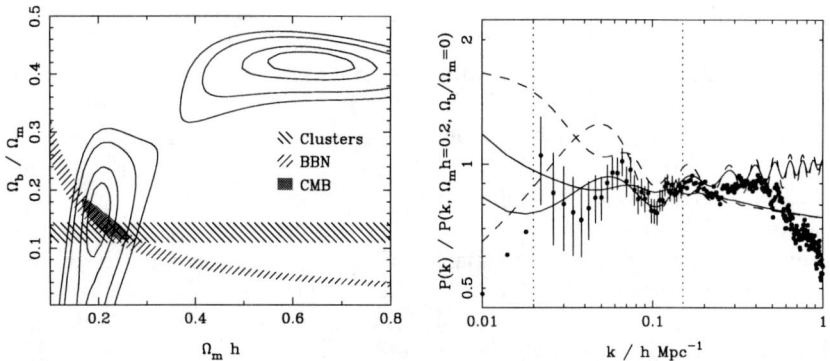

FIGURE 8. Likelihood contours for the best-fit linear power spectrum over the region $0.02 < k < 0.15$. The normalization is a free parameter to account for the unknown large scale biasing. Contours are plotted at the usual positions for one-parameter confidence of 68%, and two-parameter confidence of 68%, 95% and 99% (i.e. $-2\ln(\mathscr{L}/\mathscr{L}_{\max}) = 1, 2.3, 6.0, 9.2$). We have marginalized over the missing free parameters (h and the power spectrum amplitude). A prior on h of $h = 0.7 \pm 10\%$ was assumed. This result is compared to estimates from X-ray cluster analysis (Evrard 1997), big-bang nucleosynthesis (Burles et al. 2001) and CMB results (Netterfield et al. 2001; Pryke et al. 2002). The CMB results assume that $\Omega_b h^2$ and $\Omega_{cdm} h^2$ were independently determined from the data. The second panel shows the 2dFGRS data compared with the two preferred models from the Maximum Likelihood fits convolved with the window function (solid lines). Error bars show the diagonal elements of the covariance matrix, for the fitted data that lie between the dotted vertical lines. The unconvolved models are also shown (dashed lines). The $\Omega_m h \simeq 0.6$, $\Omega_b/\Omega_m = 0.42$, $h = 0.7$ model has the higher bump at $k \simeq 0.05 h\,\mathrm{Mpc}^{-1}$. The smoother $\Omega_m h \simeq 0.20$, $\Omega_b/\Omega_m = 0.15$, $h = 0.7$ model is a better fit to the data because of the overall shape. A preliminary analysis of the complete final 2dFGRS sample yields a slightly smoother spectrum than the results shown here (from Percival et al. 2001), so that the high-baryon solution becomes disfavoured.

the window function in Figure 8. This shows where the two branches of solutions come from: the low-density model fits the overall shape of the spectrum with relatively small 'wiggles', while the solution at $\Omega_m h \simeq 0.6$ provides a better fit to the bump at $k \simeq 0.065 h\,\mathrm{Mpc}^{-1}$, but fits the overall shape less well. A preliminary analysis of $P(k)$ from the full final dataset shows that $P(k)$ becomes smoother: the high-baryon solution becomes disfavoured, and the uncertainties narrow slightly around the lower-density solution: $\Omega_m h = 0.18 \pm 0.02$; $\Omega_b/\Omega_m = 0.17 \pm 0.06$.

It is interesting to compare these conclusions with other constraints. These are shown on Figure 8, assuming $h = 0.7 \pm 10\%$. Latest estimates of the Deuterium to Hydrogen ratio in QSO spectra combined with big-bang nucleosynthesis theory predict $\Omega_b h^2 = 0.020 \pm 0.001$ (Burles et al. 2001), which translates to the shown locus of f_B vs $\Omega_m h$. X-ray cluster analysis predicts a baryon fraction $\Omega_b/\Omega_m = 0.127 \pm 0.017$ (Evrard 1997) which is within 1σ of our value. These loci intersect very close to our preferred model. Moreover, these results are in good agreement with independent estimates of the total density and baryon content from data on CMB anisotropies (e.g. Netterfield et al. 2001; Pryke et al. 2002).

Perhaps the main point to emphasise here is that the 2dFGRS results are not greatly sensitive to the assumed tilt of the primordial spectrum. We have used CMB results to motivate the choice of $n = 1$, as discussed below, but it is clear that very substantial tilts are required to alter the conclusions significantly: $n \simeq 0.8$ would be required to turn zero baryons into the preferred model.

The main residual worry about accepting the above conclusions is probably whether the assumption of linear bias can really be valid. In general, concentration towards higher-density regions both raises the amplitude of clustering, but also steepens the correlations, so that bias is largest on small scales. One way in which this issue can be studied is to use the split by colour introduced above. Figure 9 shows the power spectra for the 2dFGRS divided in this way. The shapes are almost identical (perhaps not so surprising, since the cosmic variance effects are closely correlated in these co-spatial samples). However, what is impressive is that the relative bias is almost precisely independent of scale, even though the passive subset is rather strongly biased relative to the active subset (relative $b \simeq 1.4$). This provides some reassurance that the large-scale $P(k)$ reflects the underlying properties of the dark matter, rather than depending on the particular class of galaxies used to measure it.

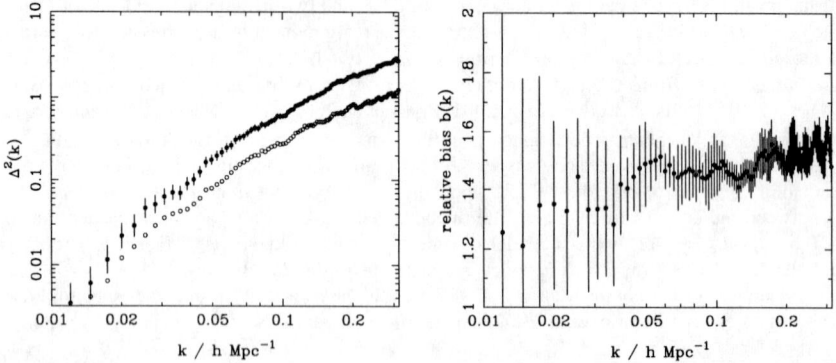

FIGURE 9. The power spectra of passive galaxies (filled circles) and active galaxies (open circles). The shapes are strikingly similar. The square root of the ratio yields the right-hand panel: the relative bias in redshift space of passive and active galaxies. The error bars are obtained by a jack-knife analysis. The relative bias is consistent with a constant value of 1.4 over the range used for fitting of the power-spectrum data ($0.015 < k < 0.15 h \, \mathrm{Mpc}^{-1}$).

7. COMBINATION WITH THE CMB AND COSMOLOGICAL PARAMETERS

The 2dFGRS power spectrum contains important information about the key parameters of the cosmological model, but we have seen that additional assumptions are needed, in particular the values of n and h. Observations of CMB anisotropies can in principle measure most of the cosmological parameters, and combination with the 2dFGRS can lift most of the degeneracies inherent in the CMB-only analysis. It is therefore of interest to see what emerges from a joint analysis.

These issues are discussed in Efstathiou et al. (2002). The CMB data alone contain two important degeneracies: the 'geometrical' and 'tensor' degeneracies. In the former case, one can evade the commonly-stated CMB conclusion that the universe is flat, by adjusting both Λ and h to extreme values. In the latter case, a model with a large tensor component can be made to resemble a zero-tensor model with large blue tilt ($n > 1$) and high baryon content. Efstathiou et al. (2002) show that adding the 2dFGRS data removes the first degeneracy, but not the second. This is reasonable: if we take the view that the CMB determines the physical density $\Omega_m h^2$, then a measurement of $\Omega_m h$ from 2dFGRS gives both Ω_m and h separately in principle, removing one of the degrees of freedom on which the geometrical degeneracy depends. On the other hand, the 2dFGRS alone constrains the baryon content weakly, so this does not remove the scope for the tensor degeneracy.

On the basis of this analysis, we can therefore be confident that the universe is very nearly flat ($|\Omega - 1| < 0.05$), so it is defensible to assume hereafter that this is exactly true. The importance of tensors will of course be one of the key questions for cosmology over the next several years, but it is interesting to consider the limit in which these are negligible. In this case, the standard model for structure formation contains a vector of only 6 parameters:

$$\mathbf{p} = (n_s, \Omega_m, \Omega_b, h, Q, \tau). \tag{4}$$

Of these, the optical depth to last scattering, τ, is almost entirely degenerate with the normalization, Q – and indeed with the bias parameter; we discuss this below. The remaining four parameters are pinned down very precisely: using the latest CMB data plus the 2dFRGS power spectrum, we obtain

$$(n_s, \Omega_c h^2, \Omega_b h^2, h) = (0.963 \pm 0.042, 0.115 \pm 0.009, 0.021 \pm 0.002, 0.665 \pm 0.047), \tag{5}$$

or an overall density parameter of $\Omega_m = 0.31 \pm 0.05$.

It is remarkable how well these figures agree with completely independent determinations: $h = 0.72 \pm 0.08$ from the HST key project (Mould et al. 2000; Freedman et al. 2001); $\Omega_b h^2 = 0.020 \pm 0.001$ (Burles et al. 2001). This gives confidence that the tensor component must indeed be sub-dominant. For further details of this analysis, see Percival et al. (2002).

8. MATTER FLUCTUATION AMPLITUDE AND BIAS

The above conclusions were obtained by considering the shapes of the CMB and galaxy power spectra. However, it is also of great interest to consider the amplitude of mass fluctuations, since a comparison with the galaxy power spectrum allows us to infer the degree of bias directly. This analysis was performed by Lahav et al. (2002). Given assumed values for the cosmological parameters, the present-day linear normalization of the mass spectrum (e.g. σ_8) can be inferred. It is convenient to define a corresponding measure for the galaxies, σ_{8g}, such that we can express the bias parameter as

$$b = \frac{\sigma_{8g}}{\sigma_{8m}}. \tag{6}$$

In practice, we define σ_{8g} to be the value required to fit a CDM model to the power-spectrum data on linear scales ($0.02 < k < 0.15 h\,\mathrm{Mpc}^{-1}$). A final necessary complication of the notation is that we need to distinguish between the apparent values of σ_{8g} as measured in redshift space (σ_{8g}^S) and the real-space value that would be measured in the absence of redshift-space distortions (σ_{8g}^R). It is the latter value that is required in order to estimate the bias.

A model grid covering the range $0.1 < \Omega_m h < 0.3$, $0.0 < \Omega_b/\Omega_m < 0.4$, $0.4 < h < 0.9$ and $0.75 < \sigma_{8g}^S < 1.14$ was considered. The primordial index was assumed to be $n = 1$ initially, and the dependence on n studied separately. For fixed 'concordance model' parameters $n = 1$, $k = 0$, $\Omega_m = 0.3$, $\Omega_b h^2 = 0.02$ and a Hubble constant $h = 0.70$, we find that the amplitude of 2dFGRS galaxies in redshift space is $\sigma_{8g}^S(L_s, z_s) = 0.94$. Correcting for redshift-space distortions as detailed above reduces this to 0.86 in real space. Applying a correction for a mean luminosity of $1.9L^*$ using the recipe of Norberg et al. (2001), we obtain an estimate of $\sigma_{8g}^R(L^*, z_s) = 0.76$, with a negligibly small random error. In order to obtain present-day bias figures, we need to know the evolution of galaxy clustering to $z = 0$. Existing data on clustering evolution reveals very slow changes: higher bias at early times largely cancels the evolution of the dark matter. We therefore assume no evolution in σ_{8g}.

The value of σ_8 for the dark matter can be deduced from the CMB fits:

$$\sigma_8 = (0.72 \pm 0.04) \exp \tau, \tag{7}$$

where the error bar includes both data errors and theory uncertainty. The unsatisfactory feature is the degeneracy with the optical depth to last scattering. For reionization at redshift 8, we would have $\tau \simeq 0.05$; it is unlikely that τ can be hugely larger (e.g. Loeb & Barkana 2001). Although direct removal of this theoretical prejudice is desirable (and will be possible with future CMB data), it seems reasonable to assume that the true value of σ_8 must be very close to 0.76. Within the errors, this agrees perfectly with our $\sigma_{8g}^R(L^*, 0) = 0.76$, implying that L^* galaxies are very nearly exactly unbiased. As we have seen, there are large variations in the clustering amplitude with type, so that this outcome must be something of a coincidence.

Finally, this conclusion of near-unity bias was reinforced in a completely independent way, by using the measurements of the bispectrum of galaxies in the 2dFGRS (Verde et al. 2002). As it is based on three-point correlations, this statistic is sensitive to the filamentary nature of the galaxy distribution – which is a signature of nonlinear evolution. One can therefore split the degeneracy between the amplitude of dark-matter fluctuations and the amount of bias. At the effective redshift and luminosity of their sample ($z_s = 0.17$ and $L = 1.9L^*$), Verde et al. found $b = 1.04 \pm 0.11$. Although the corrections to zero redshift and to luminosity L^* are probably significant, this reinforces the point that on large scales there is no substantial difference in clustering between typical galaxies and the dark matter (small scales, of course, are another matter entirely).

9. CONCLUSIONS

The 2dFGRS is the first 3D survey of the local universe to achieve 100,000 redshifts, almost an order of magnitude improvement on previous work. The final database should yield definitive results on a number of key issues relating to galaxy clustering. For details of the current status of the 2dFGRS, see http://www.mso.anu.edu.au/2dFGRS. In particular, this site gives details of the 2dFGRS public release policy, in which approximately the first half of the survey data were made available in June 2001, with the complete survey database to be made public by mid-2003. Some key results of the survey to date may be summarized as follows:

(1) The galaxy luminosity function has been measured precisely as a function of spectral type (Folkes et al. 1999; Madgwick et al. 2002).
(2) The amplitude of galaxy clustering has been shown to depend on luminosity (Norberg et al. 2001). The relative bias is $b/b^* = 0.85 + 0.15(L/L^*)$.
(3) The redshift-space correlation function has been measured out to $30\,h^{-1}$ Mpc. Redshift-space distortions imply $\beta \equiv \Omega_m^{0.6}/b = 0.49 \pm 0.09$, for galaxies with $L \simeq 1.4L^*$.
(4) The galaxy power spectrum has been measured to high accuracy (10–15% rms) over about a decade in scale at $k < 0.15\,h\,\text{Mpc}^{-1}$. The results are very well matched by an $n = 1$ CDM model with $\Omega_m h = 0.18$ and 16% baryons.
(5) Combining the power spectrum results with current CMB data, very tight constraints are obtained on cosmological parameters. For a scalar-dominated flat model, we obtain $\Omega_m = 0.31 \pm 0.05$, and $h = 0.68 \pm 0.04$, independent of external data.
(6) Results from the CMB comparison imply a large-scale bias parameter consistent with unity. This conclusion is also reached in a completely independent way via the bispectrum analysis of Verde et al. (2002).

Overall, these results provide precise support for a cosmological model that is flat, with $(\Omega_b, \Omega_c, \Omega_v) \simeq (0.04, 0.25, 0.71)$, to a tolerance of 10% in each figure. Although the ΛCDM model has been claimed to have problems in matching galaxy-scale observations, it clearly works extremely well on large scales, and any proposed replacement for CDM will have to maintain this agreement. So far, there has been no need to invoke either tilt of the scalar spectrum, or a tensor component in the CMB. If this situation is to change, the most likely route will be via new CMB data, combined with the key complementary information that the large-scale structure in the 2dFGRS can provide.

ACKNOWLEDGEMENTS

The 2dF Galaxy Redshift Survey was made possible by the dedicated efforts of the staff of the Anglo-Australian Observatory, both in creating the 2dF instrument, and in supporting it on the telescope.

REFERENCES

Ballinger W.E., Peacock J.A., Heavens A.F., 1996, MNRAS, 282, 877
Baugh C.M., Efstathiou G., 1993, MNRAS, 265, 145
Baugh C.M., Efstathiou G., 1994, MNRAS, 267, 323

Benoist C., Maurogordato S., da Costa L.N., Cappi A., Schaeffer R., 1996, ApJ, 472, 452
Benson, A.J., Frenk, C.S., Baugh, C.M., Cole, S., Lacey, C.G., 2001, MNRAS, 327, 1041
Burles S., Nollett K.M., Turner M.S., 2001, ApJ, 552, L1
Cole S., Kaiser N., 1989, MNRAS, 237, 1127
Cole S., Hatton S., Weinberg D.H., Frenk C.S., 1998, MNRAS, 300, 945
Colless M. et al., 2001, MNRAS, 328, 1039
Davis M., Geller M.J., 1976, ApJ, 208, 13
Davis M., Peebles, P.J.E., 1983, ApJ, 267, 465
De Propris R. et al., 2002, astro-ph/0212562
Efstathiou G. et al., 2002, MNRAS, 330, L29
Evrard A., 1997, MNRAS, 292, 289
Folkes S.J. et al., 1999, MNRAS, 308, 459
Feldman H.A., Kaiser N., Peacock J.A., 1994, ApJ, 426, 23
Freedman W.L. et al., 2001, ApJ, 553, 47
Hambly N.C., Irwin M.J., MacGillivray H.T., 2001, MNRAS 326 1295
Hawkins E.J. et al., 2002, astro-ph/0212375
Kaiser N., 1987, MNRAS, 227, 1
Kauffmann G., Nusser A., Steinmetz M., 1997, MNRAS, 286, 795
Lahav O. et al., 2002, MNRAS, 333, 961
Lewis I.J. et al., 2002, MNRAS, 333, 279
Loeb A., Barkana R., 2001, ARAA, 39, 19
Loveday J., Maddox S.J., Efstathiou G., Peterson B.A., 1995, ApJ, 442, 457
Maddox S.J., Efstathiou G., Sutherland W.J., Loveday J., 1990a, MNRAS, 242, 43P
Maddox S.J., Sutherland W.J., Efstathiou G., Loveday J., 1990b, MNRAS, 243, 692
Maddox S.J., Efstathiou G., Sutherland W.J., 1990c, MNRAS, 246, 433
Madgwick D. et al., 2002, 333, 133
Mo H.J., White S.D.M., 1996, MNRAS, 282, 347
Mould J.R. et al., 2000, ApJ, 529, 786
Netterfield C.B. et al., 2001, astro-ph/0104460
Norberg P. et al., 2001, MNRAS, 328, 64
Norberg P. et al., 2002, MNRAS, 332, 827
Peacock J.A. et al., 2001, Nature, 410, 169
Percival W.J. et al., 2001, MNRAS, 327, 1297
Percival W.J. et al., 2002, MNRAS, 337, 1068
Pryke C. et al., 2002, ApJ, 568, 46
Schlegel D.J., Finkbeiner D.P., Davis M., 1998, ApJ, 500, 525
Stoughton C.L. et al., 2002, AJ, 123, 485
Verde L. et al., 2002, MNRAS, 335, 432

Testing Cosmological Models with Clusters of Galaxies

Hans Böhringer* and Peter Schuecker*

Max-Planck-Institut für extraterestrische Physik, D85748 Garching, Germany

Abstract. Galaxy clusters are ideal probes for the large-scale structure of the Universe and for the tests of cosmological models. We use, REFLEX, the currently largest and best defined cluster X-ray survey to illustrate this application of galaxy cluster studies. Based on this survey of X-ray selected clusters of galaxies we determine statistical properties of the galaxy cluster population, their spatial correlation, and the density fluctuation power spectrum of the cosmic matter distribution on large scales up to about 1 Gpc. Comparing these results with predictions of cosmological models we obtain tight constrains for the matter density parameter of the Universe, consistent with the combined results from observations of the microwave background anisotropies and distant type Ia supernovae. The only difference between the present results and the "concordance model" is a low value for the σ_8-normalization. Exploring the parameter space of the cosmic matter density and the equation of state parameter of dark energy most favoured by the combined observations of REFLEX clusters and distant type Ia supernovae we find that the conventional cosmological constant model is best consistent with the observational data.

INTRODUCTION

Galaxy clusters with masses from about 10^{14} to over 10^{15} $M_\odot$ are the largest clearly defined building blocks of our Universe. Their formation and evolution is tightly connected to the evolution of the cosmic large-scale structure. Clusters are therefore ideal probes for the study of the large-scale matter distribution.

The hot, intracluster plasma with temperatures of several 10 Million degrees makes galaxy clusters the most luminous X-ray emitters next only to quasars. The hot gas and its X-ray emission allows us to obtain mass estimates of the clusters and to detect clusters as gravitationally bound entities out to very large distances [18]. Systematic studies show that clusters have within a first order description, a quite standardized, self-similar appearance. This is reflected in the tight correlation of X-ray luminosity and cluster mass enabling us to construct interesting cosmological samples of clusters above a certain mass limit based on the detection of their X-ray luminosities. Fig. 1 shows the X-ray luminosity-mass relation constructed from the observation of the X-ray brightest clusters [16] found in the ROSAT All-Sky Survey (RASS, [25]).

We have used these brightest RASS clusters to construct the first empirical mass function of clusters based on X-ray data. Fig. 2 shows the cumulative mass density of the clusters as a function of the lower mass limit in units of the critical density of the Universe. Only about 2% of the critical density and about 6% of the matter density (for $\Omega_m \sim 0.3$) of the Universe is found in galaxy clusters and groups of galaxies with a mass above about $3 \cdot 10^{13} h_{50}^{-1}$ $M_\odot$ [16] . Nevertheless this small mass fraction provides tight constraints on the large-scale structure parameters of the matter distribution in the Universe.

THE REFLEX SURVEY AND THE CLUSTER X-RAY LUMINOSITY FUNCTION

For the present illustration of the use of galaxy clusters as cosmological probes we use the REFLEX cluster survey, the currently largest, most homogeneous, and best documented X-ray cluster survey. The construction of the REFLEX cluster sample is described in detail by Böhringer et al. [5]. The survey area covers the southern sky up to declination $\delta = +2.5^o$, avoiding the band of the Milky Way ($|b_{II}| \leq 20^o$) and the regions of the Magellanic clouds. The total survey area is 13924 deg^2 or 4.24 sr. The X-ray detection of the clusters is based on the second processing of the RASS [23] and a reanalysis of the X-ray properties of the sources by means of the growth curve analysis (GCA) method [4]. The current sample has a a flux-limit of $F_x \geq 3 \cdot 10^{-12}$ erg s^{-1} cm^{-2} and comprises 452 objects (3 without certain identification and redshifts). An extension of the sample termed REFLEX II to a lower flux limit is almost completed. The cluster candidates for this study were found using a machine based correlation of the X-ray sources with galaxy density enhancements in the COSMOS optical data base. The candidate list was carefully screened based on X-ray and optical information, literature data, and results from the optical follow-up observation program (ESO key program) which also provided the missing redshifts. The final cluster catalogue is estimated to be

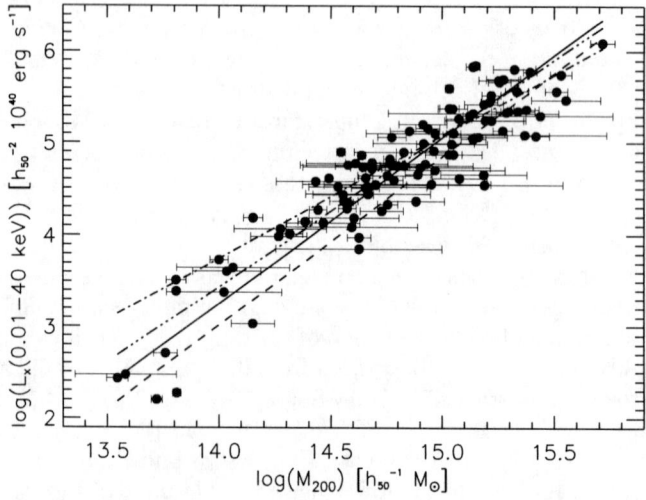

FIGURE 1. Mass-X-ray luminosity relation determined from the 106 brightest galaxy clusters found in the ROSAT All-Sky Survey by Reiprich & Böhringer (2002)

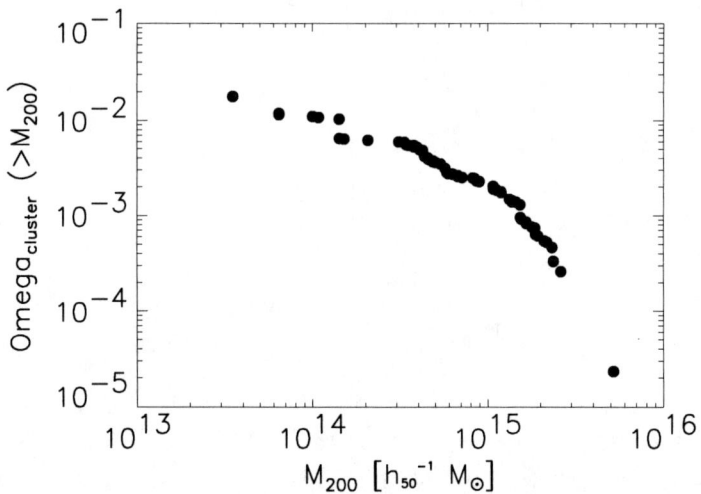

FIGURE 2. Mass density of galaxy clusters in the Universe as a function of the lower mass cut in units of the critical density of the Universe (Reiprich & Böhringer 2002).

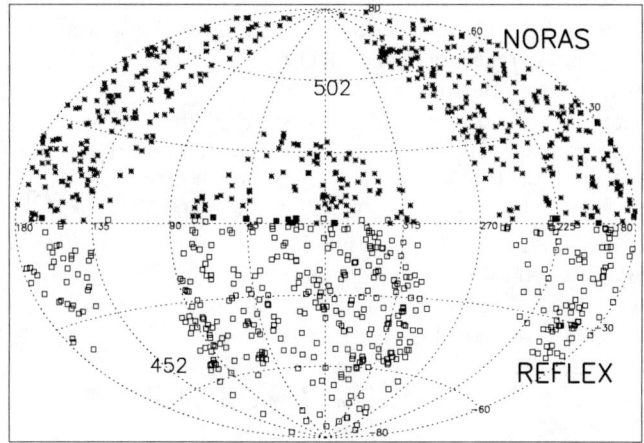

FIGURE 3. Sky distribution of the brightest galaxy clusters found in the ROSAT All-Sky Survey investigated in the NORAS (Böhringer et al. 2000) and REFLEX (Böhringer et al. 2001) cluster redshift surveys.

better than 90% complete with a contamination from X-ray luminous AGN less than 9%. This high completeness of the cluster identification of the RASS sources ensures that the selection effects introduced by the optical identification process is minimized

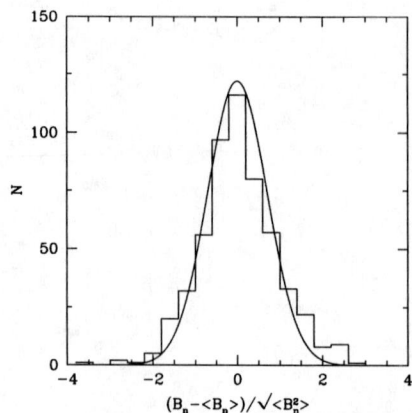

FIGURE 4. Histogram of the normalized KL eigenvalues used to analyse the large-scale structure statistics as described in section III (from Schuecker et al. 2002a); a fitted Gaussian function is superposed on the data. This distribution is effectively a cluster count in cells distribution showing that the large-scale structure is approximately Guassian as far as can be measured with the low number statistics of REFLEX.

and negligible for our purpose [5]. A series of tests demonstrate the high quality of the sample. Fig. 4 shows for example the Gaussianity of the cluster distribution of a study similar to counts in cells conducted in connection with the KL-structure analysis described below (see also Schuecker et al. [20]).

The X-ray luminosity function of the sample was determined by taking the selection function of the survey into account. The selection function describes the flux limit of the survey as a function of the sky position and is determined from the local survey exposure time and the interstellar absorbing column density of hydrogen in the line of sight. The selection function is almost uniform over about 87% of the sky. A binned representation of the luminosity function (with 20 objects per bin) is shown in Fig. 5. Due to the large sample size the uncertainty of the amplitude of the luminosity function around $L^\star$ is characterized by a pure statistical uncertainty of only about 5%. A Schechter function can be quite well fitted to the data as shown in the Figure. The best fitting parameters obtained with an ML method including the uncertainties in the flux measurement are: $L^\star = 6.5 \ 10^{44} (\pm 0.6) h_{50}^{-2}$ erg s^{-1}, $\alpha = 1.69 \ (\pm 0.045)$, $n_0 = 1.07 \ (\pm 0.4) \ 10^{-7} h_{50}^5 (10^{44}$ erg s$^{-1})^{-1}$ Mpc^3. Note that the amplitude error includes here the fitting with all other parameters left free, therefore this error is larger than the above quoted 5%.

MODELING THE REFLEX LUMINOSITY FUNCTION IN THE FRAME OF COSMOLOGICAL SCENARIOS

Due to the close correlation of the X-ray luminosity with cluster mass [16] the luminosity function provides a very good census of the galaxy cluster population. It can in particular

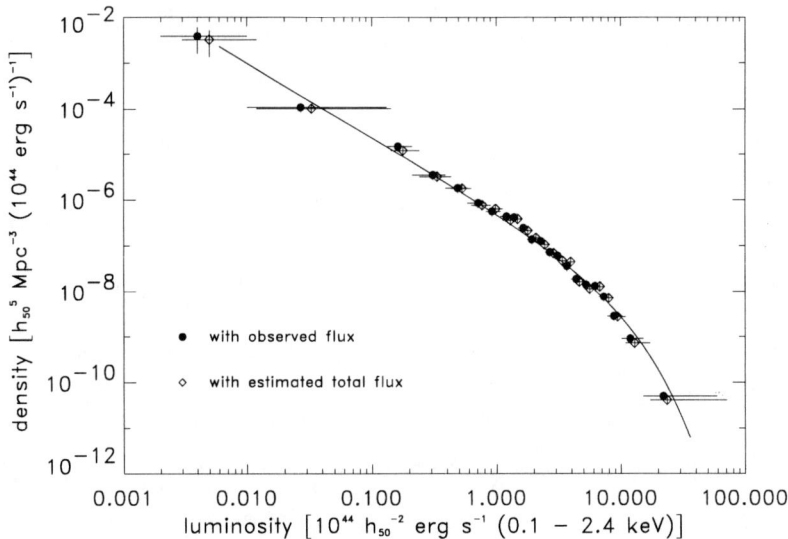

FIGURE 5. X-ray luminosity function of the REFLEX cluster Survey. The filled and open data point refer to observed and corrected total luminosities, respectively. A Schechter function (solid line) is fitted to the data by an ML method including the flux uncertainties (Böhringer et al. 2002)

be used to obtain constraints on the amplitude of the primordial density fluctuations of the matter distribution in the Universe (e.g. [24, 1]). We are using analytical methods and the results of large N-body simulations to model the X-ray luminosity function. Comprehensive modeling with a maximum likelihood analysis is still ongoing and we are still testing the final results for the best fitting models. Therefore we discuss the implications of our results on a more qualitative basis.

The modeling of the X-ray luminosity function in the frame of cosmic structure formation models involves six major steps: (i) we adopt a ΛCDM cosmological model with the following parameters fixed, Hubble parameter, $H_0 = 70$ km s^{-1} Mpc^{-1}, baryon fraction, $\Omega_b h^2 = 0.02$, and index of the primordial power spectrum, $n = 1$, (ii) we use the transfer function by Eisenstein & Hu [9] to calculate the dark matter power spectrum after recombination, (iii) we normalize this power spectrum by the free parameter σ_8 and calculate from the power spectrum the variance $\sigma^2(M)$ of the density fluctuation field as a function of the applied "top-hat" filter radius and filter mass, (iv) we use the N-body simulation results and the analytical fitting formulae by Evrard et al. [10] to derive the cluster mass function (at an overdensity of 200 over the critical density) from $\sigma(M)$, (v) we use the mass X-ray luminosity relation of Reiprich & Böhringer [16] which is based on the same mass definition (for the same overdensity) to convert the mass function into the X-ray luminosity function, and (vi) we include the scatter of the mass luminosity relation.

In the following illustrations we use a reasonably good fitting model with $\Omega_m = 0.3$, $\Omega_\Lambda = 0.7$ and $\sigma_8 = 0.8$ as a reference to show the effect of varying different

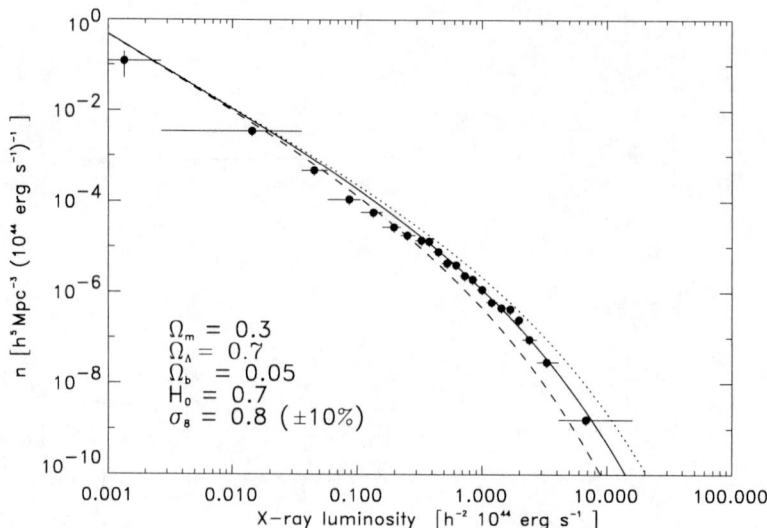

FIGURE 6. Effect of varying the normalization, σ_8, of the primordial pow+ spectrum on the predicted X-ray luminosity function. The reference model (solid line) is given in the text. For comparison we show models with $\sigma_8 = 0.7$ (dashed line) and 0.9 (dotted line)

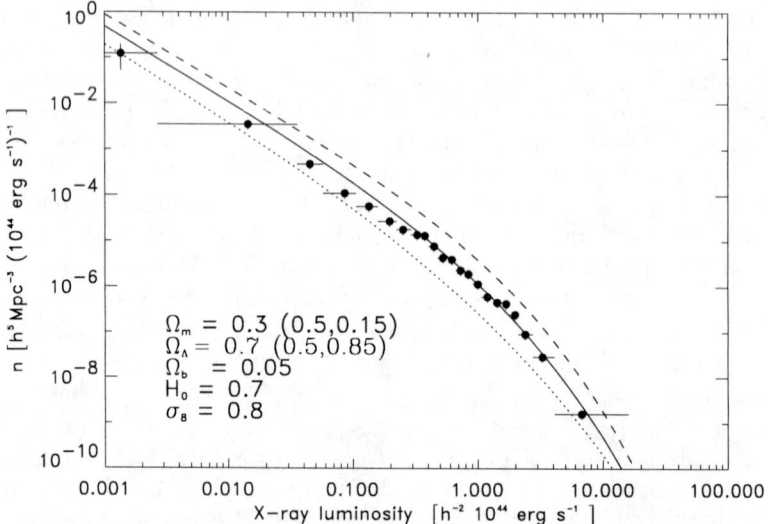

FIGURE 7. Effect of varying the parameter Ω_m. The reference model is the same as in Fig. 4. We also show models with $\Omega_m = 0.5$ (dashed line) and 0.15 (dotted line) while adjusting Ω_Λ to keep the geometry of the Universe flat.

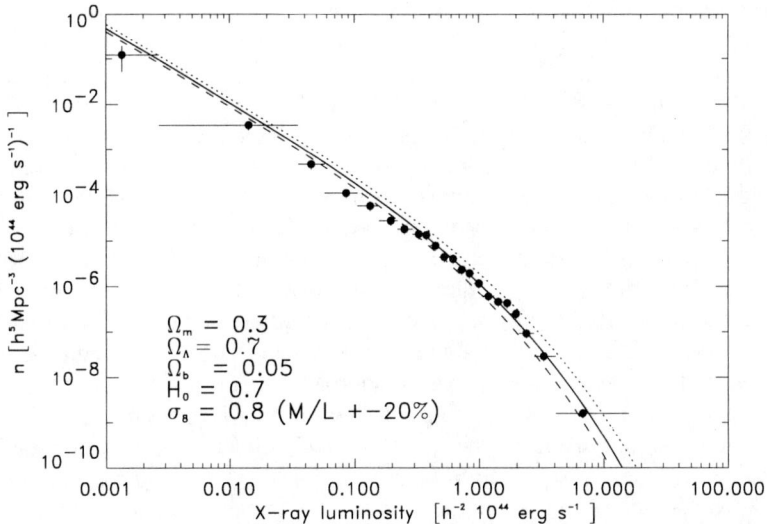

FIGURE 8. Effect of changing the normalization of the the X-ray luminosity mass relation by +20% (dashed line) and -20% (dotted line) on the model predicted X-ray luminosity function.

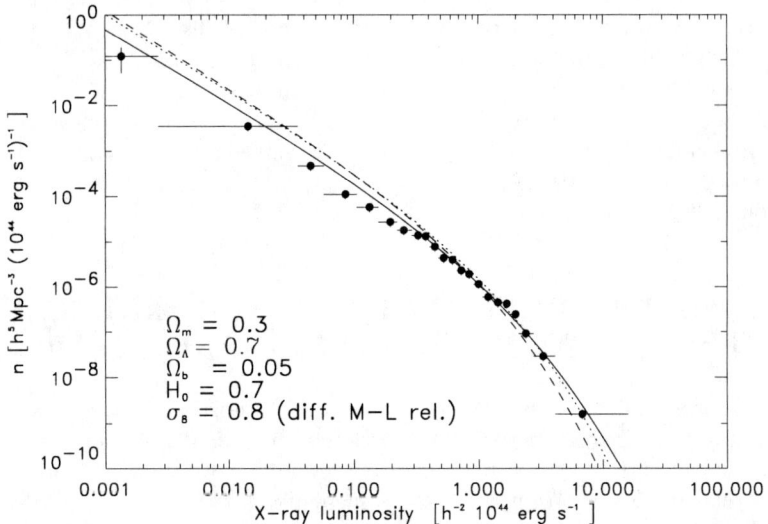

FIGURE 9. Effect of changing the slope of the the X-ray luminosity mass relation on the model predicted X-ray luminosity function. The (L|M) relation by Reiprich & Böhringer (2002) is used for the extended sample (solid line), the flux limited sample of 63 clusters (dashed line), and an even shallower relation (dotted line).

model parameters. Note that in this model we have not included the scatter in the mass luminosity relation (step vi above); its inclusion would push the best fitting model to a lower value of σ_8, closer to 0.7 in line with our findings from large scale structure [21] described below. Fig. 6 shows the effect of different normalizations of the input power spectrum. The data constrain this parameter if used as the only fitting parameter to a few percent. We note - as is well known - that the best constraint comes from the high luminosity end of the luminosity function. Fig. 7 shows the variation of the second most important parameter of the model, Ω_m. Even though the effect of varying σ_8 can partly be compensated by changing Ω_m, leading to the well known degeneracy in the determination of the two parameters for cluster abundance constraints, we can clearly see a different effect in Figs. 6 and 7. Thus with a good established shape of the luminosity function over a large enough luminosity scale as given here, this degeneracy can be broken. This was also shown e.g. by Pierpaoli et al.[15] and by Ikebe et al. [13] for the temperature function and by Reiprich & Böhringer [16] for the mass function.

One of the important results of the recent determinations of the cluster abundance is the low value obtained for σ_8 for fixed Ω_m (see e.g. [1, 15, 16]) compared to the cluster normalization adopted earlier and in the so-called concordance model (e.g. [2, 24]). One of the possible sources of this discrepancy is the difference between the theoretically predicted and observed mass luminosity and mass temperature relations (e.g. [11]) used. In Fig. 8 we test how the results for the X-ray luminosity function and the best fitting σ_8 and Ω_m depend on the precise knowledge of the mass luminosity relation. We see indeed that the variation is such that a lower normalization will imply a lower value of σ_8. However, the effect is not large enough to explain the discrepancy. Further fitting tests show that a change in the mass luminosity correlation normalization by about 35% is necessary to change σ_8 only by an amount of about 0.1. Much smaller uncertainties are estimated, however. Therefore with finding best fitting results for σ_8 of around 0.7 to 0.8 at maximum (for $\Omega_m = 0.3$) we can rule out values as large as 0.9. In Fig. 9 we further show that also a change of the slope of the mass luminosity relation has not a large enough effect.

MEASURING THE STATISTICS OF THE LARGE-SCALE STRUCTURE WITH THE REFLEX CLUSTER SAMPLE

The essential goal of this survey is the assessment of the statistics of the large-scale structure. The most fundamental statistical description of the spatial structure is based on the second moments on the distribution, characterized either by the two-point-correlation function or its Fourier transform, the density fluctuation power spectrum. The two-point correlation function of REFLEX shows a power law shaped correlation function with a slope of 1.83, a correlation length of $18.8 h_{100}^{-1}$ Mpc and a possible zero crossing at $\sim 45 h_{100}^{-1}$ Mpc [7]. The density fluctuation power spectrum (Fig. 10, [19]) is characterized by a power law at large values of the wave vector, k, with a slope of $\propto k^{-2}$ for $k \leq 0.1 h$ Mpc^{-1} and a maximum around $k \sim 0.03 h$ Mpc^{-1} (corresponding to a wavelength of about $200 h^{-1}$ Mpc). This maximum reflects the size of the horizon when the Universe featured equal energy density in radiation and matter and is a sensitive mea-

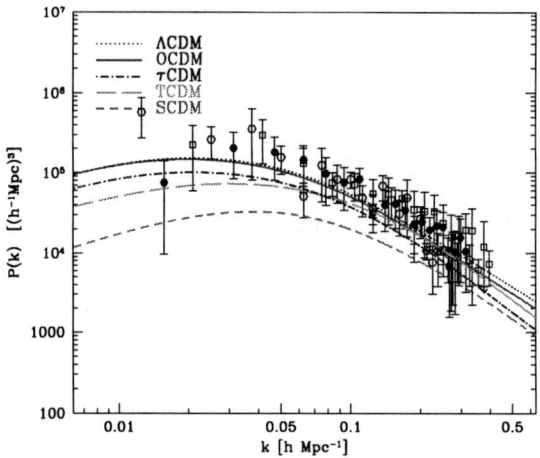

FIGURE 10. Power spectra of the density fluctuations in the REFLEX cluster sample together with predictions from various popular cosmological models taken from the literature. For details see Schuecker et al. (2001).

sure of the mean density of the Universe, Ω_0, providing approximately the following constraints $h\Omega_0 = 0.12$ to 0.26 [19].

COSMOLOGICAL TESTS WITH GALAXY CLUSTERS

For a more stringent cosmological test we are combining the results of the measurement of the cluster abundance and the statistics of the large-scale cluster distribution. One method that allows this combination in an ideal way is an analysis based on an eigenmode decomposition of the spatial structure of the REFLEX cluster distribution with the Karhunen-Loéve method in comparison with semi-analytic cosmological models of structure formation [20, 21]. With this method the density fluctuations of the distribution as well as the redshift dependent cluster number density has been taken into account simultaneously for the first time. This breaks the partly degenerate behavior of the fitting parameters, σ_8 and Ω_m, which is experienced if both analyses are conducted separately. The constraints obtained by this combined method applied by Schuecker et al. [20, 21] considering in a first step only random errors is shown in Fig.11. A careful analysis of the random and systematic errors leads to the final result: $\Omega_m = 0.341 \left(^{+0.031}_{-0.028}\right)_{\text{random}} \left(^{+0.087}_{-0.071}\right)_{\text{systematic}}$ and $\sigma_8 = 0.711 \left(^{+0.039}_{-0.031}\right)_{\text{random}} \left(^{+0.120}_{-0.162}\right)_{\text{systematic}}$. The systematic errors come mostly from the assumed uncertainty in our knowledge of the mass X-ray luminosity relation and its scatter similar to the problems discussed in section II.

These results are to be compared to the results obtained from observations of the cosmic microwave anisotropies (e.g. [8]) and of the study of distant SN Ia (e.g. [14]) providing combined constraints that encircle a region in the model parameter space spanned by the cosmological parameters Ω_0 and Ω_Λ around values of $\Omega_0 = 0.3$ and $\Omega_\Lambda = 0.7$. The galaxy cluster results provide a different cut through this parameter space crossing the other two results at their intersection. Thus, the evidence for a low density universe is solidifying.

We can carry this one step further and look for constraints of cosmological models including more complex forms of dark energy. Fig. 12 shows for example the results for constraints on the cosmological density parameter, Ω_m, and the equation of state parameter of the dark energy, w_x, for the REFLEX cluster data combined with results from distant supernovae [14, 17] as derived in the work by Schuecker et al. [22]. Note that here the parameter w_x is not constraint to values greater -1 done in most analyses. As a result we note that a model family including the "simple" Λ-cosmology models are best consistent with the data. For more details of the implications of this result see Schuecker et al. [22]

CONCLUSION

Cluster abundance measurements and the study of the large-scale distribution of clusters can provide very important cosmological tests. What may be most important is that these tests are complementary in many ways to other current methods of determining cosmological parameters. Further dramatic improvements in the use of clusters as cosmolog-

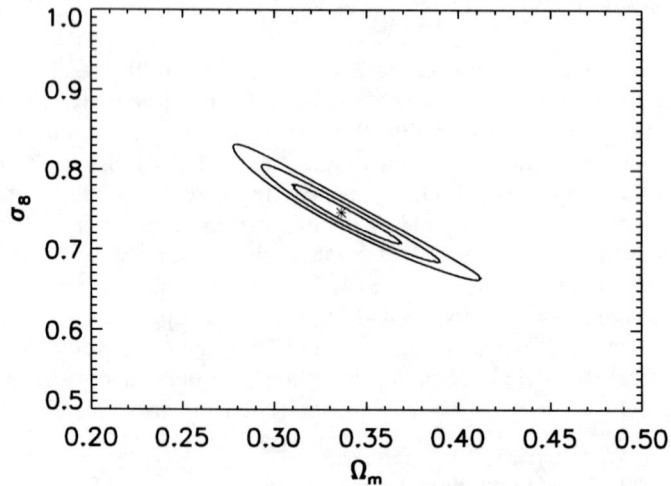

FIGURE 11. Constraints on the cosmological density parameter, Ω_m, and the amplitude of the matter density fluctuations on a scale of $8h^{-1}$ Mpc, σ_8, obtained from the comparison of the density fluctuation power spectrum and the cluster abundance as a function of redshift in a Karhunen-Loéve statistical analysis of the REFLEX Survey data (Schuecker et al. 2002b). The likelihood contours give the 1,2, and 3σ limits.

ical probes will be possible if new and deeper X-ray all-sky surveys become available as have been proposed as the DUET mission to NASA and the ROSITA mission to the German space agency. At some time such a survey will be conducted and allow similar but more precise tests with on the order of several 10,000 clusters.

REFERENCES

1. Bahcall, N. A., Ostriker, J. P., Perlmutter, S., Steinhardt, P. J., *Science*, **284**, 1481 (1999)
2. Bahcall, N.A., Dong, F., Bode, P., et al., *astro-ph/0205490*, 2002
3. Böhringer, H., Guzzo, L., Collins, C.A., et al., *The Messenger*, **94**, 21 (1998)
4. Böhringer, H., Voges, W., Huchra, J.P., et al., *ApJS*, **129**, 435 (2000)
5. Böhringer, H., Schuecker, P., Guzzo, L., et al., *A&A*, **369**, 826 (2001)
6. Böhringer, H., Collins, C.A., Guzzo, L., et al., *ApJ*, **566**, 1 (2002)
7. Collins, C.A., Guzzo, L., Böhringer, H., et al. *MNRAS*, **319**, 939 (2000)
8. De Bernardis et al., *ApJ*, **564**, 556 (2002)
9. Eisenstein, D., Hu, W., *ApJ*, **498**, 137 (1998)
10. Evrard, A.E., et al. *ApJ*, **573**, 7 (2002)
11. Finoguenov, A., Reiprich, T. H., Böhringer, H., *A&A*, **368**, 749 (2001)
12. Guzzo, L., Böhringer, H., Schuecker, P., et al., *The Messenger*, **95**, 27 (1999)
13. Ikebe, Y., Reiprich, T. H., Böhringer, H., Tanaka, Y., Kitayama, T., *A&A*, **383**, 773 (2002)
14. Perlmutter, N., et al., *ApJ*, **517**, 565 (1999)
15. Pierpaoli, E., Borgani, S., Scott, D., White, M., *astro-ph/0210567* (2002)
16. Reiprich, T.H. & Böhringer, H., *ApJ*, **567**, 716 (2002)
17. Riess, et al., *AJ*, **116**, 1009 (1998)

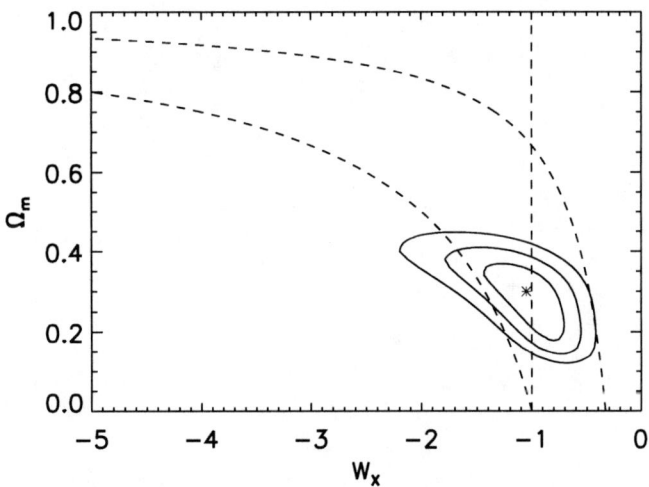

FIGURE 12. Constraints on the cosmological density parameter, Ω_m, and the equation of state parameter of the dark energy, w_x, for the REFLEX cluster data combined with results from distant supernovae from Perlmutter et al. (1999) and Riess et al. (1998) obtained by Schuecker et al. (2002c). The likelihood contours give the 1,2, and 3σ limits.

18. Sarazin, C.L., 1986, *Rev. Mod. Phys.*, **58**, 1 (1986)
19. Schuecker, P., Böhringer, H., Guzzo, et al., *A&A*, **368**, 86 (2001)
20. Schuecker, P., Guzzo, L., Collins, C.A., Böhringer, H., *MNRAS*, **335**, 807 (2002a)
21. Schuecker, P., Böhringer, H., Collins, C.A., & Guzzo, L., *A&A (in press), astro-ph/0208251* (2002b)
22. Schuecker, P., Caldwell, R.R., Böhringer, H., Collins, C.A., Guzzo, L., *A&A (in press), astro-ph/0211480* (2002c)
23. Voges, W., Aschenbach, B., Boller, T., et al., *A&A*, **349**, 389 (1999)
24. White, S.D.M., Efstathiou, G., & Frenk, C.S., *MNRAS*, **262**, 1023 (1993)
25. Trümper, J., *Science*, **260**, 1769 (1993)

SHUFFLE: A New Statistical Bootstrap Method: Applied to Cosmological Filaments

Suketu P. Bhavsar*, Somnath Bharadwaj[†] and Jatush V. Sheth**

University of Kentucky, Lexington, KY 40506, USA
[†]*IIT Kharagpur, Kharagpur, India*
**IUCAA, Pune, India*

Abstract. We introduce Shuffle, a powerful statistical procedure devised by Bhavsar and Ling [1] to determine the true physical extent of the filaments in the Las Campanas Redshift Survey [LCRS]. At its heart, Shuffle falls in the category of bootstrap like methods [2]. We find that the longest physical filamentary structures in 5 of the 6 LCRS slices are longer than 50 h^{-1} Mpc but not quite extending to 70 h^{-1} Mpc. The -3 degree slice contains filamentary structure longer than 70 h^{-1} Mpc.

INTRODUCTION

A traditional measure of clustering of point data is the two point correlation function $\xi(r)$ [3]. For the LCRS, on scales larger than 30-40 h^{-1} Mpc, $\xi(r)$ fluctuates closely around zero [4], indicating a high level of uniformity in the galaxy distribution at and beyond these scales. This observation raises the question whether the visual very long filamentary structure is real and on what scale true filaments exist. The presence of large scale filamentarity above that measured for a Poisson distribution has been established using Shapefinder [5], a technique to quantify filamentarity. But this excess of filaments, at some scale, could merely be the result of chance alignment, the probability of which is enhanced by the strong clustering of galaxies [6].

SHUFFLE

Can we determine the scale at which real structure exists and beyond which the structure is merely due to chance? Here we show how this can be achieved by a method: Shuffle. Shuffle generates fake data sets, practically identical in their clumping properties to the original data up to a scale length L, but in which all structures longer than L, both real and chance, of the original data, have been eliminated. In these Shuffled data, structures longer than L are visually evident, even identified by objective methods like Shapefinders; but all visual structure longer than L has formed accidentally. Quantitative comparison of the identified features in the original data with those in the Shuffled data gives us a statistical estimate of the level at which *real* features longer than L occur in the original data. The excess of features in the original data over Shuffled is obtained

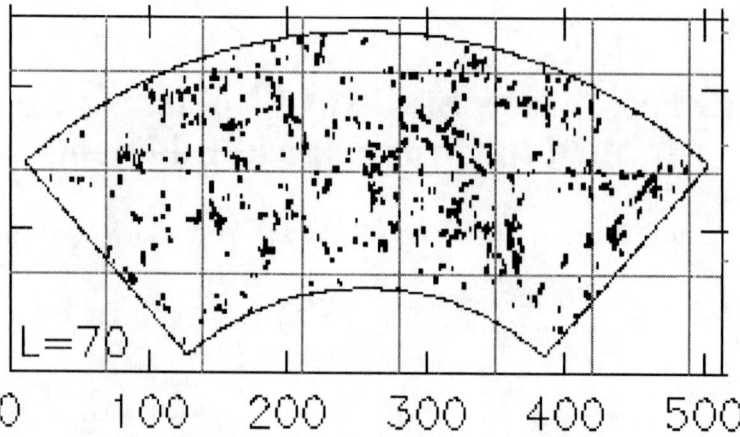

FIGURE 1. Shuffle divides the region of data into sub regions that are randomly swapped with rotation

through a range of L, Where L ranges from values of the order of the correlation length up to values that approach the scale sizes of the entire region.

We have done this for the LCRS [7]. If there is an excess in the number of filamentary features in the original data over the Shuffled data sets, this indicates that real features are present up to this length scale in the survey slices. If beyond some scale L the number of features in the Shuffled data occur just as often as in the original, the conclusion is that these features even in the original data are due to chance and not true filaments.

METHOD

- Embed a 1 h^{-1} Mpc x 1 h^{-1} Mpc rectangular grid on each slice.
- Generate "coarse grained" map by filling cells of occupied neighbors. This creates larger structures as a function of the filling factor, FF, in a slice.
- Use "friends of friends" to define features for at each value of the FF.
- Use Shapefinders to obtain average filamentarity, F_2, of the features as a function of FF.
- Shuffle LxL square regions in each slice to form new fake features of length greater than L as shown in Figure 1.
- Repeat first four steps.
- Plot F_2 versus FF for the original data and Shuffled slices as shown in Figure 2.
- The excess of $F2$ above its values for Shuffled slices gives the REAL filamentarity through the range of FF for each slice.

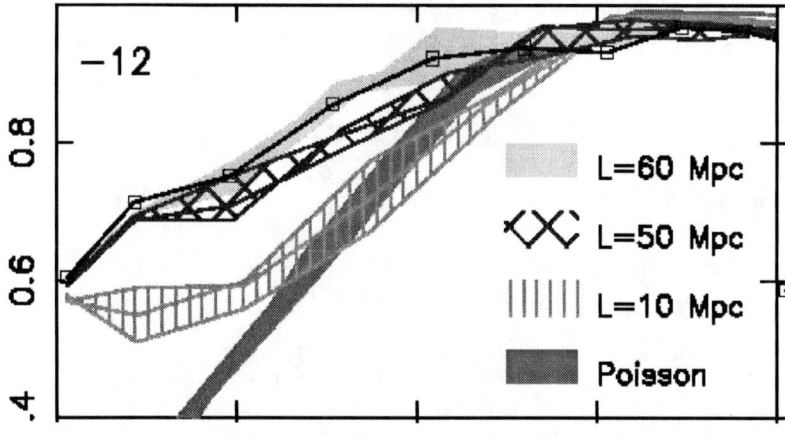

FIGURE 2. A plot of F_2 versus FF

RESULTS

Figure 2 shows results for a map of galaxies in the LCRS. The black line with the square data points graphs the average filamentarity F_2 for the galaxies in the -12 degree slice of the LCRS as a function of filling factor FF. The dark filled region graphs the filamentarity detected for a Poisson distribution; such filaments appear completely by chance in a non-clustered distribution of points. The vertical hatched region shows the filamentarity F_2 when Shuffling is performed at $L = 10 \, h^{-1}$ Mpc on the data. The cross hatched region for $L = 50$ Mpc, and the light filled region for $L = 60 \, h^{-1}$ Mpc. Note the transition from the shuffling at $L = 50 \, h^{-1}$ Mpc to $L = 60 \, h^{-1}$ Mpc. The Shuffled $L = 60 \, h^{-1}$ Mpc slice has the same filamentarity as the original data, indicating the extent that real coherent structure is present in the -12 degree slice.

CONCLUSION

Each slice is Shuffled at successively larger L-values until Shuffling no longer suppresses filamentarity. When this transition occurs the shuffled slice is statistically indistinguishable from the original data slice. The L-value of the transition point identifies the largest length scale at which real filaments exist. Any filamentary feature longer than this L-value in the original data is the result of chance. For the -3 degree slice the transition occurs when L is increased from $70 \, h^{-1}$ Mpc to $80 \, h^{-1}$ Mpc. In other words, features larger than $80 \, h^{-1}$ Mpc are not real, and features around $70 \, h^{-1}$ Mpc are. The transition occurs at a scale greater than $50 \, h^{-1}$ Mpc but less than $70 \, h^{-1}$ Mpc in the other LCRS slices.

Features on these scales imply a power spectrum peaking at a wavelength between 100-$140 \, h^{-1}$ Mpc, consistent with power spectrum measurements for the LCRS. While filaments are a natural outcome of gravitational instability, any model claiming to accurately describe formation of large scale structure in the universe must produce coherence on scales that match these observations.

ACKNOWLEDGMENTS

We thank the LCRS team for making the data public

REFERENCES

1. Bhavsar, S. P. & Ling E. N. 1988, Astrophys.Jour. lett., 331, L63.
2. Barrow, J. D., Bhavsar, S. P., Sonoda, D. H., 1984, Mon. Not. R. astr. Soc.,210, 19.
3. Peebles, P. J. E. 1993, Principles of Physical Cosmology (princeton U. Press)
4. Tucker, D. L. et al, 1997, Mon. Not. R. astr. Soc., 285, L5.
5. Bharadwaj, S., Sahni, V., Sathyaprakash, B. S., Shandarin, S. F., & Yess, C., 2000' Astrophys. Jour., 528, 21.
6. Barrow, J. D., & Bhavsar, S. P., 1987, Quart. Jour. R. astr. Soc.,28, 109.
7. Shectman, S. A., et al. 1996, Astrophys. Jour., 470, 172-188.

A National Virtual Observatory (NVO) Science Case: Properties of Very Luminous IR Galaxies (VLIRGs)

Kirk D. Borne*, S. Arribas[†], H. Bushouse[†], L. Colina** and R. Lucas[†]

Raytheon Information Technology & Scientific Services
[†]*STScI, Baltimore, Maryland*
***CSIC, Madrid, Spain**

Abstract. The class of Very Luminous IR Galaxies (VLIRGs) is defined by those galaxies whose IR luminosities are between 10^{11} and $10^{12} L_o$. They represent a key class of objects in the study of galaxy formation and evolution. For example, they offer the opportunity to study how the fundamental astrophysical and morphological properties of galaxies vary with IR luminosity, providing the link between the dynamically diverse Ultra-Luminous IR Galaxies (ULIRGs: $L_{IR} > 10^{12} L_o$) and *normal* galaxies. Furthermore, VLIRGs may be closely related to several recently identified cosmological populations in the sense that VLIRGs may be either their low-redshift analogs or the direct result of their evolution: EROs (Extremely Red Objects), GRB host galaxies, IR-selected AGN, faint SCUBA submm sources, or galaxies comprising the IR background. We describe a prototype analysis of the multi-wavelength properties of VLIRGs. Our research plan utilizes data and catalogs from a variety of distributed astronomical information sources. This manually intensive research exercise represents one of many possible NVO science scenarios that will be greatly facilitated and made much more efficient through the application of planned NVO resources. Support for this work was provided in part by NASA (NAS5-99226) and NSF (AST0122449).

INTRODUCTION

The NVO (National Virtual Observatory) will enable seamless user access to astronomical data from diverse geographically distributed data sources. As the volume of astronomical data resources grows astronomically, one must resort to sophisticated data mining techniques. *What is data mining?* Data mining is defined as "an information extraction activity whose goal is to discover hidden knowledge in large databases." We envision data mining as a key component of future astronomical research projects involving large distributed databases. To support this "new science", we have begun to explore various data mining techniques for astronomy, and we have compiled a comprehensive Scientific Data Mining Resource Guide, available from http://nvo.gsfc.nasa.gov/.

As a demonstration of some of the key aspects of distributed data mining that can be applied to astrophysics research problems, we present a case study of one particularly interesting class of galaxies that has not been well studied to-date: the Very-Luminous IR Galaxies (VLIRGs). Data mining in a variety of multi-wavelength multi-mission multi-modal (imaging, spectroscopic, catalog) databases will eventually yield interesting new scientific properties, knowledge, and understanding for the VLIRGs as well as for many other objects of astrophysical significance.

VERY-LUMINOUS IR GALAXIES (VLIRGS)

Very-Luminous IR Galaxies (VLIRGs: $10^{11} <$ L[8-400 μ] $< 10^{12} L_{\odot}$) are key in the study of galaxy formation and evolution. On one hand they offer the opportunity to study how the fundamental physical and structural properties of galaxies vary with (infrared) luminosity, providing the link between Ultra-Luminous IR Galaxies (ULIRGs: L[8-400μ] $> 10^{12} L_{\odot}$) and *normal* galaxies. On the other hand, they are believed to be closely related with recently identified cosmological populations in the sense that VLIRGs should be their low-redshifted analogs, or the direct result of their evolution.

PROPERTIES OF VLIRGS: THEIR RELATION WITH ULIRGS AND *NORMAL* GALAXIES

VLIRGs represent an atypical subset of the galaxy population, not represented among the nearby Shapley-Ames galaxies. In fact in a sample of 165 Shapley-Ames galaxies, IRAS detected nearly all late-type spirals (Sb-Sd) and Irr-Am galaxies, approximately half of the early type (S0-Sa) galaxies, and none of the ellipticals (de Jong et al. 1984). It is remarkable that no objects were found with L[8-400μ] $> 10^{11}$ (i.e., VLIRGs and ULIRGs). Whereas ULIRGs have been the subject of numerous studies over the past decade (e.g., Sanders & Mirabel 1996; Borne *et al.* 2000; Colina *et al.* 2001; Farrah *et al.* 2001; Bushouse *et al.* 2002), their slightly less luminous cousins, the VLIRGs, have not been the subject of anywhere near as much scrutiny — for example, in a preliminary restricted search of the *HST* science data archive we found observations for only 2 out of the 927 potential candidates for this study. The situation is similar for ground-based studies. This is remarkable especially if we take into account that locally the density of VLIRGs is ∼2 orders of magnitude higher than the density of ULIRGs; see for example Soifer *et al.* 1987; Saunders *et al.* 1990). A major objective of our NVO data mining research program on VLIRGs is to solve this situation, allowing a complete analysis of IR galaxies through a broader range of luminosities.

HIGH-Z POPULATIONS - VLIRGS RELATIONSHIP

Several cosmological interesting populations have been recently identified including: (a) the galaxies that comprise the far-IR background (Dwek *et al.* 1998); (b) the high-redshift submm sources (Smail *et al.* 1998; Barger *et al.* 1999; Blain *et al.* 1999); (c) the infrared-selected AGN (a new population of AGN identified in the 2MASS survey; Beichman *et al.* 1998, Nelson *et al.* 1998a, 1998b); and (d) the "Extremely Red Objects" found in HST-NICMOS images and in other recent deep surveys (Smail *et al.* 1999, Soifer *et al.* 1999, Thompson *et al.* 1999, Yan *et al.* 2000). The primary questions pertaining to these cosmological sources are: What are they? Are they dusty AGN?

or dusty starbursts? or massive old stellar systems? Some of them are undoubtedly ULIRGs, while the majority are probably related to the significantly more numerous class of galaxies: the VLIRGs. Our NVO data mining project will enable us to identify signatures of VLIRGs that are indicative of interactions and mergers, to determine the distribution of star–formation rates among the galaxies in this class, and to identify starburst or AGN phenomena that are otherwise obscured by the characteristic large dust opacities in these galaxies (heated dust being the source of the IR emission). The results will therefore be applicable to understanding and interpreting the properties of the 4 cosmological samples listed above. If, as believed, both ULIRGs and VLIRGs represent low-redshift analogs to the high-redshift galaxies, a comprehensive analysis of the VLIRGs is urgently needed. This is clear, for instance, when interpreting the diffuse IR background (Hauser et al. 1998). Deep submm surveys have confirmed that IR-luminous galaxies are a significant, if not the dominant, contributor to the IR background (Blain et al. 1999; Lilly et al. 1999). However the recent submm observations have been able to identify only ULIRGs and the most luminous VLIRGs. Taking into account their significant larger density it appears likely that VLIRGs could contribute a significant fraction of the infrared radiation at high redshifts. Further advance depends on acquiring detailed knowledge of the properties of VLIRGs and of their relation to ULIRGs and normal galaxies. A thorough data mining exercise within the NVO across multiple databases will enable a significant research advantage toward resolving these scientific questions.

A PROOF-OF-CONCEPT NVO-STYLE ARCHIVAL STUDY

We initiated a proof-of-concept NVO search scenario by attempting to identify potential candidate contributors to the CIB (Cosmic Infrared Background; Hauser et al. 1988). Our approach (Borne 2000) is similar to that of Haarsma & Partridge (1998), except that we applied the full power of on-line databases and linkages between these databases, archives, and published literature. Our search scenario involved finding object cross-identifications among the IRAS Faint Source Catalog and FIRST survey catalog, and then attempting to find those commonly identified objects also within other databases, such as the HST observation log. In a very limited sample of targets that we investigated to test our NVO data mining approach to the problem, we did find one object in common among the HST-IRAS-FIRST databases: a known hyperluminous infrared galaxy (HyLIRG) at z=0.780 harboring an AGN, which was specifically imaged by HST because of its known HyLIRG characteristics. In this extremely limited test scenario, we did in fact find what we were searching for: a distant IR-luminous galaxy that is either a likely contributor to the CIB, or else it is an object similar in characteristics to the more distant objects that likely comprise the CIB.

REFERENCES

1. Barger, A. J., Cowie, L. L., & Sanders, D. B., 1999, ApJ, 518, L5
2. Beichman, C. A. et al., 1998, PASP, 110, 367

3. Blain, A. W., Kneib, J.-P., Ivison, R. J., & Smail, I., 1999, ApJ, 512, L87
4. Borne, K. D., 2000, in "Virtual Observatories of the Future", p. 333
5. Borne, K. D., Bushouse, H., Lucas, R. A., & Colina, L., 2000, ApJ, 529, L77
6. Bushouse, H., Borne, K., Colina, L., Lucas, R., et al., 2002, ApJS, 138, 1
7. Colina, L., Borne, K., Bushouse, H., Lucas, R., et al., 2001, ApJ, 563, 546
8. de Jong et al., 1984, ApJ, 278, L67
9. Dwek, E. et al., 1998, ApJ, 508, 106
10. Farrah, D., Rowan-Robinson, M., Oliver, S., Serjeant, S., Borne, K., et al., 2001, MNRAS, 326, 1333
11. Haarsma, D. B., & Partridge, R. B., 1998, ApJ, 503, L5
12. Hauser, M. G. et al., 1998, ApJ, 508, 25
13. Lilly, S. D. et al., 1999, ApJ, 518, 641
14. Nelson et al., 1998a, BAAS, 192, 55.07
15. Nelson et al., 1998b, BAAS, 193, 81.04
16. Sanders, D. B., & Mirabel, I. F., 1996, Ann.Rev.Astr.Ap., 34, 749
17. Saunders, W. et al., 1990, MNRAS, 242, 318
18. Smail, I., Ivison, R. J., Blain, A. W., & Kneib, J.-P., 1998, ApJ, 507, L21
19. Smail, I. et al., 1999, MNRAS, 308, 1061
20. Soifer, B. T. it et al., 1987, ApJ, 320, 238
21. Soifer, B. T. it et al., 1999, AJ, 118, 2065
22. Thompson, D. et al., 1999, ApJ, 523, 100
23. Yan, L. et al., 2000, AJ, 120, 575

X-Ray Analysis of the MS0302 Supercluster

Donald J. Horner* and Megan Donahue*

*Space Telescope Science Institute, 3700 San Martin Drive, Baltimore MD 21218 USA

Abstract. The MS0302 supercluster is comprised of three massive clusters at $z = 0.42$ (GHO 0303+170, MS 0302.7+1658, and MS0302.5+1717). While this supercluster has been the subject of deep photometric and lensing studies, it has been rather poorly observed in the X-rays. We are conducting an analysis of existing ROSAT HRI, ASCA, and Chandra data of supercluster. Here, we present a new mosaic of the region using existing ASCA observations of the region and X-ray temperatures for GHO 0303+170 and MS0302.5+1717.

INTRODUCTION

Superclusters are the most massive and rarest objects in the hierarchy of structure in the universe. They usually contain several massive clusters over a region $\sim 10 h_{50}$ Mpc in size. Since the crossing time of superclusters is greater than a Hubble time, they also retain an imprint of conditions that existed in the early universe. Studies of these systems have been used to measure the Ω_0, to constrain variations of the mass-to-light ratio, and to test theories of the formation and evolution of galaxies and clusters. In addition to massive clusters, observations and numerical simulations suggest that superclusters are not isolated entities but contain a hierarchy of constituents. Between the massive galaxy clusters are galaxy groups in higher density regions and filaments in lower density regions.

The MS0302 supercluster is one of the few high redshift supercluster known. It is comprised of at least three massive clusters (see Table 1 for a summary of their properties and Figure 1 for images) of similar flux and X-ray luminosity. The first of the clusters to be discovered, GHO 0303+1706, was found during the optical cluster survey of Gunn et al. (1986) [1]. An Einstein IPC observation of the cluster revealed two additional clusters in the field. These became part of the EMSS flux limited sample as MS 0302.5+1717 and MS 0302.7+1658 [2]. All were found to be at redshifts $z \approx 0.42$.

Spectroscopic studies have brought the total of spectroscopically identified galaxies to ~ 50–100 in the field around each cluster [3, 4, 5]. Kaiser et al. (1998) [6] used deep V and I imaging of the supercluster to conduct a photometric and weak lensing survey. All three clusters have also been targets of HST observations (using WFPC2 and NICMOS). However, the X-ray data so far presented in the literature has been rather poor. Besides the original Einstein IPC observation of GHO 0303+1706 (which led to the discovery of the other two clusters), the only X-ray results in the literature are a ROSAT HRI surface brightness profile and an ASCA temperature for MS 0302.7+1658 [7, 8]. However, the MS0302 region has been observed several times by ASCA as well as by Chandra (an ACIS-I snapshot of MS 0302.7+1658).

TABLE 1. ASCA Spectral Fitting Results. Units of right ascension and declination are degrees. Units of temperature are keV. Units of flux are 10^{-13} ergs cm^{-2} s^{-1} in the 0.5–2.0 keV band. Units of bolometric luminosity are 10^{44} ergs s^{-1}.

Cluster	R.A.	Dec.	z	T_x	f_x	L_{bol}	χ_r^2	dof
CL–N	46.3291	17.4773	0.4241	$5.73^{+2.52}_{-1.45}$	2.76	7.41	0.43	244
CL–S	46.3913	17.1682	0.4248	$5.31^{+1.94}_{-1.29}$	2.68	6.92	0.51	159
CL–E	46.5779	17.3009	0.4195	$7.14^{+14.8}_{-3.32}$	2.41	7.24	0.38	101

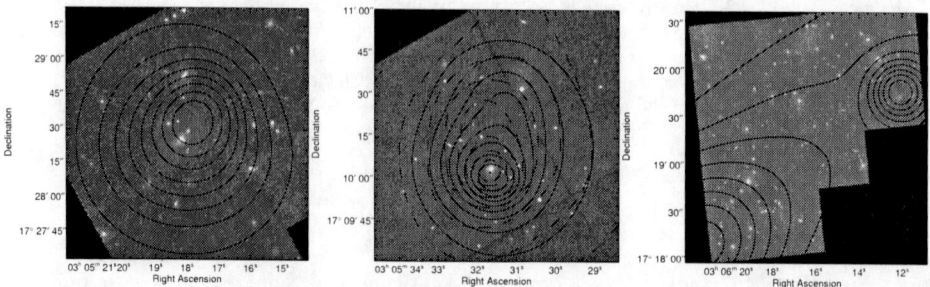

FIGURE 1. Adaptively smoothed ROSAT HRI X-ray contours overlaid on HST image of CL–N (left), CL–S (middle), and CL–E (right). Dashed contours for CL–S indicate adaptively smoothed Chandra ACIS-I contours

In this contribution, we present an analysis of the available X-ray data for the MS0302 supercluster. We use $H_0 = 50$ km s^{-1} Mpc^{-1} and $q_0 = 0.5$, and quoted errors are 90% confidence levels unless otherwise stated. For simplicity we will hereafter refer to to MS 0302.5+1717 as CL–N, MS 0302.7+1658 as CL–S, and GHO 0303+1706 as CL–E.

ASCA ANALYSIS

We created a mosaic of the ASCA GIS observations of the MS0302. An exposure map was generated for each GIS observation using the FTOOLS *ascaexpo* and *ascaeffmap* weighted by the spectrum of CL–N. The FTOOL *fmosaic* was used to create a combined exposure map and background image. We then subtracted the background and divided by the exposure map to create the final image (see Figure 2).

For the spectral analysis, we started with the standard screened events file provided by the HEASARC. We extracted spectra around each cluster using a $5'$ radius for the GIS data and $3'$ for the SIS data. We used the standard redistribution matrix files (RMFs) for the GIS. Due to the time evolution of the SIS response, we generated the RMFs for each SIS observation. We then created ancillary response files (ARFs) using a weighted-average effective area with the weighting factor being the number of events at each pixel position. We added together the spectra from each observation for each detector to increase the signal-to-noise using the *addascaspec* tool. The resulting spectrum files

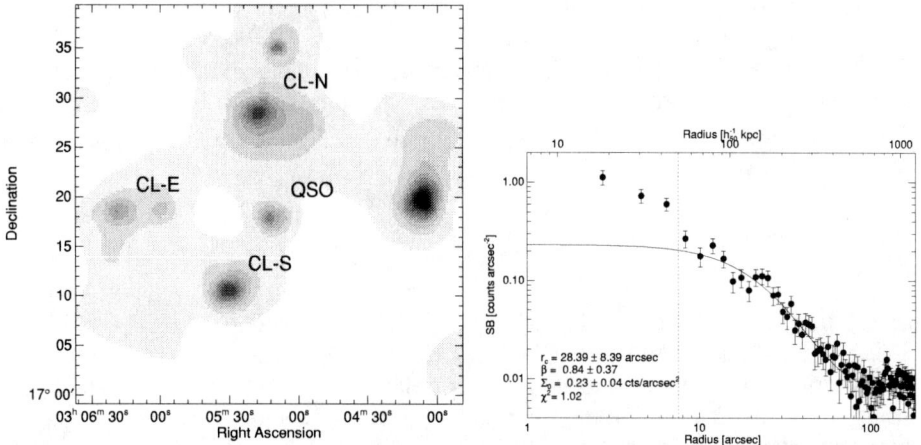

FIGURE 2. (left) Adaptively smoothed ASCA GIS mosaic of the MS0302 supercluster. (right) Radial profile of CL–S in ACIS-I. The solid line is a β–profile fit for points $r > r_{lim}$ as indicated by the dashed line.

were grouped to a minimum of 25 counts per channel in order to use χ^2 statistics.

Using *XSPEC V11.1.0*, we fit the processed spectra to a single temperature MEKAL spectral model [9] over the energy range 0.8–8 keV. The spectra from each instrument were fit simultaneously, allowing the relative normalizations to vary as necessary. The metal abundance was fixed at $0.3Z_\odot$. The temperatures and luminosities are presented in Table 1 with their 90% confidence limits. Obviously, we cannot well constrain the temperatures of the clusters, especially CL-E, although we can say that all the clusters must be fairly massive.

The temperatures of CL-N and CL-E have never been previously measured. Henry (2000) [8] reports an X-ray temperature of $4.4^{+0.8}_{-0.6}$ (68% confidence limits) keV for CL–S. Although about 1 keV lower than our derived temperature $5.3^{+2.0}_{-1.3}$ (90% confidence limits), it is well within our 90% confidence limits.

CHANDRA ANALYSIS

Chandra observed CL–S with ACIS-I for 10 kiloseconds in October 2000. This exposure is too short to obtain a reliable temperature for the cluster (with only a few hundred counts from the cluster). CL–N and CL–E were off the ACIS-I chips and so not observed.

We reprocessed the data using updated gain files and create a new event file for events with energies between 0.3 and 8.0 keV. The light curve of the I3 chip was extracted to construct a new GTI file for the observation. There were no strong flares during the observation.

Figure 2 shows the radial profile of the CL–S. We fit a standard β-model profile to the data, excluding the central 50 kpc which is likely affected by a cooling flow. Using the ASCA temperature for CL–S and the Chandra surface brightness values we can estimate the mass of CL–S using the isothermal β-model. We find a $M(< 1.19 Mpc) = 5.84^{+2.94}_{-2.74} \times 10^{14}$ M$_\odot$ (1σ errors). This is about a factor of 2 higher than the optical viral mass of [10], although the large error bars mean that they are just barely consistent at the 1σ level. However, it is over a factor of four higher than the weak lensing mass of Kaiser et al. (1998) [6].

FUTURE WORK

In the future we intend to analyze the archival HRI data to obtain radial profiles for CL–E and CL–N and estimate their masses, gas masses, baryon fractions, and mass-to-light ratios. We would also to study the correlation of the X-ray data with the weak lensing mass distribution. Future X-ray and optical observations will be necessary to better understand the system. For example, the ASCA mosaic reveals another bright source to the west of the three clusters. The nature of this source is unclear. It is always out of the field-of-view of the X-ray instruments or near the edge of their field-of-view where they are poorly calibrated, and there are no deep optical observations of this region either.

REFERENCES

1. J. E. Gunn, J. G. Hoessel, and J. B. Oke, ApJ, **306**, 30 (1986).
2. I. M. Gioia et al., ApJS, **72**, 567 (1990).
3. D. G. Fabricant, M. W. Bautz, and J. E. McClintock, AJ, **107**, 8 (1994).
4. E. Ellingson et al., ApJS, **113**, 1 (1997).
5. A. Dressler et al., ApJS, **122**, 51 (1999).
6. N. Kaiser et al., preprint, **astro-ph/9809268** (1998).
7. A. D. Lewis, E. Ellingson, S. L. Morris, and R. G. Carlberg, ApJ, **517**, 587 (1999).
8. J. P. Henry, ApJ, **534**, 565 (2000).
9. R. Mewe, D. S. Kaastra, and D. A. Liedahl, Legacy **6**, 16 (1995).
10. M. Girardi and M. Mezzetti, ApJ, **548**, 79 (2001).

Recent Results from the NOAO Fundamental Plane Survey

J. E. Nelan*, G. A. Wegner*, R. J. Smith[†], M. J. Hudson[†], J. J. Malecki[†], J. R. Lucey**, S. A. W. Moore**, S. J. Quinney**, D. Schade[‡], N. B. Suntzeff[§] and R. L. Davies[¶]

Dept. of Physics and Astronomy, Dartmouth College, Hanover, NH
[†]*Dept. of Physics and Astronomy, University of Waterloo, Canada*
**Dept. of Physics, Durham University, UK*
[‡]*CADC/Dominion Astrophysical Observatory, Victoria BC, Canada*
[§]*Cerro Tololo Inter-American Observatory, Chile*
[¶]*Dept. of Astrophysics, Oxford University, UK*

Abstract. The NOAO Fundamental Plane Survey (NFPS) is a deep, homogeneous, all-sky spectroscopic and photometric study of 100 X-ray selected clusters within 200 h^{-1} Mpc (z < 0.06) with the principle goal of determining the large-scale flow of clusters with respect to the CMB. The photometry of tens of thousands of galaxies in two colors (B and R) will yield quantitative morphological information; spectroscopic parameters of early-type galaxies will include velocity dispersions, line indices, and redshifts. When combined with high-quality Fundamental Plane parameters, these measurements will shed light on the evolution of galaxies in the cluster and environmental dependences of spectroscopic parameters. Early results include detection of a significant dependence of the Mgb index on cluster-centric radius for Abell 3558, which can be interpreted as a gradient in metallicity or abundance ratios, but is not compatible with an age gradient. In addition, we have used the photometric catalogues to construct maps of the color-magnitude scatter as a function of position. Substructures identified in these maps often correspond with known X-ray/velocity subsystems.

INTRODUCTION

The primary goal of the NFPS is to find the large-scale bulk flow out to 20,000 km/s. The analysis of peculiar velocities provides the means for studying the total mass distribution in the nearby universe [1]; NFPS aims to measure this distribution on the largest scale possible for these types of surveys (> 100 h^{-1}). Similar studies have revealed that about 420 ± 180 km/s of the Local Group's 600 km/s peculiar velocity with respect to the Cosmic Microwave Background is produced by structures on scales of 100 h^{-1} Mpc [2]; however, this is not entirely consistent with the ΛCDM model and therefore better data is required for more accurate results. With data for ∼50 early-type galaxies per cluster, fundamental plane distances can be calculated with random errors of < 3 percent and then used to find the bulk flow of the sample. Another major goal of NFPS is to understand how cluster environment contributes to the evolution of early-type galaxies.

Each of the ∼5000 galaxies in approximately 100 X-ray selected clusters have been imaged in both the Northern and Southern hemispheres at CTIO, KPNO, MDM and CFHT observatories in the R and B bands and color-magnitude relations have been obtained. Early-type galaxies are identified as red-sequence cluster members and are

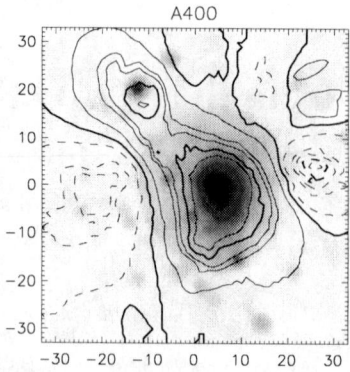

FIGURE 1. Locally-measured color-magnitude scatter, compared to X-ray emission in Abell 400. The contours represent the confidence levels at which the local scatter is lower than that of the color-scatter in the 'field': 50% (bold), 68%, 90%, 99% (bold) and 99.9%. The axis scales are in arcminutes.

used for spectroscopic follow-up. All bright galaxies will be measured to find overall morphological parameters and will be fit with a bulge/disk ratio.

Spectroscopic follow-up was accomplished using the Hydra multi-fiber spectrographs at the WIYN 3.5m and CTIO 4.0m telescopes with a 60 arcmin and 40 arcmin field of view, respectively. 40 to 70 galaxies in each cluster with $R < 17$ ('red' relative to the cluster color-magnitude relation) were observed resulting in redshifts, internal velocity dispersions, and metal and Balmer-series absorption lines over the spectral range of 4000-6000 Å.

PHOTOMETRIC RESULTS

We define a red sequence for each cluster by fitting the $B - R$ vs R color-magnitude relation with constant slope and zero-point to galaxies with $R < 17$. The zero-point variation of this sequence is very stable, with a cluster-to-cluster variation of < 0.04 magnitudes, after correcting for galactic extinction and k-correction [3]. We then construct a smoothed map of the local mean and variance of the CMR residuals which we call an Excess Color and HOmogeneity (ECHO) map. ECHO maps allow us to identify "cluster" regions by their very small CMR scatter in contrast to the high-scatter background of the field [4]. We can also use the mean residuals as a rough distance indicator which leads to an estimate of the three-dimensional structure of the region and indentification of background clusters. ECHO maps can also be used to reveal cluster substructure which in many cases replicate those from redshift samples and/or X-ray images.

Figure 1 shows the ECHO contours for cluster Abell 400 with ROSAT/PSPC X-ray data. Two significant low-variance substructures are detected; both are coincident with X-ray emission and galaxy distribution peaks.

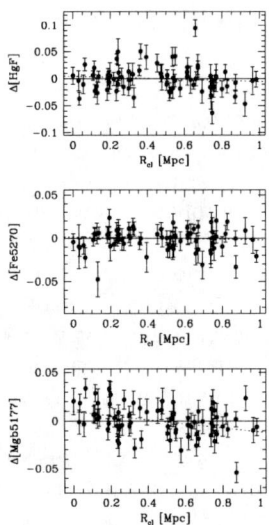

FIGURE 2. Residuals from the Mgb - σ, Fe5270 - σ, and Hγ_F - σ relations, plotted against cluster-centric distance R_{cl} for galaxies in the rich cluster Abell 3558.

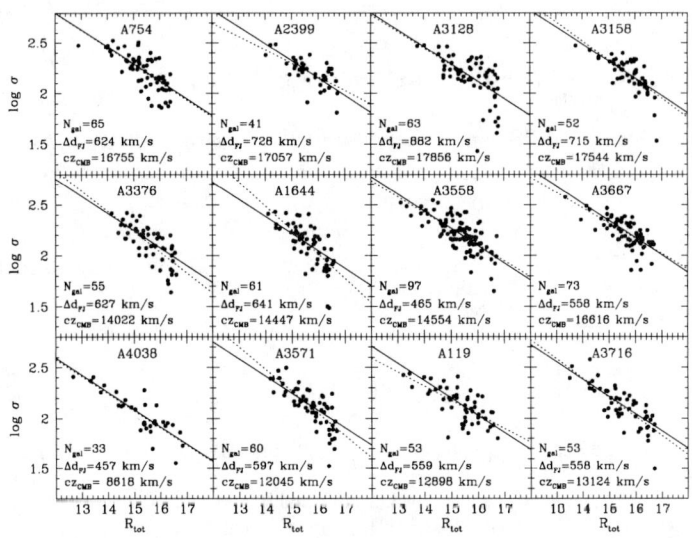

FIGURE 3. R-band Faber-Jackson relations for selected clusters with complementary imaging and spectroscopic data. The solid and dotted lines indicate fixed-slope and free-slope fits, respectively. With ~50 galaxies per cluster, even this very crude distance-indicator yields errors of only ~600 km/s. We anticipate a factor-of-two improvement in distance precision when the Fundamental Plane is employed.

SPECTROSCOPIC RESULTS

Spectroscopic data of 40-70 galaxies per cluster has been taken and the signal-to-noise and spectral resolution are sufficient to measure linestrength indices with high precision, ultimately yielding stellar-population parameters for the brighter $\sim$50 % of the sample. Using these data, we aim to charaterize the formation and enrichment history of cluster galaxies, as a function of mass, morphology, cluster-location, host-cluster properties, etc. As suggested in many other studies, the Mgb line at 5177Å increases strongly with velocity dispersion σ, most likely due to an increase of metallicity with mass; however, the age of a cluster may also play a role in this relation. For one of the NFPS clusters, Abell 3558, Mgb is stronger at given sigma for galaxies in the center of the cluster than galaxies at larger radii, indicating that galaxies in the core of the cluster are more metal-rich, at given mass, than those farther out (Figure 2). Since the Hγ_F-σ, residuals do not vary significantly with the radius of A3558, we can rule out an age effect as an explanation for the metal-rich galaxies at small radii (Smith et al., in preparation).

This effect of stronger Mgb emission at given mass for galaxies centrally located in the cluster has been shown for several NFPS clusters. Previous studies have shown that the same effect is present in the Coma cluster. However, initial indications are that this effect is not ubiquitous in the entire NFPS sample. It is possible that these gradients might correlate with the absence of substructure since it is likely that gradients in metallicity or age could be erased by mixing in cluster mergers.

Faber-Jackson Relation

Data analysis for the NFPS sample is still underway, but Figure 3 shows examples of Faber-Jackson relations for 12 Abell clusters. The FJ relation is the correlation between galaxy luminosity, or radius, and its velocity dispersion, and can be used as a distance indicator. Once the Fundamental Plane parameters are added, the errors in ΔFJ, the error in peculiar velocity if the FJ is employed as a distance indicator, should improve by a factor of two.

REFERENCES

1. Strauss, M. A. and Willick, J. A., "The Density and Peculiar Velocity Fields of Nearby Galaxies," *Physics Reports,* v.261, p. 271-431, 1995.
2. Hudson, M. J., in "Where's the Matter? Tracing Dark and Bright Matter with the New Generation of Large Scale Surveys," eds. Treyer & Tresse, 2001, Frontier Group.
3. Hudson, M. J., et al., "Early-Type Galaxies in the NOAO Fundamental Pla ne Survey," in *Galaxy Evolution: Theory and Observations (2002)*, RevMe xAA.
4. Smith, R. J., et al., "A Spatial Analysis of Colour-Magnitude Residuals and Variance in Wide-Field Images of Low-Redshift Galaxy Clusters," submitte d to *MNRAS*, December 2001.

Cosmic Structure Traced by Precision Measurements of the X-Ray Brightest Galaxy Clusters in the Sky

Thomas H. Reiprich*[†], Craig L. Sarazin*, Joshua C. Kempner**, Michael F. Skrutskie*, Gregory R. Sivakoff*, Hans Böhringer[‡] and Jörg Retzlaff[‡]

*Department of Astronomy, University of Virginia, PO Box 3818, Charlottesville, VA 22903-0818
[†]thomas@reiprich.net, http://www.reiprich.net
**Harvard-Smithsonian Center for Astrophysics, 60 Garden Street, MS-67 Cambridge, MA 02138
[‡]Max-Planck-Institut für extraterrestrische Physik, PO Box 1312, 85741 Garching, Germany

Abstract. The current status of our efforts to trace cosmic structure with 10^6 galaxies (2MASS), 10^3 galaxy clusters (NORAS II cluster survey), and precision measurements for 10^2 galaxy clusters (*HIFLUGCS*) is given. The latter is illustrated in more detail with results on the gas temperature and metal abundance structure for 10^0 cluster (A1644) obtained with XMM-Newton.

Galaxy clusters have been very important tools to study cosmic structure and determine cosmological parameters. Even moderately sized samples yield competitive statistical constraints, e.g., *HIFLUGCS* (Fig. 1), [1]. However, accuracy is currently limited by systematic uncertainties. Luckily major improvements are now feasible, e.g., by taking advantage of new multiwavelength data and especially Chandra and XMM-Newton X-ray observations. We've started a project to tackle systematic uncertainties in flux, temperature, gas and gravitational mass estimates by detailed analyses of the 63 X-ray brightest galaxy clusters in the sky (*HIFLUGCS*). Many of these clusters have already been observed by Chandra and XMM-Newton. Data are accumulating in the archives. The first out of further nine approved Chandra observations proposed by us is being carried out today (2002-12-16).

The NORAS II cluster survey contains about 800 X-ray selected galaxy clusters (Böhringer, Retzlaff et al., in prep.). In order to test for systematic effects caused by different selection techniques — which must be well understood to obtain precise cosmological constraints — we started correlating the NORAS II clusters with the 2MASS extended source catalog (the latter shown in Fig. 1, [2]) and color selected point source catalog. Furthermore galaxy overdensities at cluster positions will be determined in order to estimate richness–mass relations and mass-to-light ratios.

In the following we illustrate possible improvements from the X-ray side on an especially tough example. XMM-Newton observations of A1644 show a very complicated surface brightness distribution on all scales (Reiprich et al., in prep.). A main clump and a smaller sub clump are easily identified. The emission surrounding both core regions is highly non-spherical. And the core region of the main clump itself contains a displaced core-within-a-core. How much do X-ray flux, gas, and gravitational mass estimates based on precision measurements differ from simple estimates where, e.g., the

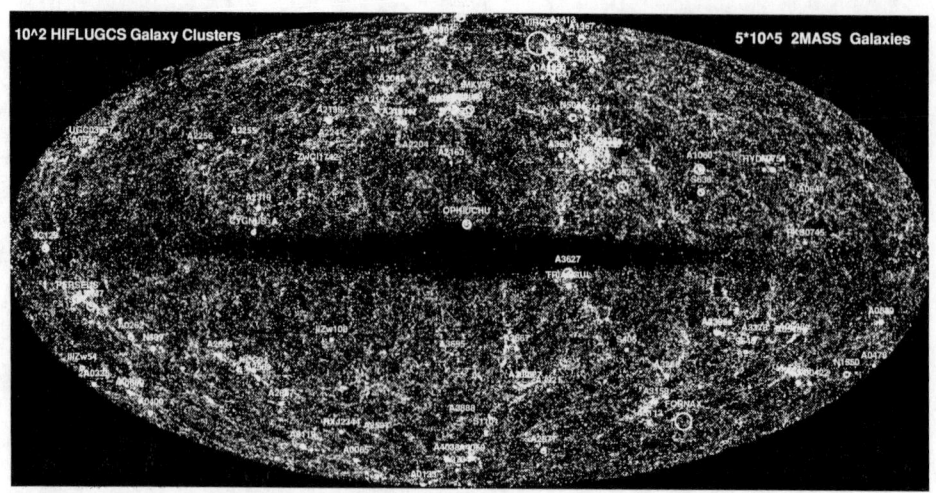

FIGURE 1. *HIFLUGCS* and other local X-ray galaxy clusters shown in Galactic coordinates on top of the distribution of half a million 2MASS galaxies. Qualitatively both populations seem to nicely trace an underlying matter distribution with clusters being located at the peaks.

whole cluster is treated as a spherical cow as might be done if only ROSAT All-Sky Survey data were available? How much do mass estimates differ when only a broad beam overall gas temperature estimate is available?

The X-ray flux (f_X) ratio between the main and sub clump is about 3:1. This means instead of one cluster in a flux-limited sample with $f_X \approx 4 \times 10^{-11}$ erg/s/cm^2 one actually has two clusters with about $f_X \approx 3$ and 1×10^{-11} erg/s/cm^2 each. This is quite an important difference, especially for construction of luminosity and mass functions!

The intraclucter gas temperature is a fundamental observable for a mass determination based on the hydrostatic assumption. One first step of refinement compared to broad beam temperature estimates is the construction of radial temperature profiles for the main and sub clump. To our surprise we found that each of the two temperature profiles appears very similar to temperature profiles of relaxed, apparently undisturbed clusters: a drop in the center to about 1/2 to 1/3 of the ambient gas temperature, an isothermal structure in the outer parts, and weak indications for a slight temperature drop in the very outermost regions accessible. This appears to be good news: this cluster may not be as complicated as it seemed, the temperature structure may not be affected by the interaction of the two sub clumps (which are located at about the same redshift). Note, however, that simple broad beam temperature estimates would still be biased low compared to the ambient gas temperature due to cool emission in the dense (high emissivity) centers. For instance, the temperature estimate in a large annulus around the main clump — taken as the ambient temperature — is a factor of 1.1–1.15 higher than a broad beam temperature estimate including both clumps. Since $M \propto T^{1.5-2.0}$ [e.g., 3] this factor translates into a factor of 1.15–1.32 in a mass estimate!

The next step we took is the subdivision of annuli into regular segments. Figure 2 shows selected regions as well as surface brightness contours overlaid onto a hardness

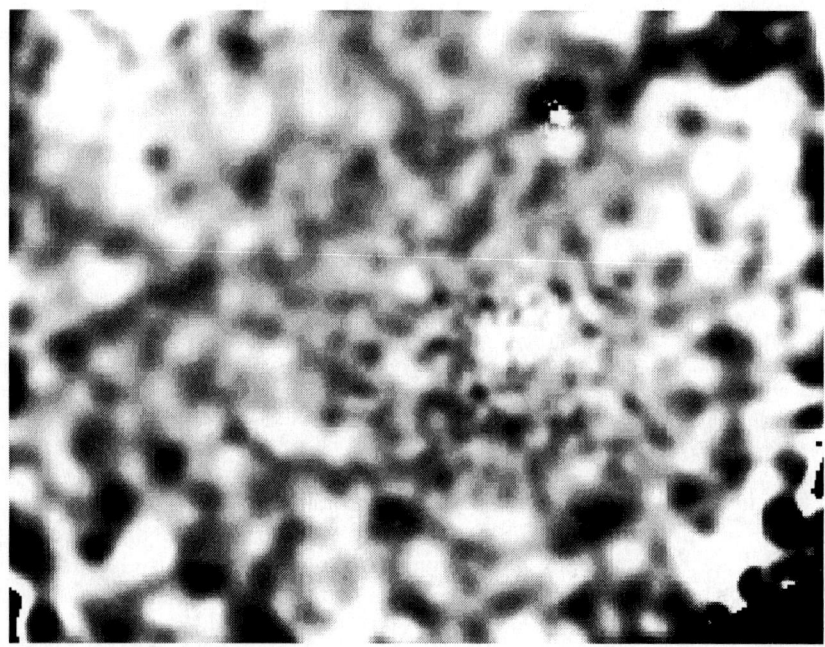

FIGURE 2. Hardness ratio map of the galaxy cluster A1644. Background, exposure, and vignetting corrected, adaptively smoothed, combined MOS1-MOS2-pn count rate image ratio for the energy bands (0.3–2.0 keV)/(2.0–10.0 keV). Soft emission appears bright and hard emission dark. Also shown are surface brightness contours for the (0.3–2.0 keV) image and regions selected for spectral analysis.

ratio (HR) map. Preliminary temperature (metal abundance) estimates for the segments based on standard spectral model fits are shown in the left (right) hand side of Fig. 3. Now it becomes obvious that the apparently regular temperature profiles are misleading and only due to averaging over (too) large regions. The temperatures of the segments to the East of the main clump (R22, R24, and R26) are all significantly lower than almost all regions to the West (R27–R32). This may indicate that the gas in the regions between the two clumps has been heated up by adiabatic compression or even shocks.

Having found irregularities in the temperature structure a further step is to select regions based on the HR map to directly search for cool/metal rich trails (bright) or hot spots (dark). The significance of brightness fluctuations is again evaluated by direct spectral fits. Note that not all artifacts, e.g., inexact exposure correction close to CCD chip boundaries, have been removed in the HR map in Fig. 2. The spectral analysis, however, is not affected by this. The region to the South of the sub clump (R18) appears fairly bright and therefore soft, whereas a small region to the Southeast (R20) appears hard. The spectral fit results (Fig. 3) reveal that indeed the temperature estimate for R18 is significantly lower than the estimate for the rest of this annulus (R19) and especially than that for R20. Furthermore the metal abundance estimate for R18 appears enhanced. These findings together with the surface brightness structure might imply that to the South we see gas that has been removed from the cooler center of the sub clump possibly by the combined effect of some energy source related to the central galaxy of the sub

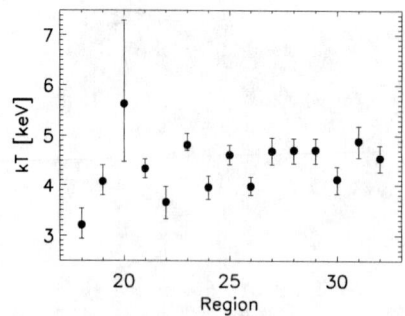 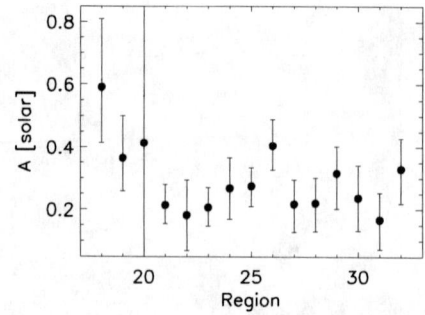

FIGURE 3. Gas temperature (left) and metal abundance (right) estimates of selected regions based on simultaneous spectral fits to background and vignetting corrected spectra from the MOS1, MOS2, and pn detectors aboard XMM-Newton. Note that R21=R27+R30, R23=R28+R31, R25=R29+32, and R19=full annulus–R18; see Fig. 2. Statistical error bars show the 90% confidence level for one interesting parameter.

clump (a radio source) and the movement through intracluster gas of the main clump. Should more detailed modeling confirm the significance of the abundance excess then other possibilities like cooling of intracluster gas onto the moving sub clump (as seen in A1795 on much smaller scales, [4]) would be more difficult to reconcile with the data.

In summary the temperature structure of A1644 shows that this system is quite complicated as indicated by the surface brightness structure. Interestingly no evidence for substructure has been found in optical and near-infrared observations [5]. The next step is to attempt an improved mass estimate and compare it to more simple estimates as generally applied to clusters at higher redshift and in larger samples. Note that this detailed study required only 12 ks of good data. Chandra and XMM-Newton observations of the complete *HIFLUGCS* are quite cheap but offer a great opportunity for clusters in the era of precision cosmology. Comparison of cluster X-ray data to the wealth of new multiwavelength data, e.g., 2MASS, will also help to reduce systematic uncertainties.

We acknowledge the benefit from the dedicated work of the NORAS II and 2MASS teams. This work was supported by NASA XMM-Newton Grant NAG5-10075. THR acknowledges support by the Celerity Foundation through a Post-Doctoral Fellowship. The XMM-Newton project is an ESA Science Mission with instruments and contributions directly funded by ESA Member States and the USA (NASA). This publication makes use of data products from 2MASS, which is a joint project of the University of Massachusetts and IPAC/Caltech, funded by NASA and NSF.

REFERENCES

1. Reiprich, T. H., and Böhringer, H., *ApJ*, **567**, 716–740 (2002).
2. Jarrett, T. H., Chester, T., Cutri, R., Schneider, S., Skrutskie, M., and Huchra, J. P., *AJ*, **119**, 2498–2531 (2000).
3. Finoguenov, A., Reiprich, T. H., and Böhringer, H., *A&A*, **368**, 749–759 (2001).
4. Fabian, A. C., Sanders, J. S., Ettori, S., Taylor, G. B., Allen, S. W., Crawford, C. S., Iwasawa, K., and Johnstone, R. M., *MNRAS*, **321**, L33–L36 (2001).
5. Tustin, A. W., Geller, M. J., Kenyon, S. J., and Diaferio, A., *AJ*, **122**, 1289–1297 (2001).

FUTURE PROSPECTS

The Future of Microwave Background Physics

Arthur Kosowsky

Department of Physics and Astronomy, Rutgers University, 136 Frelinghuysen Road, Piscataway, NJ 08854-8019

Abstract. The cosmic microwave background is now fulfilling its promise of determining the basic cosmological parameters describing our Universe. Future study of the microwave background will mostly be directed towards two basic questions: a complete characterization of the initial perturbations, and probes of the nonlinear evolution of structure in the Universe. The basic scientific issues in both of these areas are reviewed here, along with possibilities for addressing them with further microwave background measurements at higher sensitivities and smaller angular scales. The proposed ACT experiment, which will map 200 square degrees of sky at arcminute resolution and micro-Kelvin sensitivity in three microwave frequency bands, is briefly described as an example of rapidly advancing experimental technique.

WHERE WE ARE NOW

The cosmic microwave background is one of the best-studied sources of cosmological information (see [1, 2, 3] for reviews). It is nearly isotropic on the sky, with small temperature fluctuations at the level of one part in 10^5. These fluctuations arise from several basic physical mechanisms in the early Universe at a redshift $z \simeq 1100$: gravitational redshift, temperature fluctuations at the last scattering surface, Doppler shifts from peculiar velocities at the last scattering surface, and diffusion damping through the thickness of the last scattering surface [4, 5]. In addition, further temperature fluctuations and spectral distortions arise from gravitational and scattering effects at comparatively recent epochs, with $z < 3$. Simple arguments show that the microwave background should also have small polarization fluctuations, roughly an order of magnitude smaller than the temperature fluctuations [6].

Power Spectra

The temperature fluctuations have been measured and studied in detail for the past decade, beginning with the watershed COBE detection [7], while the first measurement of polarization fluctuations by Kovac and collaborators has been discussed at this conference. So far, on angular scales down to about a degree, the temperature fluctuations appear to possess a Gaussian random distribution [8, 9]. To the extent that the fluctuations are Gaussian, they are described completely by their power spectra. A parity-invariant distribution of primordial fluctuations will result in four non-zero power spectra: temperature C_l^T, two polarization C_l^E and C_l^B, and the cross-correlation C_l^{TE} [10, 11, 12] (where l is the multipole moment, inversely proportional to angular scale with $l = 200$ corresponding to one degree). These are the observables we will primarily consider here, although additional non-Gaussian temperature structure in the maps will also be very interesting to probe.

Theoretically, we can model how well a given measurement will probe the power spectrum, given its sky coverage, angular resolution, and sensitivity [13, 10, 11]. Experimentally, the MAP satellite will map the full sky with an angular resolution of around 12 arcminutes and an effective sensitivity of around 25 μK, in five frequency bands ranging from 30 GHz to 150 GHz. A few ground-based experiments have displayed higher angular resolution and sensitivity over far smaller regions of the sky; see Max Tegmark's contribution to these proceedings for a current compilation of power spectrum measurements.

Cosmological Parameters

The intense interest in microwave background temperature fluctuations has been fueled largely by the realization that the power spectrum contains much information about cosmological parameters which describe the fundamental properties of the Universe [14, 15, 16]: the Hubble parameter h; the densities of baryons Ω_b, dark matter Ω_{cdm}, and "dark energy" Ω_Λ; the primordial power spectrum of scalar perturbations parameterized by an amplitude A_s and power law index n_s; the same for tensor fluctuations A_t and n_t; and the optical depth to the surface of last scattering τ.

The ability of the microwave background to constrain these parameters hinges on the existence of acoustic oscillations of the primordial plasma at the last scattering surface. These oscillations are phase-coherent in simple cosmological models, since each k-mode has a characteristic cosmological time at which it enters the horizon, and result in the well-known series of "acoustic peaks" in the power spectrum. The amplitudes and angular scales of these peaks in turn depend on the entire set of cosmological parameters. Precision measurements of the power spectrum constrain the parameters, with only one essential degeneracy. The microwave background provides, by far, the single most powerful set of constraints on the basic properties of the Universe, and indeed is on the verge of giving definitive answers to most of the historically most important and vexing questions of classical observational cosmology.

Parameter Constraints

Extracting parameter constraints from microwave background power spectrum measurements is conceptually simple, but somewhat difficult in practice. Finding the best-fit cosmological model for a given power spectrum is not hard, but evaluating an error region requires looking around in a multi-dimensional parameter space. Approximations based on linear extrapolations of the likelihood in the above parameters give the right qualitative answer [14], but are inadequate to support the currently available data. Brute-force analyses on parameter space grids have been performed [17], but are not in general sufficiently accurate or flexible for upcoming data.

Monte Carlo techniques are much better, provided that the power spectra for a given cosmological model can be evaluated efficiently enough [18, 19, 20]. To this end, it is convenient to use a different set of cosmological parameters which better reflect the physical effects determining the acoustic peak structure in the power spectra. The following set of physical parameters has several advantages, including being largely uncorrelated and having nearly linear power spectrum dependence [21]:

- $\mathscr{A} \equiv r_s(a_*)/D_A(a_*)$, where $r_s(a_*)$ is the sound horizon at the time of last scattering, and $D_A(a_*)$ is the angular diameter distance to the surface of last scattering. This parameter determines the angular scale of the acoustic peaks.
- $\mathscr{B} \equiv \Omega_b h^2$, the baryon density.
- $\mathscr{V} \equiv \Omega_\Lambda h^2$, the vacuum energy density.
- $\mathscr{R} \equiv a_* \Omega_{\mathrm{mat}}/\Omega_{\mathrm{rad}}$, the ratio of matter to radiation energy density at last scattering.
- $\mathscr{M} \equiv (\Omega_{\mathrm{mat}}^2 + a_*^{-2}\Omega_{\mathrm{rad}}^2)^{1/2} h^2$, which is approximately a degenerate direction in the space of physical parameters. It is fixed if the number of neutrino species is assumed known.
- $\mathscr{L} \equiv e^{-2\tau}$, which parameterizes the effect of reionization.
- n, the primordial scalar perturbation power spectrum index.
- $\mathscr{S}$, the amplitude of the CMB power spectrum due to scalar perturbations at large l for $n = 1$; for fixed $\mathscr{S}$, varying $\mathscr{L}$ changes the power spectrum only at large scales.
- $\mathscr{T}$ and n_T, the tensor perturbation amplitude and spectral index. Note it is far more convenient to use $\mathscr{S}$ and $\mathscr{T}$ separately, rather than their sum and ratio as is customary.

These parameters allow an extremely efficient evaluation of the power spectrum for a given set of cosmological parameters over a large region of parameter space via simple functional approximations; they make Monte Carlo evaluations of parameter space error regions far less demanding computationally [21]. Also, the parameters give insight into the fundamental physical effects the cosmological parameters have on the power spectrum. Some conclusions are immediate; for example, with these parameters, it is clear that power spectrum measurements at angular scales $l > 1000$ will give very little additional constraint on any physical parameters besides n and $\mathscr{S}$, because the change in the power spectrum for $l > 1000$ when the other parameters are varied is negligible. A full sky map with MAP's angular resolution and sensitivity will provide 1-σ constraints of approximately 0.5% for $\mathscr{A}$, 2% for $\mathscr{S}$, 3% to 5% for $\mathscr{B}$, $\mathscr{R}$, and n, 30% for $\mathscr{M}$; $\mathscr{V}$ is essentially undetermined by the CMB alone [21].

THE MICROWAVE BACKGROUND AND FUNDAMENTAL PHYSICS

Beyond simple cosmological parameter estimation, the microwave background can provide other data with potential impact on fundamental physics.

Primordial Power Spectrum

Generally, the primordial power spectra of scalar and tensor perturbations have been parameterized as power laws. This approximation appears to be fairly good in the case of scalar perturbations, and slow-roll models of inflation predict power law spectra. However, more complicated inflation models generically predict departures from exact power laws [22], and the microwave background fluctuations have considerable power to measure the primordial power spectrum without prior assumptions about its shape.

The angular scales between $l = 1000$ and $l = 3500$ are largely unaffected by variations in the physical parameters in the previous section, and thus reflect the primordial power spectrum directly. However, this is only a factor of 3 in angular scale. To constrain the primordial perturbations over a significantly wider range of scales, the region between $l = 2$ and $l = 1000$ must be probed, but here the microwave background fluctuations vary greatly with the other cosmological parameters. The temperature fluctuations alone exhibit a virtual degeneracy between the primordial power spectrum and the effect of the other cosmological parameters on this range of scales [23]. The power spectrum in this range cannot be probed effectively by temperature fluctuations alone, without further assumptions or other measurements of the cosmological parameters. However, the acoustic peaks in the polarization power spectrum are generally out of phase with those in the temperature power spectrum, which breaks this degeneracy to a large extent. Accurate measurement of both the temperature and polarization power spectrum between $l = 2$ and $l = 3500$ has the potential to measure directly the primordial power spectrum of scalar perturbations over a significant range in wavenumber [24].

Initial Conditions

Usually, "adiabatic" initial conditions are assumed, which means that the fractional density fluctuations in each particle species are identical. Again, this is the natural prediction of the simplest inflationary models, but in general other fluctuations are possible. Efforts have been made to classify all such "isocurvature" fluctuations [25], although no mathematically rigorous classification has yet been obtained. Arbitrary initial conditions greatly expand the parameter space of possible models and reduce the ability to determine the cosmological parameters [26, 27]. The current CMB measurements show that the primordial perturbations are not far from adiabatic; power spectrum measurements of both temperature and polarization have the potential to put fairly sharp limits on the contributions from any other isocurvature components, which in turn could constrain the number of dynamical fields in inflation and their couplings to each other.

Gaussianity

The statistical distribution of the temperature fluctuations on the sky, and not just their power spectrum, is another way to probe the characteristics of the primordial perturbation. Simple inflation models predict the primordial perturbations should be Gaussian random distributed; departures from this prediction would signify a more complicated inflation mechanism or other new physics [28]. Gaussianity is a highly special case of all possible fluctuation patterns; no single definitive test for non-Gaussianity exists. Various techniques have been developed in the context of the microwave background (e.g. [29, 30, 31]), but so far all measurements of the temperature fluctuations down to sub-degree angular scales show no statistically significant departure from Gaussianity [8, 9, 32].

Gravitational Waves

Inflation generically predicts primordial tensor perturbations, or gravitational waves, with an amplitude proportional to the energy scale of inflation. Tensor perturbations can be cleanly separated from scalar perturbations by observing a B-polarization signal, which is produced by tensor but not by scalar perturbations [33, 34, 11]. Current measurements of the temperature power spectrum limit the amplitude of tensor perturbations to be no larger than around 20% of the scalar perturbation amplitude. Recently, Knox and collaborators have shown that the B-polarization induced by gravitational lensing provides a lower limit to the amplitude of tensor perturbations which can be detected via microwave background polarization, corresponding to an inflation energy scale of around 3×10^{15} GeV [35, 36]. If inflation occurred at energy scales higher than this, we can eventually expect direct confirmation of the inflationary scenario via detection of the microwave background polarization signal produced by the inevitable inflationary gravitational waves. Note this energy scale is generally below the coupling-constant unification scale in GUT models of particle physics.

Topology

A novel application of full-sky microwave background maps is a strong test for non-trivial large-scale topology of the Universe. While the assumption of local homogeneity and isotropy determines the Friedmann equation governing the expansion of the Universe, it says nothing about large-scale topology (see [37] for a review). Cornish, Spergel, and Starkman [38] have proposed an elegant topological test using the cosmic microwave background. They noticed that since the last scattering surface is a sphere, any non-trivial topology, which translates into intersecting the last scattering surface with multiple copies of itself, will lead to circles on the sky with identical temperature patterns. (This is actually only approximately true, because part of the observed temperature fluctuation comes from Doppler shifts at the last scattering surface instead of temperature fluctuations; see simulated maps in Ref. [39].) The number of these circles and their relative directions and orientations can be used to reconstruct the topology of the Universe [40]. Only a small number of topologies are possible in Universes with zero spatial curvature, but curved ones can support a rich variety of topologies, including some relatively small compared to the curvature scale [41, 42]. The MAP satellite has sufficient resolution to perform this test. Related ideas have also been studied [43, 44];

see [45] for a recent review.

What Cosmology Offers High-Energy Theorists

Here is a brief list of cosmological information that is relevant to fundamental physics, probed mostly by the microwave background:

- The primordial power spectrum of fluctuations over three decades in wavelength.
- Limits on or characterization of non-Gaussianity in the primordial fluctuations.
- Limits on or characterization of any isocurvature components in the primordial fluctuations.
- Detection of primordial gravitational waves, or a limit on their amplitude.
- Limits on or detection of global topology.
- Expansion rate of the Universe at the epoch of nucleosynthesis [46].
- Expansion rate of the Universe at recent epochs, either from direct observation of standard candles like SNIa [47, 48] or from nonlinear fluctuations in the microwave background (see below).
- Constraints on dark matter properties. These come from galaxies and clusters, but are less clean than microwave background conclusions, and are currently in an unsettled state (see, e.g., [49]).
- Direct detection of gravitational radiation from the early Universe. This is the other potential direct source of information about the early Universe besides the microwave background, and could potentially probe the electroweak phase transition and similar epochs [50, 51].

This is the ground on which high-energy theorists must meet cosmologists. A general and largely unaddressed question is in what ways these sources of information can constrain fundamental theories of matter and its interactions. Optimistically, eventually we will have a candidate "theory of everything" with unavoidable cosmological predictions, and the above data sources will serve as a strong test of any such theory.

SMALL-SCALE NONLINEAR FLUCTUATIONS

At small angular scales, $l > 3500$, the power spectrum of microwave background fluctuations becomes dominated by secondary effects from nonlinear structures at recent epochs, not by the primary fluctuations from linear perturbations in the early Universe. MAP will produce the definitive measurement of microwave background temperature fluctuations out to $l = 800$, and the upcoming Planck satellite will measure the temperature (and polarization) fluctuations out to $l = 3000$ in many frequency bands before the end of this decade. At that point, observations of the primary temperature anisotropies will largely be exhausted. Presently, attention is shifting to the small-scale, non-linear fluctuations, and particularly the fluctuations induced by clusters of galaxies, the largest gravitationally bound objects in the Universe. Clusters are potentially powerful tracers of the growth of cosmic structure, and thus have the potential for further probes of the fundamental properties of the Universe, though these conclusions will necessarily be accompanied by more severe systematic challenges than for probes based on the primary, linear CMB fluctuations. Below, several interesting aspects of nonlinear CMB fluctuations are sketched.

Thermal Sunyaev-Zeldovich Effect

The largest non-linear signal is that of the thermal Sunyaev-Zeldovich effect [52, 53], the spectral distortion that occurs when microwave background photons, with an initially blackbody spectrum, are Compton scattered by hot electrons, generally in clusters of galaxies. Lower energy photons are boosted to higher energies; the spectrum amplitude at the "null" of the effect, around 218 GHz (depending slightly on the density and temperature of the electrons), remains constant. In the direction of galaxy clusters, therefore, the microwave radiation appears cooler for frequencies below the null, hotter for frequencies above. The amplitude of this effect can be as large as 1 mK for large clusters of galaxies, much larger than the primary temperature fluctuations. The total spectral distortion is proportional to the product of the electron density and the electron temperature, integrated along the line of sight.

The thermal SZ effect has now been detected for numerous clusters of galaxies, at roughly arcminute angular resolution (e.g. [54, 55]). Besides providing direct information about the density and temperature of the gas in galaxy clusters, the great utility of the SZ effect is its independence of cluster redshift: since the induced spectral distortion remains as the microwave background radiation propagates, a given galaxy cluster will produce the same observed SZ distortion independent of its distance from the observer. This is in marked contrast to other methods of observing clusters from their direct emission of radiation. SZ observations hold the promise of cluster catalogs with relatively simple and complete selection functions, which in turn are necessary for any use of galaxy clusters as precision cosmological probes.

Kinematic Sunyaev-Zeldovich Effect

A related but smaller nonlinear effect results from the Doppler shift experienced by photons scattering from electrons moving with a coherent peculiar velocity. This bulk velocity induces a blackbody temperature shift of the microwave photons proportional to the radial component of the electrons' peculiar velocity. (In the mildly nonlinear regime of structure, this effect is known as the Ostriker-Vishniac effect [56].) In galaxy clusters, the typical amplitude of this temperature shift is a few μK, much smaller than the thermal effect. The two can be separated by their spectral dependences, in principle. Even though galaxy clusters are highly dynamic objects with significant internal bulk flows due to mergers, the average kinematic SZ signal provides a largely unbiased measure of the cluster's peculiar radial velocity [57]. If the kinematic SZ distortion can be extracted reliably for individual galaxy clusters, then galaxy clusters can be used as tracers of the cosmic peculiar velocity field out to redshifts beyond $z = 1$. Current peculiar velocity surveys extract velocities by estimating the distance to galaxies and then subtracting the inferred Hubble velocity from the observed redshift velocity to obtain a peculiar velocity [58]. Such a procedure quickly becomes dominated by systematic errors at cosmologically modest distances due to the difficulty of accurate distance estimation. The great advantage of the kinematic SZ effect, in comparison, is that it provides a direct peculiar velocity estimate without requiring a distance determination. A peculiar velocity map over a substantial portion of the observable Universe will provide a sharp test of the gravitational instability paradigm of structure growth [59] and a strong consistency check with surveys of the cosmic density field [60, 61].

Weak Gravitational Lensing

As the microwave background radiation propagates from the last scattering surface to the observer, its geodesics will be altered by the presence of intervening matter inhomogeneities. This gravitational lensing has only small effects on the power spectrum of the microwave background fluctuations [62], but does change the pattern of the radiation, inducing a specific form of non-Gaussianity [63]. Lensing creates a correlation between two-point correlations on degree scales and four-point correlations on smaller scales. Recently, algorithms have been developed to reconstruct the lensing mass distribution given a lensed temperature map. Temperature information alone can give the correct qualitative structure, while a high-sensitivity polarization map on scales of a few arcminutes can determine the projected lensing mass distribution to good accuracy on sub-degree scales [64]. The lensing mass distribution can also be determined from shear measurements of background galaxies on comparable angular scales, providing a valuable cross-check. Microwave background lensing has the advantage of a well-defined source redshift, along with completely different systematic errors for these challenging observations.

Strong Gravitational Lensing

In the sky regions of galaxy clusters, the large mass concentration significantly distorts the microwave background temperature pattern. This strong gravitational lensing signal has several distinct features. Most notably, it produces a double-lobe distortion aligned with the temperature gradient of the background fluctuations [65]. This distortion can be used to reconstruct the cluster mass profile, in principle: while observations of background galaxy shapes yield only information about the relative shear field of the mass distribution, lensing of the microwave background provides direct information about the displacement field induced by the mass distribution. The characteristic angular scale of the strong lensing distortion is an arcminute; on this scale, the primary CMB fluctuations have essentially no power. Thus, in certain regions of the sky where the primary fluctuations are especially regular, the lensing displacement field from a cluster can likely be modelled with good precision and used to estimate cluster masses. Since the displacement is estimated directly, this method has no mass sheet degeneracy, like optical galaxy lensing estimates of cluster masses, and has no uncertainties related to the background source redshift. Ultimately, the accuracy of this cluster mass determination method will depend on how well the background microwave temperature distribution can be modelled, and on how well the blackbody lensing distortion can be separated from the blackbody kinematic SZ distortion. Accurate cluster mass determination is crucial for directly measuring $N(M,z)$, the number density of clusters at a given mass and redshift. This function is a highly sensitive probe of the recent growth of structure and constrains the cosmological constant or other "dark energy" contributions, as well as small neutrino masses.

FUTURE EXPERIMENTAL PROSPECTS: ACT

Experimental techniques for measuring the microwave background radiation are, remarkably, advancing at an accelerating pace. The various kinds of observations outlined in the previous section require, roughly, arcminute resolution maps with micro-Kelvin temperature sensitivity. Such measurements are on the menu for the coming half-decade. As an example, I provide a brief overview of the Atacama Cosmology Telescope (ACT), an experimental collaboration between Princeton, U. Pennsylvania, Rutgers, NASA Goddard, NIST, and several other smaller partners. This collaboration exhibits many technological and organizational characteristics which we anticipate will come to dominate the microwave background field over the coming decade. Experiments of comparable scope and ambition are also being planned by other groups.

The ACT collaboration plans to construct a custom-designed 6-meter off-axis telescope which is optimized to minimize the systematic errors which can easily dominate any precision measurement of the microwave background radiation. It will scan the sky at fixed elevation by rotating on a turntable; the entire telescope, including ground screens, will rotate as a unit to eliminate any change in side lobes. Constant-elevation scans greatly aid in controlling systematic errors, since the atmospheric emission changes by many hundreds of μK for elevation changes smaller than a degree. A similar design philosophy was used successfully in the Saskatoon [66] and MAT [67] experiments. The telescope will be cited on Cerro Toco in the Atacama Desert of the Chilean Andes, near the site for the ALMA interferometer. Site studies show that the atmospheric signal at microwave frequencies will generally be smaller than our detector noise for all scales $l > 100$. At this latitude, scans at a constant 45° elevation with an amplidude of $\pm 1.5°$ can be combined over 24 hours to produce a map of an annular region of sky approximately 2° wide by 120° degrees around. About half the scan will cover the galactic plane and will be useful for galactic astronomy; the other half will be of cosmological quality.

ACT's novel detector technology will enable it to produce maps of unprecedented sensitivity and angular resolution. We plan to build a "camera" composed of bolometer arrays at three different frequencies, each array containing 1024 square-millimeter transition edge sensor detectors [68, 69]. These bolometers are very fast, enabling rapid scanning of the sky. Each individual bolometer's sensitivity is within a factor of two of the bolometers planned for the Planck satellite; packing them into a large array gives ACT the raw detector sensitivity required for achieving a nominal angular resolution of 1.7' (at 150 GHz) with a nominal sensitivity per pixel of near 1 μK for a three-month observing campaign. The largest technological challenge of the experiment is fabricating the detector arrays with their associated SQUID multiplexors [70] and other back-end electronics. Current state-of-the-art bolometer arrays employ tens of bolometers; we aim to increase this by a factor of a hundred. The anticipated time scale for building the experiment and observing for two seasons is five years.

The microwave observations from ACT will be combined with an integrated optical and X-ray observing campaign aimed at the galaxy clusters discovered via their SZ signals. Optical observations with the Southern African Large Telescope (currently under construction) and other large telescopes will provide redshifts and galaxy velocity dispersions for hundreds of clusters out to redshift unity, while X-ray observations will give information about gas temperature and density. Through this combination of observations, we aim to construct the most complete and useful cosmological cluster sample to date. We will begin to probe all of the small-scale signals outlined in the previous section.

As spectacular as the advances in microwave background physics have been over the past decade, we can reasonably expect another decade of comparable discovery, with remarkable implications for cosmology, fundamental physics, and astrophysics.

ACKNOWLEDGMENTS

At the time of this writing, the Atacama Cosmology Telescope is a currently pending NSF proposal; Lyman Page (Princeton University) is the Principal Investigator. The author thanks all of the collaborators on this proposal, who are too numerous to list here. This work has been supported in part by NASA's Space Astrophysics Research and Analysis Program, and by the Cottrell Scholar program of the Research Corporation.

REFERENCES

1. Kamionkowski, M., and Kosowsky, A., *Ann. Rev. Nuc. Part. Sci.*, **49**, 77 (1999).
2. Durrer, R., *J. Phys. Stud.*, **5**, 177 (2001).
3. Hu, W., and Dodelson, S., *Ann. Rev. Astron. Astrophys.*, **40**, 171 (2002).
4. Kosowsky, A., "The Cosmic Microwave Background," in *Modern Cosmology*, edited by S. Bonometto, V. Gorini, and U. Moschella, IOP Publishing, Bristol and Philadelphia, 2002, p. 219.
5. Hu, W., *Ann. Phys.*, p. in press (2003).
6. Kosowsky, A., *New Astron. Rev.*, **43**, 157 (1999).
7. Gorski, K. M., et al., *Astrophys. J. Lett.*, **430**, 89 (1994).
8. Park, C.-G., Park, C., Ratra, B., and Tegmark, M., *Astrophys. J.*, **556**, 582 (2001).
9. Shandarin, S., Feldman, H., Xu, Y., and Tegmark, M., *Astrophys. J. Suppl.*, **141**, 1 (2002).
10. Kamionkowski, M., Kosowsky, A., and Stebbins, A., *Phys. Rev. D*, **55**, 7368 (1997).
11. Zaldarriaga, M., and Seljak, U., *Phys. Rev. D*, **55**, 1830 (1997).
12. Lue, A., Wang, L., and Kamionkowski, M., *Phys. Rev. Lett.*, **83**, 1506 (1999).
13. Knox, L., *Phys. Rev. D*, **52**, 4307 (1995).
14. Jungman, G., Kamionkowski, M., Kosowsky, A., and Spergel, D., *Phys. Rev. D*, **54**, 1332 (1996).

15. Zaldarriaga, M., Spergel, D., and Seljak, U., *Astrophys. J.*, **488**, 1 (1997).
16. Bond, J., Efstathiou, G., and Tegmark, M., *Mon. Not. R. Ast. Soc.*, **291**, 33 (1997).
17. Wang, X., Tegmark, M., and Zaldarriaga, M., *Phys. Rev. D*, **65**, 123001 (2002).
18. Christensen, N., Meyer, R., Knox, L., , and Luey, B., *Class. Quant. Grav.*, **18**, 2677 (2001).
19. Knox, L., Christensen, N., and Skordis, C., *Astrophys. J. Lett.*, **563**, 95 (2001).
20. Lewis, A., and Bridle, S., *Phys. Rev. D*, **66**, 103511 (2002).
21. Kosowsky, A., Milosavljevic, M., and Jimenez, R., *Phys. Rev. D*, **66**, 063007 (2002).
22. Salopek, D. S., Bond, J. R., and Bardeen, J. M., *Phys. Rev. D*, **40**, 1753 (1989).
23. Wang, Y., Spergel, D., and Strauss, M., *Astrophys. J.*, **510**, 20 (1999).
24. Tegmark, M., and Zaldarriaga, M., *Phys. Rev. D*, **66**, 103508 (2002).
25. Bucher, M., Moodley, K., and Turok, N., *Phys. Rev. D*, **62**, 083508 (2000).
26. Bucher, M., Moodley, K., and Turok, N., *Phys. Rev. D*, **66**, 023528 (2002).
27. Trotta, R., Riazuelo, A., and Durrer, R., *Phys. Rev. Lett.*, **87**, 231301 (2001).
28. Bernardeau, F., and Uzan, J.-P., *Phys. Rev. D*, **66**, 103506 (2002).
29. Rocha, G., Magueijo, J., Hobson, M., and Lasenby, A., *Phys. Rev. D*, **64**, 063512 (2001).
30. Winitzki, S., and Kosowsky, A., *New Astron.*, **3**, 75 (1997).
31. Ferreira, P. G., and Magueijo, J., *Phys. Rev. D*, **55**, 3358 (1997).
32. Kunz, M., et al., *Astrophys. J. Lett.*, **563**, 99 (2001).
33. Kamionkowski, M., Kosowsky, A., and Stebbins, A., *Phys. Rev. Lett.*, **78**, 2058 (1997).
34. Kamionkowski, M., and Kosowsky, A., *Phys. Rev. D*, **57**, 685 (1998).
35. Knox, L., and Song, Y., *Phys. Rev. Lett.*, **89**, 011303 (2002).
36. Kesden, M., Cooray, A., and Kamionkowski, M., *Phys. Rev. Lett.*, **89**, 011304 (2002).
37. Lachieze-Rey, M., and Luminet, J., *Phys. Rep.*, **254**, 135 (1995).
38. Cornish, N., Spergel, D., and Starkman, G., *Class. Quant. Grav.*, **15**, 2657 (1998).
39. Riazuelo, A., Uzan, J.-P., Lehoucq, R., and Weeks, J., *Phys. Rev. D*, p. submitted (2002).
40. Weeks, J., *Class. Quant. Grav.*, **15**, 2599 (1998).
41. Hodgson, C., and Weeks, J., *Experimental Math.*, **3**, 261 (1994).
42. Gausmann, E., et al., *Class. Quant. Grav.*, **18**, 5155 (2001).
43. Levin, J., et al., *Phys. Rev. D*, **58**, 123006 (1998).
44. Bond, J., Pogosyan, D., and Souradeep, T., *Phys. Rev. D*, **62**, 043005 (2000).
45. Levin, J., *Phys. Rep.*, **365**, 261 (2002).
46. Carroll, S., and Kaplinghat, M., *Phys. Rev. D*, **65**, 063507 (2002).
47. Perlmutter, S., et al., *Astrophys. J.*, **517**, 565 (1999).
48. Garnavich, P. M., et al., *Astrophys. J.*, **509**, 74 (1998).
49. de Blok, W. J. G., McGaugh, S. S., and Rubin, V. C., *Astron. J.*, **122**, 2396 (2001).
50. Kamionkowski, M., Kosowsky, A., and Turner, M., *Phys. Rev. D*, **49**, 2837 (1994).
51. Kosowsky, A., Mack, A., and Kahniahsvili, T., *Phys. Rev. D*, **66**, 024030 (2002).
52. Birkinshaw, M., *Phys. Rep.*, **310**, 97 (1999).
53. Carlstrom, J. E., Holder, G. P., and Reese, E. D., *Ann. Rev. Astron. Astrophys.*, **40**, 643 (2002).
54. Mauskopf, P. D., et al., *Astrophys. J.*, **538**, 505 (2000).
55. Reese, E. D., et al., *Astrophys. J.*, **581**, 53 (2002).
56. Jaffe, A., and Kamionkowski, M., *Phys. Rev. D*, **58**, 043001 (1998).
57. Nagai, D., Kravtsov, A., and Kosowsky, A., *Astrophys. J.*, p. in press (2003).
58. Willick, J. A., et al., *Astrophys. J. Suppl.*, **109**, 333 (1997).
59. Juszkiewicz, R., Springel, V., and Durrer, R., *Astrophys. J. Lett.*, **518**, 25 (1999).
60. Dodelson, S., et al., *Astrophys. J.*, **572**, 140 (2002).
61. Percival, W. J., et al., *Mon. Not. R. Ast. Soc.*, **327**, 1297 (2001).
62. Seljak, U., *Astrophys. J.*, **463**, 1 (1996).
63. Bernardeau, F., *Astron. Astrophys.*, **324**, 15 (1997).
64. Hu, W., and Okamoto, T., *Astrophys. J.*, **574**, 566 (2002).
65. Seljak, U., and Zaldarriaga, M., *Astrophys. J.*, **538**, 57 (2000).
66. Wollack, E. J., et al., *Astrophys. J. Lett.*, **419**, 49 (1993).
67. Miller, A., et al., *Astrophys. J. Suppl.*, **140**, 115 (2002).
68. Irwin, K., Hilton, G., Wollman, D., and Martinis, J., *Appl. Phys. Lett.*, **69**, 1945 (1996).
69. Lee, A., Richards, P., Nam, S., Cabrera, B., and Irwin, K., *Appl. Phys. Lett.*, **69**, 1801 (1996).
70. Chervenak, J., et al., *Appl. Phys. Lett.*, **74**, 4043 (1999).

What can we learn about cosmic structure from gravitational waves?

Joan M. Centrella

Laboratory for High Energy Astrophysics, NASA Goddard Space Flight Center, Greenbelt, MD 20771 USA

Abstract. Observations of low frequency gravitational waves by the space-based LISA mission will open a new observational window on the early universe and the emergence of structure. LISA will observe the dynamical coalescence of massive black hole binaries at high redshifts, giving an unprecedented look at the merger history of galaxies and the reionization epoch. LISA will also observe gravitational waves from the collapse of supermassive stars to form black holes, and will map the spacetime in the central regions of galaxy cusps at high precision.

INTRODUCTION

Most of the information we have about the universe in general, and cosmic structure in particular, has been gleaned through detections of electromagnetic radiation. For example, observations with optical telescopes have revealed galaxy morphologies and clustering patterns; studies of galaxy spectra uncovered the universal recession as well as the presence of dark matter. Radio telescopes have shown a variety of active galaxies with vast cosmic jets emanating from their centers. Observations of the microwave sky revealed the 3K background radiation and the imprints of structure at early epochs. And high energy detectors unveiled a variety of exotic phenomena, ranging from hot gas in galaxy clusters to accretion disks around massive black holes (MBHs) at the centers of galaxies. This wealth of this electromagnetic information comes generally from excitations in gases in the outer regions of stars, within galaxies and clusers of galaxies, and in accretion flows around compact objects such as black holes.

As the 21st century begins, a new observational window on the universe is being opened by detectors designed to measure gravitational waves from astrophysical sources. Gravitational radiation is produced through the bulk motion of massive objects such as black holes, and reveals their dynamical behavior directly. This article begins with a short introduction to gravitational radiation, and then encompasses a brief tour of several gravitational wave sources that can contribute to our understanding of cosmic structure.

A GRAVITATIONAL RADIATION PRIMER

Gravitational waves are ripples in spacetime curvature that travel at the speed of light. They are typically generated by compact matter distributions that have time-changing quadrupole moments[1] such as binary systems and nonspherical collapses, and carry information about the bulk motion of the sources. Since gravitational waves couple very weakly to matter, they easily escape from the centers of galaxies or collapsed objects, bringing information about these deep, hidden regions. Gravitational waves carry both energy and momentum. For a binary system, this means that both the binary separation and the orbital period will decrease due to the emission of gravitational radiation. Such behavior has been observed in the binary pulsar PSR B1913+16, providing strong confirmation of the existence of gravitational radiation as predicted by general relativity [2].

The characteristic dimensionless gravitational wave amplitude for a source of mass M located at a distance r can be estimated as (e.g., [3])

$$h \sim \frac{G}{c^4}\frac{1}{r}\frac{d^2Q}{dt^2} \sim \frac{R_{\mathrm{Schw}}}{r}\frac{v_{\mathrm{ns}}^2}{c^2}, \quad R_{\mathrm{Schw}} = \frac{2GM}{c^2}. \tag{1}$$

Here, v_{ns} is the characteristic nonspherical velocity in the source, Q is its (trace-free) quadrupole moment, and R_{Schw} its Schwarzschild radius. Thus, the strongest sources have large masses moving with $v_{\mathrm{ns}} \sim c$. For a stellar black hole system having mass $M = 10 M_\odot$, this gives $h \sim 10^{-16}$ at distance $r = 15\mathrm{kpc}$, and $h \sim 10^{-21}$ at $r = 3000\mathrm{Mpc}$. The gravitational waves from a MBH system with $M = 2.5 \times 10^6 M_\odot$ at a distance $r = 3000\mathrm{Mpc}$ typically have amplitudes $h \sim 10^{-16}$.

The observed gravitational wave frequency of a source depends on its mass, physical scale, and redshift. For example, the nonspherical collapse of an object at redshift z will produce a burst of gravitational waves with a characteristic frequency

$$f_{\mathrm{burst}} \sim \frac{1}{2\pi\, t_{\mathrm{dyn}}}\frac{1}{(1+z)} \sim (1.1 \times 10^{-2}\mathrm{Hz})\frac{10^6 M_\odot}{(1+z)M}\left(\frac{R_{\mathrm{Schw}}}{R}\right)^{3/2}, \tag{2}$$

where $t_{\mathrm{dyn}} = \sqrt{GM/R^3}$ is the dynamical timescale. The nonspherical collapse of a star at $z \sim 0$ with mass $M \sim 10 M_\odot$ near its Schwarzschild radius $R \sim R_{\mathrm{Schw}}$ can produce a burst of high frequency gravitational radiation at $f_{\mathrm{burst}} \sim 1\mathrm{kHZ}$. For the collapse of a supermassive star with $M \sim 10^6 M_\odot$ at $z \sim 2$, the radiation is produced at low frequency, $f_{\mathrm{burst}} \sim 3.7\mathrm{mHz}$.

A distorted black hole, such as might be formed from nonspherical collapse or the coalescence of two black holes or neutron stars, will emit gravitational waves as it settles down to a quiescent axisymmetric Kerr hole. These so-called ringdown waves have frequencies in the range [4, 5]

$$f_{\mathrm{ring}} \sim (1.2 - 3.2) \times 10^{-2}\mathrm{Hz}\left(\frac{10^6 M_\odot}{(1+z)M}\right), \tag{3}$$

where the lower end of the frequency range corresponds to a nonrotating Schwarzschild hole, and the upper end to a maximally rotating Kerr hole. Thus, a distorted stellar

---

[1] Conservation of mass and momentum guarantee that there can be no monopolar or dipolar gravitational radiation, respectively. [1]

black hole with mass $M \sim 20M_\odot$ in a nearby galaxy with $z \sim 0$ emits high frequency gravitational waves with $f_{\text{ring}} \sim 1\text{kHz}$. In contrast, a distorted MBH with mass $M \sim 2.5 \times 10^6 M_\odot$ rings down at low frequencies, with $f_{\text{ring}} \sim 4\text{mHz}$ at $z \sim 1$ and $f_{\text{ring}} \sim 1\text{mHz}$ at $z \sim 5$.

Another important class of gravitational wave sources consists of binary systems. The waves generated by a binary system whose components are on circular orbits and have comparable masses have frequency [6]

$$f_{\text{binary}} = \frac{2f_{\text{orbital}}}{1+z} = \frac{1}{\pi}\frac{1}{(1+z)}\left(\frac{GM}{a^3}\right)^{1/2}, \qquad (4)$$

where M is the total mass of the binary and a is the separation of the binary components. As the gravitational waves are emitted, the binary separation a decreases until it reaches $a \sim 3R_{\text{Schw}}$, by which point the components have begun to merge. This gives a frequency range for these inspiral waves [7]

$$f_{\text{inspiral}} \lesssim 400\text{Hz}\left(\frac{10M_\odot}{(1+z)M}\right). \qquad (5)$$

Again we see that stellar black hole systems produce high frequency waves, so that $f_{\text{inspiral}} \lesssim 200\text{Hz}$ for a binary with total mass $M \sim 20M_\odot$ at $z \sim 0$. MBH binaries emit waves at lower frequencies; a binary with total mass $M \sim 2.5 \times 10^6 M_\odot$ produces gravitational waves at frequencies $f_{\text{inspiral}} \lesssim 1\text{mHz}$ at $z \sim 1$ and $f_{\text{inspiral}} \lesssim 0.3\text{mHz}$ at $z \sim 5$.

DETECTING GRAVITATIONAL WAVES

Broad-band gravitational wave detectors use laser interferometry to monitor the separation between test masses located at a distance L from each other. When a gravitational wave impinges on a detector, the separation between the masses changes by ΔL, resulting in a strain amplitude $h(t) = \Delta L/L$. Since gravitational wave detectors measure displacement $h \sim 1/r$ (see Eq. (1)), improving the sensitivity by a factor of 2 increases the regions of the universe from which sources can be detected by a factor of 8. Electromagnetic detectors generally measure energy flux $\sim 1/r^2$, so a similar increase in the observable volume would require improving the detector sensitivity by a factor of 4.

Ground-based kilometer-scale interferometers, such as LIGO, VIRGO, and GEO600, are sensitive to high frequency gravitational radiation in the range $10\text{Hz} \lesssim f_{\text{GW}} \lesssim 10^4\text{Hz}$. For example, the LIGO (Laser Interferometric Gravitational-wave Observatory) project [8] has 2 interferometers with arm length $L = 4\text{km}$, one in Livingston, LA and and the other in Hanford, WA, and a third with $L = 2\text{km}$ at Hanford. LIGO has recently begun to take scientific data, and should be able to observe the dynamics of black holes with masses up to $\sim$ few $\times 10M_\odot$ as well as other high frequency sources. The first generation interferometers are expected to reach their best strain sensitivities $h \sim 10^{-21}$ at frequencies near $f \sim 200\text{Hz}$. Second generation detectors can lower the broad-band sensitivity to $h \sim 8 \times 10^{-23}$ and achieve even better sensitivity with "tunable" narrow band techniques [9].

The planned space-based LISA (Laser Interferometric Space Antenna) mission [10] is a joint NASA/ESA collaboration designed to detect low frequency gravitational waves in the range $10^{-5}\text{Hz} \lesssim f_{\text{GW}} \lesssim 1\text{Hz}$. LISA consists of 3 spacecraft at the vertices of an equilateral triangle with arm lengths $L \simeq 5 \times 10^6 \text{km}$. The detector will follow 20° behind the Earth in its orbit about the Sun, and will observe gravitational waves from the dynamics of MBHs and other low frequency sources. LISA will reach its best strain sensitivity $h \sim 10^{-23}$ around frequency $f \sim 5\text{mHz}$ and should be capable of detecting MBH binaries to redshifts $z > 10$.

The timing of millisecond pulsars provides a means of probing the very low frequency gravitational wave band, $10^{-9}\text{Hz} \lesssim f \lesssim 10^{-7}\text{Hz}$. The rotation period of a millisecond pulsar is determined to high precision by measuring the arrival times of the pulses. Perturbations in the arrival times arise when the electromagnetic waves from the pulsar pass through distortions in spacetime caused by a stochastic background of gravitational waves [11], such as that from the ensemble of MBH binary coalescences across the universe. Current pulsar data sets reach a minimum characteristic strain $h \sim 4 \times 10^{-15}$ at frequencies $f \sim$ few nHz; observations with a planned Pulsar Timing Array could lower this sensitivity to $h \sim 10^{-15}$ [12].

GRAVITATIONAL WAVES FROM MBH BINARIES

MBH binaries are expected to result from the merger of galaxies [13, 14] and, in hierarchical structure formation scenarios, the merger of halos containing seed black holes [15, 16]. A variety of astrophysical mechanisms have been proposed to bring a MBH binary from an initally wide orbital separation $a \gtrsim 1\text{kpc}$ to the regime in which gravitational radiation reaction will cause it to coalesce in less than a Hubble time. These include both dynamical friction from background stars [13, 17] and gas dynamical effects [18]. The time needed for a binary of total mass M and separation a to coalesce once gravitational radiation reaction becomes the dominant energy loss mechanism is [6]

$$t_{\text{GR}} = \frac{5}{64} \frac{c^5}{G^3} \frac{a^4}{M^3}, \tag{6}$$

assuming circular orbits and equal mass components. For example, a MBH binary of total mass $M \sim 2 \times 10^7 M_\odot$ starting from a separation $a \sim 10^{-2}\text{pc}$ will coalesce in a Hubble time $\sim 10^{10}$ years.

The final gravitational radiation-driven coalescence can be thought of as proceeding in 3 stages: an adiabatic inspiral, followed by a dynamical merger, and a final ringdown. As the binary evolves from inspiral through merger and ringdown, the gravitational wave frequency increases. As long as any accompanying accretion disks are dynamically negligible, black hole binaries coalescing through the emission of gravitational radiation are solutions to the vacuum (i.e., source-free) equations of general relativity. As such, their timescales are proportional to the total mass and, after time scaling, all other properties of the dynamics and waveforms depend only on ratios involving the masses and spins of the components. Thus, calcuations of the gravitational waveforms from all three stages of black hole binary coalescence can be easily scaled to apply to any masses and spins [3].

During the slow inspiral, the black holes are well-separated and spiral together adiabatically. The resulting gravitational waveform is a "chirp," which is a sinusoid that increases in frequency and amplitude as the orbital period shrinks; see Eq. (4). The gravitational waveforms from this inspiral phase are computed analytically using higher order post-Newtonian techniques in which the black holes are approximated as point masses [19]. These waveforms can be used as templates for data analysis algorithms based on matched filtering [20].

Typically, MBH binaries will be within the LISA frequency band for periods of roughly several months to years. If a sufficient number of cycles of the inspiral waveform are thus observed, both the so-called chirp mass $\mathcal{M}_c = (M_1 M_2)^{3/5}/(M_1+M_2)^{1/5}$ and, with less accuracy, the reduced mass $m_{\rm red} = M_1 M_2/(M_1+M_2)$ can be measured, as well as some information on the spins [21].

LISA's ability to detect a binary black hole inspiral depends on both the mass and redshift of the system. In particular, the gravitational wave frequencies of a binary of masses M_1 and M_2 at redshift z are the same as those emitted by a binary local to the detector but having masses $(1+z)M_1$ and $(1+z)M_2$ [21]. Increasing the mass and/or the redshift lowers the frequency; see Eqs. (3) and (5). For binaries with mass $(1+z)M \sim 10^5 M_\odot$, most of the inspiral signal occurs in the frequency band in which LISA has very good sensitivity; this results in high signal-to-noise ratios (SNR) $\sim 10^2 - 10^3$ for such systems [21]. Overall, LISA should be able to detect the inspiral phase at SNR $\gtrsim 10$ for equal mass binaries of total mass $M \sim 10^3 M_\odot - 10^6 M_\odot$ out to $z \sim 9$, and higher mass inspirals $M \sim 10^7 M_\odot$ at lower redshifts $z \sim 1$ [21].

If MBH binaries proceed directly to the regime in which they coalesce under gravitational radiation reaction, recent calculations predict ~ 10 or more events per year will occur within LISA's frequency band [22, 23, 24]. Since many of these may originate at redshifts $z \gtrsim 7$, LISA can provide an outstanding probe of the high redshift universe, including the reionization epoch and the merger history of galaxies [24]. Observations in the nHz frequency range by pulsar timing arrays will detect the stochastic gravitational wave background from MBH binaries at lower redshifts, providing important constraints on that population [12].

The relatively slow inspiral is followed by the dynamical merger phase, in which the black holes plunge toward each other and merge into a single remnant, emitting a burst of gravitational waves. This stage is governed by strong nonlinear gravitational fields, and the resulting waveforms can only be calculated using full 3-D numerical relativity simulations [25, 26]. Typically, LISA will be able to observe gravitational wave bursts from this dynamical merger phase on timescales of roughly minutes to hours, providing an outstanding probe of a regime that is expected to be phenomenologically rich. For example, the merger of two black holes may cause a substantial change in the direction of the spin axis of the more massive hole [27]. The jets emanating from the center of an active galaxy are believed to be directed along the spin axis of the central MBH [28]. Thus, if such a hole suffered a change in spin direction as a result of merging with another MBH, the change in direction of the jet would cause a new radio lobe to be

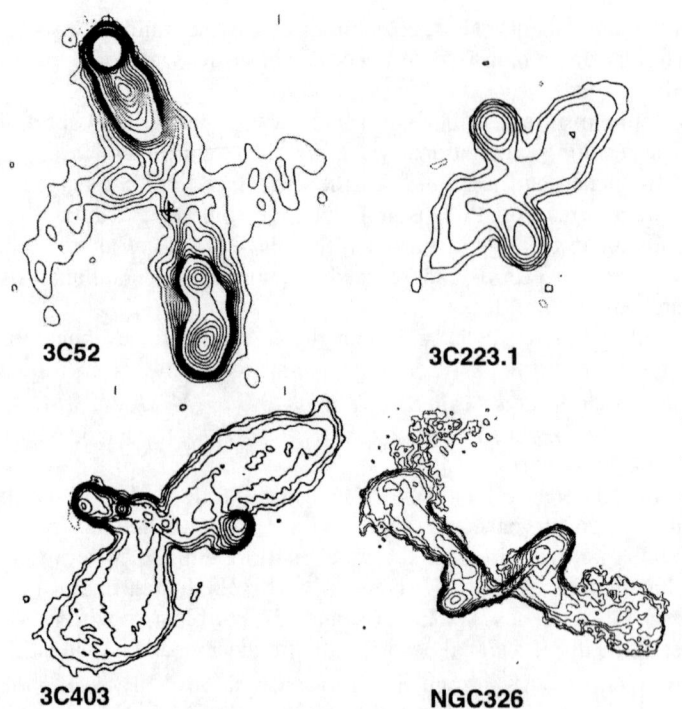

FIGURE 1. VLA radio observations of 4 radio galaxies showing the X-type morphology [30, 31, 32]. Reprinted with permission from Merritt, D. and Ekers, R. D., *Science*, **297**, 1310 (2002). Copyright 2002 American Association for the Advancement of Science.

generated at some angle to the original one. This has been proposed as an explanation for the "winged" or "X-type" radio sources shown in Fig. 1 [29].

After the dynamical merger is complete and a common event horizon forms, the distorted remnant will ring down to a quiescent Kerr hole through the emission of gravitational radiation in quasi-normal modes. The ringdown waveforms are known analytically from calculations of black hole perturbations, and take the form of damped sinusoids [4]. LISA will observe these bursts on time scales of typically minutes to hours. Detections at SNR $\gtrsim 10^2$ should be possible for masses $M \sim 10^5 - 10^7 M_\odot$ out to $z \sim 9$ for any value of the spin [21]. For $M \sim 10^8 M_\odot$ at $z \sim 1$, the ringdown occurs at the edge of LISA's low frequency sensitivity, and detection of the ringdown with large SNR requires large spin; c.f. Eq. (3) [33]. Observations of the gravitational waves from MBH ringdown will allow the identification of the mass and spin of the final Kerr black hole.

SUPERMASSIVE STARS AND MBH FORMATION

The origin of MBHs is a topic of considerable interest for understanding the high redshift universe and the emergence of structure [34]. Various scenarios have been proposed, including the collapse of baryonic gas clouds (c.f. [35]) and dark matter halos [36]. MBHs may form hierarchically through the coalescence of smaller "seed" black holes. They may also form through direct collapse of large gas clumps; in fact, a recent study reveals the intriguing possibility that such MBHs may *form* in binaries [37].

A large gas cloud undergoing direct collapse may first form a rotating supermassive star as an intermediate state. Such an object is dominated by radiation pressure and is subject to a general relativistic radial instability leading to further collapse [6]. Recent numerical simulations have demonstrated that a uniformly rotating supermassive star collapses coherently to a black hole containing $\sim 90\%$ of the original mass; the rest of the gas forms a rotating disk outside the hole [38, 39]. A burst of gravitational waves from the collapse itself will be followed by ringdown radiation as the newly formed black hole settles into a quiescent state. For an interesting range of masses and redshifts, f_{burst} and f_{ring} lie within LISA's sensitive range; see Eqs. (2) and (3). If the supermassive star is differentially rotating, a non-axisymmetric instability could trigger bar formation before the catastrophic collapse; the rotation of this bar could generate quasi-periodic gravitational waves in the LISA band even if a MBH does not form [40].

MAPPING MBH SPACETIMES

The final class of gravitational waves sources that we consider here consists of compact remnants of evolved stars – white dwarfs, neutron stars, or black holes – orbiting close to the central MBH in a galaxy [41, 42]. Compact remnants are expected to comprise a significant fraction of the total stellar population in the central cusp surrounding the MBH. If such an object is scattered into an orbit around the MBH, it will spiral in through the emission of gravitational radiation, eventually being captured by the MBH. If the remnant is on a tight eccentric trajectory, roughly the final year before capture will be spent executing $\sim 10^5 - 10^6$ orbits through the very strong field region near the event horizon of the MBH [43]. The gravitational radiation emitted during this stage can be detected by LISA at high accuracy [44, 45].

Since a stellar remnant of mass μ spiralling into a MBH of mass M constitutes an extreme mass ratio binary, typically $\mu/M \lesssim 10^{-4}$, the motion of the remnant can be treated perturbatively in the MBH spacetime. The emission of gravitational radiation as the particle moves in the MBH potential causes the orbit to evolve. Technical issues arise in computing the effects of radiation reaction on the orbits and the accompanying waveforms [45]. Recent work on nonequatorial circular inspirals around a massive Kerr black hole shows that the gravitational waveforms are modulated by the effects of MBH spin and frame dragging, as shown in Fig. 2 [43]. The modulation is stronger at later times when the remnant is closer to the MBH event horizon. Complicated behavior also arises from eccentric orbits [46] and the spin of the remnant.

These trajectories and the resulting gravitational waveforms encode information about

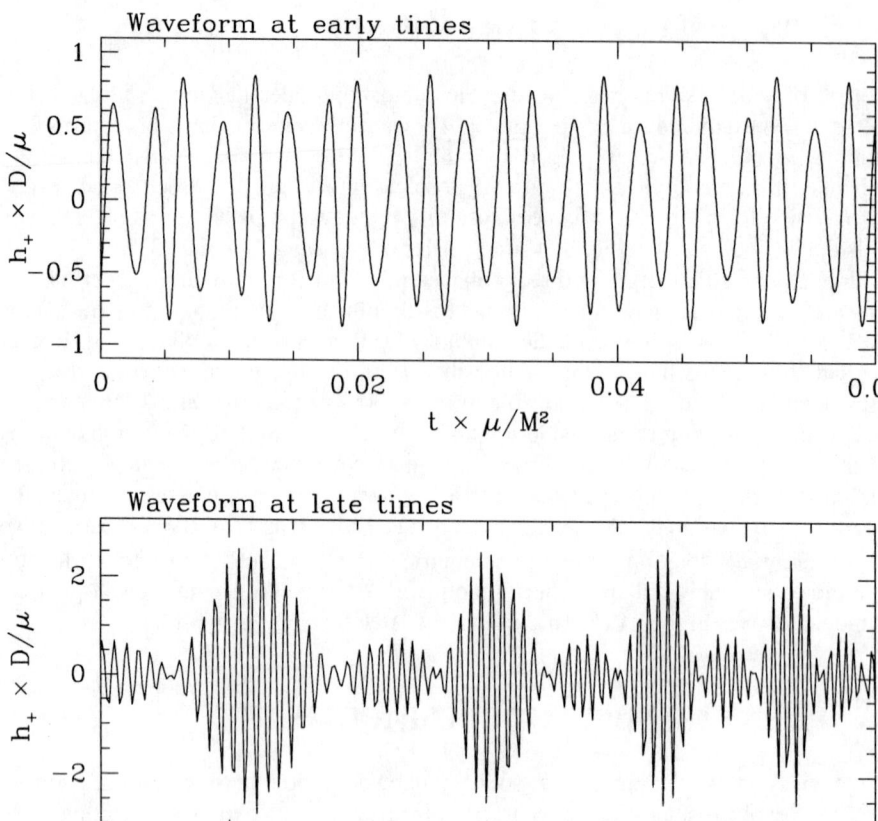

FIGURE 2. The gravitational waveform from an extreme mass ratio inspiral with $\mu/M = 10^{-4}$. The particle moves on a circular, nonequatorial trajectory around a central Kerr black hole which has nearly maximal spin, 0.998M. Reprinted from Ref. [43].

the multipolar structure of the gravitational potential of the central massive object. Measuring just three multipole moments can falsify whether this massive object is indeed a black hole [5]. LISA should be able to measure these extreme mass ratio waveforms out to a distance of a few Gpc if the remnant is a black hole with mass $\mu \sim 10 M_\odot$, and out to a few ×100Mpc if the remnant is a white dwarf or neutron star [41, 42, 45]. Conservative estimates suggest that LISA will measure at least several such events per year, and possibly many more.

SUMMARY

Gravitational wave detectors are poised to open a new observational window on the universe. Ground-based interferometers will observe the high frequency regime, $10\text{Hz} \lesssim f_{GW} \lesssim 10^4\text{Hz}$, including sources such as the dynamics of stellar black holes. The low frequency region of the spectrum, $10\text{Hz} \lesssim f_{GW} \lesssim 10^4\text{Hz}$, will be probed by the space-based LISA mission, and is especially important for understanding the emergence of structure in the universe. LISA will observe the coalescence of MBH binaries at high redshifts, yielding an unprecedented view of the merger history of galaxies and the reionization epoch. LISA will also be able to detect gravitational waves from the collapse of a supermassive star to a black hole, and will provide a high precision probe of the central regions of galaxy cusps. Pulsar timing arrays will observe the very low frequency regime, $10^{-9}\text{Hz} \lesssim f \lesssim 10^{-7}\text{Hz}$, which includes the stochastic gravitational wave background from MBH binaries at lower redshifts.

ACKNOWLEDGMENTS

It is a pleasure to thank David Merritt for supplying Fig. 1; and Scott Hughes for stimulating discussions on MBH detection with LISA, and for supplying Fig. 2.

REFERENCES

1. Misner, C. W., Thorne, K. S., and Wheeler, J. A., *Gravitation*, W. H. Freeman, New York (1973).
2. Weisberg, J. M., and Taylor, J. H., "The Relativistic Binary Pulsar B1913+16," in *Radio Pulsars*, edited by M. Bailes, D. J. Nice, and S. E. Thorsett, ASP Conference Series, 2003 (in press).
3. Thorne, K. S., "Probing Black Holes and Relativistic Stars with Gravitational Waves," in *Black Holes and Relativistic Stars*, edited by R. M. Wald, The University of Chicago Press, Chicago, 1992, pp. 41 - 78.
4. Echeverria, F., *Phys. Rev. D*, **40**, 3194 (1989).
5. Hughes, S. A., *Annals of Physics*, in press (2003).
6. Shapiro, S. L. and Teukolsky, S. A., *Black Holes, White Dwarfs, and Neutron Stars*, Wiley-Interscience, New York (1983).
7. Flanagan, É. É. and Hughes, S. A., *Phys. Rev. D*, **57**, 4535 (1998).
8. Barish, B. C., "First Generation Interferometers," in *Astrophysical Sources for Ground-Based Gravitational Wave Detectors*, edited by J. M. Centrella, AIP Conference Proceedings 575, American Institute of Physics, Melville, New York, 2001, pp. 3 - 14.
9. Fritschel, P., "The Second Generation LIGO Interferometers," in *Astrophysical Sources for Ground-Based Gravitational Wave Detectors*, edited by J. M. Centrella, AIP Conference Proceedings 575, American Institute of Physics, Melville, New York, 2001, pp. 15 - 23.
10. Bender, P., et al., *LISA, Pre-Phase A Report*, 2nd. edition, unpublished, available online at: http://lisa.jpl.nasa.gov/documents/ppa2-09.pdf (1998).
11. Phinney, S., astro-ph/0108028 (2001).
12. Jaffe, A. H., and Backer, D. C., astro-ph/0210148 (2002).
13. Begleman, M. C., Blandford, R. D., and Rees, M. J., *Nature*, **287**, 307 (1980).
14. Komossa, S., Burwitz, V., Hasinger, G., Predehl, P., Kaastra, J. S., and Ikebe, Y., *Ap. J*, in press (2003).
15. Islam, R. R., Taylor, J. E., and Silk, J., *Mon. Not. Roy. Astron. Soc.*, in press (2003).
16. Volonteri, M., Haardt, F., and Madau, P., *Ap. J.*, in press (2003).
17. Milosavljević, M. and Merritt, D., *Ap. J.*, **563**, 34 (2001).

18. Armitage, P. J. and Natarajan, P., *Ap. J.*, **567**, L9.
19. Will, C. M. and Wiseman, A. G., *Phys. Rev. D*, **54**, 4813 (1996).
20. Flanagan, É. É. and Hughes, S. A., *Phys. Rev. D*, **57**, 4566 (1998).
21. Hughes, S. A., *Mon. Not. Roy. Astron. Soc.*, **331**, 805 (2002).
22. Haehnelt, M. G., "Supermassive Black Holes as Sources for LISA," in *Laser Interferometer Space Antenna, Second International LISA Symposium on the Detection and Observation of Gravitational Waves in Space*, edited by W. M. Folkner, AIP Conference Proceedings 456, American Institute of Physics, Woodbury, New York, 1998, pp. 45 - 49.
23. Menou, K., Haiman, Z., and Narayanan, V. K., *Ap. J.*, **558**, 535 (2001).
24. Wyithe, J. S. B. and Loeb, A., astro-ph/0211556 (2002).
25. Baker, J., Campanelli, M., Lousto, C. O., and Takahashi, R., *Phys. Rev. D* **65**, 124012 (2002).
26. Alcubierre, M., Bruegmann, B., Diener, P., Koppitz, M., Pollney, D., Seidel, E., and Takahashi, R., gr-qc/0206072 (2002).
27. Biermann, P. L., Chirvasa, M., Falcke, H., Markoff, S., and Zier, C., "Single and Binary Black Holes and their Active Environment,", in *High Energy Astrophysics from and for Space*, edited by N. Sanchez and H. de Vega, Proceedings of the 7eme Colloquium Cosmologie, Paris, in press (2002).
28. Begelman, M., Blandford, R., and Rees, M., *Rev. Mod. Phys.*, **56**, 255 (1984).
29. Merritt, D. and Ekers, R. D., *Science*, **297**, 1310 (2002).
30. Murgia, M., Parma, P., Ruiter, H. R., Bondi, M., Ekers, R. D., Fanti, R., and Fomalont, E. B., *Astron. Astrophys*, **380** 102 (2001).
31. Leahy, J. P. and Williams, A. G., *Mon. Not. R. Astron. Soc.*, **210**, 929 (1984).
32. Dennett-Thorpe, J., Bridle, A. H., Laing, R. A., and Scheuer, P. A. G., *Mon. Not. R. Astron. Soc.*, **304**, 27 (1999).
33. Hughes, S. A., private communication (2003).
34. Barkana, R. and Loeb, A., *Physics Reports*, **349**, 125 (2001).
35. Loeb, A. and Rasio, F., *Ap. J.*, **432**, 52 (1994).
36. Balberg, S. and Shapiro, S. L., *Phys. Rev. Lett.*, **88**, 101301, (2002).
37. Bromm, V. and Loeb, A., astro-ph/0212400 (2002).
38. Saijo, M., Baumgarte, T. W., Shapiro, S. L., and Shibata, M., *Ap. J.*, **569**, 349 (2002).
39. Shibata, M. and Shapiro, S. L., *Ap. J.*, **572**, L39 (2002).
40. New, K. C. B. and Shapiro, S. L., *Ap. J.*, **548**, 439 (2001).
41. Sigurdsson, S., *Class. Quantum Grav.*, **14**, 1425 (1997).
42. Sigurdsson, S. and Rees, M. J., *Mon. Not. R. Astron. Soc.*, **284**, 318 (1997).
43. Hughes, S. A., *Phys. Rev. D*, **64**, 064004 (2001).
44. Ryan, F. D., *Phys. Rev. D*, **56**, 1845 (1997).
45. Finn, L. S. and Thorne, K. S., *Phys. Rev. D*, **62**, 124021 (2000).
46. Glampedakis, K. and Kennefick, D., *Phys. Rev. D*, **66**, 044002 (2002).

Prospects for Future Observations in the Mid/Far IR

John C. Mather

Infrared Astrophysics Branch, NASA's Goddard Space Flight Center, Greenbelt, MD 20771

Abstract. The rapid advancement of telescope and detector technology enables a sequence of ambitious new airborne and space missions to observe in the mid- and far-infrared. I review the scientific forces that push us to these wavelengths, and the observatories that are planned or conceived to meet the goals. This decade and several more promise to be the "age of the infrared," with marvelous discoveries yet to come. I describe the main scientific objectives and the missions conceived to meet them, beginning with the SIRTF (Space Infrared Telescope Facility) planned for launch in early 2003, to the SPECS (Submillimeter Probe of the Evolution of Cosmic Structure), which may be decades in the future.

INTRODUCTION

There are several exciting topics for study in the mid- and far-infrared, including the highly redshifted distant universe, the nuclei of galaxies, many of which are obscured by dust at shorter wavelengths, and the formation of stars and planets (also obscured by dust). For this conference we concentrate on the most distant universe, but many of the tools we desire to have will be equally valuable for these other topics.

SCIENTIFIC OBJECTIVES

Approximately half of the energy produced in the universe after the Big Bang is now measurable in the far-IR wavelength range, so it is evident that many questions of cosmic structure can not be answered without observations in this area. Many far-IR objects have no detectable counterparts at UV or optical wavelengths. Dust grains in the Milky Way and most similar spiral galaxies convert about 1/3 of their starlight to far IR, and about 0.3% of the galactic luminosity appears in a single far IR fine structure line, ionized carbon at 158 μm wavelength. This is energetically the strongest line of our Galaxy, though it is not at all easy to observe.

Another factor favoring the use of far IR observations is that the cosmic redshift favors distant objects at long wavelengths. The slope of the dust emission spectrum is steeper than thermal, because the dust grains are small compared to the emitted wavelengths, so the dust spectrum is typically a Planck form multiplied by v^2. Hence, a starburst galaxy observed at 2000 μm is about equally bright for redshifts from 1 to 10. Therefore, if observations could be made with adequate sensitivity and angular resolution, the far IR would be one of the best places to study the emergence of cosmic structure.

This means that the ALMA project (Atacama Large Millimeter Array,[1], has an excellent opportunity to do cosmology, and indeed the design of the ALMA has been significantly modified since it was first conceived, specifically to enable the study of high redshift objects. It also means that the short wavelength information about such objects must be observed from space, because the ALMA performance degrades rapidly with increasing frequency, and because only a few wavelengths shorter than 850 μm can be observed from the ground on a regular basis.

Moreover, such a space mission must span a huge size. For example, to match the angular resolution of the Hubble Space Telescope [2], a far IR system needs an effective aperture of about 1 km.

NEW TECHNOLOGY

There is rapid progress in the necessary technologies to achieve the needed sensitivity and angular resolution in the infrared, from near IR to very far IR. At all IR wavelengths, we continue to see an exponential growth of detector array sizes, and indeed the doubling time is less than it is for standard semiconductor technology. Detectors are now available over the entire range from 1 to 1000 μm that can be built in large multiplexed arrays and that are limited in sensitivity by the cosmic IR background from distant galaxies and interplanetary dust.

New mirror technologies are also progressing well. For the James Webb Space Telescope (JWST, formerly known as the Next Generation Space Telescope, NGST, [3], both beryllium and glass mirrors are still in the competition. Both can support diffraction-limited imaging at 2 μm with areal densities of 20 kg/m^2. For larger, colder telescopes there are many alternative technologies that may work better, including stretched membranes and replicated composite materials. The DART concept developed at JPL and Lockheed Martin [4] uses stretched metallic membranes to create a pair of crossed cylindrical paraboloids. Together they can function as a large parabolic primary mirror, and on 1 m scales have already achieved diffraction-limited performance at 40 μm.

Another breakthrough in infrared space missions was made by the SIRTF (Space Infrared Telescope Facility, [5] project. This was the warm launch, cold operation concept, in which the telescope is not cooled down until after launch. It avoids the need for a cryostat that is larger than the telescope, and enables the use of unlimited telescope apertures.

Deep space orbits are now also commonly considered for infrared missions. The SIRTF will use an escape orbit that carries it up to 1 AU from the Earth after 10 years, providing an extremely stable thermal environment without the need for hydrazine thrusters, but requiring serious effort to get the data back over this long distance. The MAP (Microwave Anisotropy Probe), JWST, Herschel [6], and Planck [7] missions have all chosen to orbit around the Sun-Earth Lagrange point L_2. This point is about 1.5×10^6 km from the Earth, only 0.01 AU, so it is much easier to get the data back. The L_2 orbit also provides an extremely stable thermal environment, but requires active orbit maintenance of a m/sec/yr since the equilibrium is only semi-stable. In these orbits, the Sun and Earth appear to be in about the same direction, so a sunshield that allows observation of half the sky at any time is readily built.

A key to the operation of large, cold, lightweight telescopes in space is the use of alignment and figure adjustment after launch. The JWST will have a segmented mirror with four actuators per segment, for piston, tip, tilt, and radius of curvature adjustments. Other missions might use a small deformable mirror (located at an image of the primary mirror) to correct for additional small errors. The necessary algorithms and procedures for determining the adjustments were developed for the Hubble Space Telescope spherical aberration problem and proven by the repair mission. The basic idea is that images taken over a range of focal positions are sufficient to determine the entire wavefront error, because the complex electric field is an analytic function. Therefore the intensity fields at a number of different positions are sufficient to determine the complex wavefunction at the primary mirror. The algorithm is rather complex and iterative but has been perfected.

Finally, active coolers are now sufficiently developed to serve as a basis for planning future missions. The ACTDP (Active Cooler Technology Demonstration Program) managed by JPL [8] recognizes that at least three NASA missions need this technology: the JWST, the Constellation-X X-ray observatory, and the TPF (Terrestrial Planet Finder). Funds are being applied and several technologies appear very promising. Eventually it will be possible to provide cooling all the way down to the millikelvin range without the use of stored cryogens.

PLANNED AND PROPOSED SPACE MISSIONS

There are many infrared space missions in our future! Some of the leading candidates are:

- 2003 - SIRTF - Space Infrared Telescope Facility
- 2004 - SOFIA - Stratospheric Observatory for Infrared Astronomy
- 2004 - ASTRO-F (IRIS) - Infrared Imaging Surveyor, Japan
- 2007 - NGSS - Next Generation Sky Survey
- 2007 - Herschel = Far Infrared and Submillimetre Telescope
- 2010 - JWST - James Webb Space Telescope (was NGST)
- TBD - SIRCE - Survey of Infrared Cosmic Evolution
- 2010 - SPICA - Japanese proposed mission
- TBD - SAFIR - Single Aperture Far IR telescope "sapphire"
- TBD - SPIRIT - Space Infrared Imaging Telescope
- TBD - SPECS - Submillimeter Probe of the Evolution of Cosmic Structure

SIRTF - Space Infrared Telescope Facility

The SIRTF is planned for launch in early 2003, with an 85 cm beryllium telescope operating below 5.5 K. It will carry three instruments to cover the 3 to 180 μm range, and will be diffraction-limited at 6.5 μm. It will provide imaging and photometry over the entire wavelength range, spectroscopy from 5 to 40 μm, and spectrophotometry from 50 to 100 μm. It will observe for at least 2.5 years and will be launched by a Delta 7920H rocket in to a solar orbit, escaping from the Earth. It will be available for use by general observers on a competitive basis, and the first rounds of observing teams have already been chosen.

ASTRO-F (IRIS) - Japan

The Infrared Imaging Surveyor (IRIS) being built in Japan will provide a 70 cm telescope operating at 6 K with superfluid liquid helium and Stirling cycle coolers. It will have two instruments [9]. The Far-Infrared Surveyor (FIS) will carry out an all-sky survey, over the wavelength range from 50 to 200 μm, with an angular resolution of 30 - 50 arcsec. The Infrared Camera (IRC), with a field of view of 10 arcmin and an angular resolution of 2 arcsec, will be pointed at selected targets. The planned launch is in February, 2004 on the ISAS M-V rocket into a sun-synchronous polar orbit at an altitude of 750 km. This orbit, similar to the COBE orbit, permits an excellent observing strategy in which the Sun does not enter the top of the telescope and the Earth remains always below. However, most targets can still not be seen for extended periods of time.

SOFIA - Stratospheric Observatory for Infrared Astronomy

The SOFIA [10] is a joint project of NASA and the German DLR, and the University Space Research Association (USRA) leads the US consortium. It is a 2.7 m telescope looking out a hole in the side of a specially modified Boeing 747 airplane. It provides a wide range of wavelength coverage, from visible through far IR, and can carry a wide variety of instruments. At short wavelengths, turbulence from the wind rushing by the open hole will limit the potential for diffraction-limited imaging. At long wavelengths, the telescope's thermal emission is a major limitation. Nevertheless, the SOFIA will be an extraordinarily powerful telescope for exploring the infrared, and at high spectral resolutions its thermal emission will be less important than for wide bandwidths. It is expected to have its first operations in 2004.

NGSS - Next Generation Sky Survey

The NGSS [11] will carry out an all-sky survey from 3.5 to 23 μm, up to 1000 times more sensitive than the IRAS survey and 10^6 times more sensitive than the DIRBE instrument on the COBE satellite. Its top scientific goals include measuring ultraluminous IR galaxies and finding the most luminous galaxy; measuring old, low-mass, faint cool brown dwarfs and finding the closest star to the Sun; and measuring radiometric diameters for almost all known asteroids and discovering 500,000 new asteroids.

The NGSS is sufficiently sensitive to see a 1 km diameter main-belt asteroid, a 200 K, 10^{-8} L_{Sun} brown dwarf at the distance of α Centauri, the ULIRG F15307 at a redshift of $z = 3$, and an L_* galaxy at $z = 1$ over a significant part of the sky. The main technical advance of NGSS over prior missions is in the detector technology, especially for the Si:As detectors used at 12 and 23 μm. NGSS will find interesting targets for pointed telescopes to study, and will extend the stellar photometric calibration scale to a faint enough limit to overlap with the JWST brightness range.

Herschel = Far Infrared and Submillimetre Telescope

The Herschel, formerly known as the FIRST mission, is a project of the European Space Agency with significant contributions from NASA. It will include a 3.5 m telescope made of SiC, operating at 70 K, and will be launched by an Ariane V rocket to the Lagrange point L_2 in 2007. The same vehicle will also carry the Planck Surveyor, a mission to measure the microwave and submillimeter cosmic background radiation. The Herschel will carry three instruments covering the 60 - 670 μm range. While the observing speed and sensitivity are limited by telescope emission, there is adequate observing time to reach the fundamental confusion limit set by overlapping images. This mission will include a heterodyne spectrometer as well as direct detectors.

JWST (James Webb Space Telescope, was NGST)

The JWST, formerly known as the Next Generation Space Telescope, will have a deployable hexagonal primary mirror of at least 6 m aperture, and will be launched to the Lagrange point L_2 around 2010. It is a joint project of NASA, ESA, and the Canadian Space Agency (CSA). The near IR camera (NIRCam) will have about 48 megapixels and will be built by the University of Arizona with Lockheed Martin and Canadian firms. It will be Nyquist sampled at 2 μm, and will cover the entire wavelength range from 0.6 to 5 μm. It will include an $R = 100$ tunable filter module for emission line studies. The Near IR Spectrograph (NIRSpec) will be built by ESA, and will cover a field of about 3 arcmin square, with a spectral resolution of 1000. It will be a Multi-Object Spectrograph with an object selector using a microshutter array, and will be capable of observing hundreds to thousands of galaxies simultaneously. NASA will provide the detectors and microshutters. The mid IR instrument (MIRI) will be built by a partnership of NASA's Jet Propulsion Laboratory and a consortium of European institutions led by the Royal Observatory of Edinburgh under the management of ESA. It will include a camera with a field of about 2 arcmin and a spectrograph with a resolution of about 1500. The optics module will be provided by Europe. In addition, a Fine Guider camera will be provided by the CSA.

The JWST has immense advantages over other observatories in survey speed, angular resolution (compared with other space missions), and full wavelength coverage (compared with ground based telescopes). It is capable of observing the predicted first galaxies at redshifts significantly greater than the apparent reionization of the universe at $z = 6$, and supernovae originating in Population III stars back to the first generation of such objects around $z = 20$. It will also be capable of observing star and planet formation in the Milky Way, and may be capable of obtaining direct images of Jupiter-like planets around nearby stars.

SIRCE - Survey of Infrared Cosmic Evolution

The SIRCE is proposed by a team led by Goddard Space Flight Center. It would provide an all-sky far infrared survey in three bands. With modern detectors and a helium-cooled telescope, it would be capable of reaching the confusion limit for an aperture of 2 m. The resulting maps would include a catalog of 10^7 galaxies of which perhaps 10^6 would be at $z > 1$. There are so many objects to find that almost the entire sky map will be filled with galaxies. Like NGSS, the SIRCE could find the most distant, most luminous, and most interesting objects in the early universe. It would be an excellent scientific discovery tool and would enhance other missions that could never have enough sky coverage to find these objects, such as JWST, ALMA, and SAFIR (see below).

The SIRCE would use a COBE-like orbit (nearly polar, sun-synchronous, nearly perpendicular to the Sun) to enable strong radiative cooling. The helium-cooled telescope would feed arrays of bolometers, probably of the Transition Edge Superconductor type with SQUID (superconducting quantum interference device) multiplexers. Such devices are already in production and will be used for the SCUBA-2 instrument on Mauna Kea.

SPICA - Japanese proposed mission

The SPICA mission, formerly known as the H2-L2 mission, has been under study by Japanese teams for several years [12]. It would provide a helium-cooled 3.5 m telescope at the Lagrange point L_2, with a launch around 2010 on the Japanese H-II rocket. The cooling design is quite mature, based on existing active Joule-Thompson and Stirling cycle refrigerators. The SPICA would cover the wavelength range from 5 to 200 μm with cameras and spectrometers. It would be the first opportunity for a large cold pointed far IR telescope in space and would be extremely powerful. With its lower temperature it would be extremely sensitive compared with the European Herschel mission. The Japanese teams are seeking international partnership to enable this project.

SAFIR - Single Aperture Far IR telescope "sapphire"

The SAFIR mission was recommended by the National Academy of Sciences' Decadal Survey (Astronomy and Astrophysics in the New Millennium) report in 2000 [13]. In that report it was conceived as a successor to the JWST, but much colder and operating at longer wavelengths. The longer wavelength range relaxes the tolerances on the optical elements, although the lower temperature requirements are an additional challenge. New mirror technologies might enable a significantly larger aperture than now planned for JWST, and there is sufficient time to develop these new concepts [14].

One of the great challenges for SAFIR is to show that there are reasonable mirror cooling techniques. Initial quantitative studies at Goddard Space Flight Center show that the main modification to the JWST design is an intermediate radiation baffle between the main sunshield and the telescope. The JWST radiation baffle designed by TRW already reduces the radiant heat load on the telescope to 23 mW. It seems to be a simple matter to intercept this heat and the heat conducted along the support tower and cables with an actively cooled shield, allowing the telescope itself to cool down to the needed few Kelvin with an available additional active cooler.

SPIRIT - Space Infrared Imaging Telescope

The SPIRIT is a concept for the first far IR imaging interferometer in space [15]. It would use two or more mirrors or telescopes attached to a deployable structure to achieve the maximum angular resolution for a single spacecraft. In the far IR, available detector sensitivity is already sufficient to reach the limit where galaxy images overlap, so more angular resolution is necessary for the next step after SAFIR. There are two basic approaches to Interferometry: Michelson interferometers using beamsplitters, and Fizeau using geometrical combination. Each has advantages, depending on the scientific objectives, the available detector technology, and the delay line possibilities. The SPIRIT would be the first sensitive instrument to reach arcsecond resolution in the far IR. With a 40 m boom and an interferometric resolution of $\delta\theta = \lambda/2D$, and $\lambda = 200\mu$, we would have $\delta\theta$ = 0.5 arcsec, comparable to ground-based visible images.

SPECS - Submillimeter Probe of the Evolution of Cosmic Structure

The SPECS [15] would be a much more ambitious imaging interferometer, with separated spacecraft up to 1 km apart. The clear advantage of such a system would be in its superior angular resolution, 25 times better than a 40 m SPIRIT. It would find immediate application in resolving galactic nuclei, protostars and protoplanets, and distant star forming regions. It adds significantly to the challenges of SPIRIT, requiring methods to relay optical beams from one spacecraft to another without adding serious stray radiation from warm objects. It also requires excellent formation-flying technology, but it will not be the first to require it - the planned Terrestrial Planet Finder mission would need an order of magnitude better stability and accuracy.

CONCLUSIONS

The next several decades will continue to be the "Decade of the Infrared" as technology and space missions continue to advance in power. There is every reason to expect that Moore's law can continue in the development of new detectors over the entire infrared spectral range. Advanced active coolers will enable large cold telescopes. Deployable, adjustable optics will enable continued increases in aperture. Interferometric techniques will enable increased angular resolution. It is difficult to predict what will happen when these new methods become available, but we confidently expect that miraculous discoveries will be made.

ACKNOWLEDGMENTS

I enjoy the continued support of the James Webb Space Telescope project, and the friendship of my colleagues who are working on all of the missions I have described. Most of the information is drawn from public web sites, to which I commend the reader, and was presented at the SPIE conference 4850 in Waikoloa, Hawaii in August 2002.

REFERENCES

1. http://www.alma.nrao.edu/
2. http://www.stsci.edu/hst/
3. http://jwst.gsfc.nasa.gov/
4. http://lmms.external.lmco.com/latestnews/archive/articles/dart.032001.html
5. http://sirtf.caltech.edu/
6. http://sci.esa.int/home/herschel/index.cfm
7. http://astro.estec.esa.nl/SA-general/Projects/Planck/
8. (http://acquisition.jpl.nasa.gov/rfp/actdp/ACTDP_QandA.pdf
9. http://www.ir.isas.ac.jp/ASTRO-F/index-e.html
10. http://sofia.arc.nasa.gov/
11. http://www.astro.ucla.edu/~wright/NGSS/
12. http://www.ir.isas.ac.jp/SPICA/h212_spie/h212.html
13. http://www.nap.edu/books/0309070317/html/
14. http://gsfctechnology.gsfc.nasa.gov/laisdocs/SAFIR_Technologies.pdf, http://safir.jpl.nasa.gov/
15. http://gsfctechnology.gsfc.nasa.gov/laisdocs/spie_paper.pdf

OTHER TOPICS

Prevalence of Magnetic Field Forces and Universal Symmetry of Nature

Yildirim Cinar

University of Abant Izzet Baysal, Konuralp, 81000, Duzce, Turkey

Abstract. Prevalence and equivalence of attraction and repulsion forces of magnets were searched. The repulsion between S poles was stronger by 21.6 percent than the repulsion between N poles. Attraction of N pole was greater than S pole, and inverse relation of distance for magnetic attractions were different for two types of poles. Repulsion effects between the same types of poles at equivalent rate constituted approximately distinct spheres. According to these results, it can be concluded that, inequality of strength of magnetic poles contradicts the concepts of particle-antiparticle duality and universal symmetry of nature, and the traditional continuous appearances of magnetic lines between two poles are actually constricted by two different lines of different poles, which get together at the equator plane of magnet. These findings can be used to explain why the axis of the earth is tilted 23.5 degree away from a normal ecliptic plane by the sun, and asymmetric dynamical findings of the universe might be clarified by application of these suggestions to Astrodynamics.

INTRODUCTION

Magnetic interactions of masses are associated by their shape, distance, direction, position, and type of poles. The magnetic field of the Earth is practically same as that of a bar magnet. The Sun's and the Earth's direction of rotation of axis, and direction of the magnetic poles are the same. The N pole of a compass is actually attracted by the Earth's S magnetic (approximately geographic north) pole. The repulsion between the S poles is greater than the repulsion between the N poles [1].

MATERIAL AND METHOD

A) Effects of prevalence of magnetic field strength of a bar magnet ($1.5x1.5x7cm$ loadstone) on measurement of iron cube weight ($20g$) were determined by a scale during parallel displacement of the bar magnet at a constant vertical distance of iron cube ($2cm$). **B)** Based on repulsion equal in magnitude due to adjusted distance, 3 dimensional distances between the same types of magnetic poles of two equal bar magnet were recorded. When the first bar magnet was hold on a diamagnetic spiral, the second was moved by standard magnetic interactions (repulsion power) that was measured by shrinks of the spiral and a plastic scale. **C)** Effects of a bar magnet (glued bottom face of a glass) to facilitating prevalence of pouring small pieces of iron wire on the glass plate were determined. This experiment was repeated by using bar magnets of different lengths (5, 7, and $14cm$). **D)** Two equal loadstones (planar and ring shaped) and a cellulose-acetate (diamagnetic) bar were used. Loadstones were placed on the bar by replacing it through the hole. First, the *N* poles (planar surfaces) of two magnets were placed face-to-face and the vertical distance between the two *N* surfaces resulting from the magnetic repulsion was measured using the scale on the bar (Figure 1). The magnetic properties (types) of the surfaces were determined by means of their interaction with a standard compass.

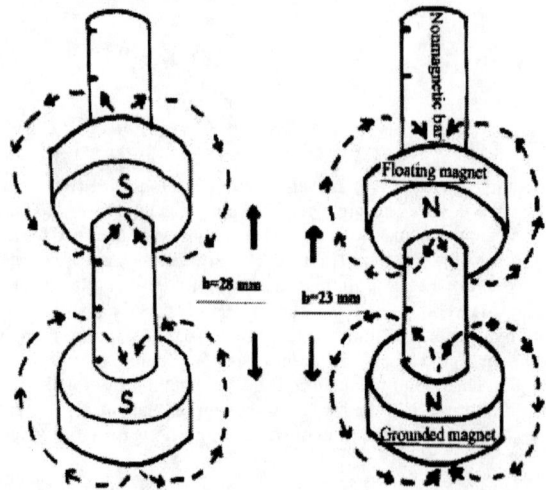

FIGURE 1. Determination of Repulsion of Magnetic Poles. The loadstones were placed on the bar by putting the bar through their holes. First, the N poles (planar surfaces) of two magnets were placed face-to-face and the vertical distance between the two *N* surfaces resulting from the magnetic repulsion was measured using the scale on the bar. The distances resulting from repulsions between the same types of poles (planar surfaces) were measured to be 28*mm* for S poles and 23*mm* for *N* poles. Therefore, it was concluded that the repulsion between S poles is stronger by 21.6% than the repulsion between *N* poles.

E) Weight of an iron ring was measured by a diamagnetic scale. According to the place of iron on the scale, the ring shaped magnet was placed on a glass shelf by center-to-center position and at the ineffective vertical distance of magnetism. Then, the distance was decreased as 0.5*cm* at each subsequent step, and magnetic effect of each pole were measured as weights of iron at every step.

RESULTS

A) Attraction of N pole was 1.6 percent greater than S pole for the same distance. According to N pole, approximately 90 percent decreased vertical attraction was found on the middle right line of bar. B) Borderline distances of repulsion effects of the poles constituted two approximately distinct spheres. C) Some of the pouring iron wires were collected on the outer borderline of bar magnet body at parallel position to surface. However, half of these wires were collected at two poles homogeny and arrows of porcupine position that directions of wires were on the centers of spherical poles. Therefore, wires constituted a drowned empty dumbbell shape. D) The distances resulting from repulsions between the same types of poles (planar surfaces) were measured to be 28 mm for S poles and 23 mm for N poles. Therefore, it was concluded that the repulsion between S poles was stronger by 21.6 percent than the one between N poles. E) Due to decreased distance (.5-1-1.5-2-2.5-3-3.5-4-4.5-5cm), and attraction power of the N pole was greater than S pole; measured weight by N pole (Cl-19.64-26.84-28.17-28.35-28.82-29.22-29.5-29.93g) were greater than S types of poles (Cl-20.42-27.61-28.72-28.79-29.19-29.5-29.56-29.93g) respectively (Cl: The mass is clutched by the magnet at this distance).

DISCUSSION

Inequality of strength of magnetic poles contradicts the concepts of particle-antiparticle duality, symmetry and equality of magnetic dipole, and universal symmetry of nature. In addition, symmetry base magnetic field equations cannot represent asymmetric dipole base nature. Traditional representations of field lines by one-direction and one type of vector could be proper to presentation of electron transfer or attraction of one magnetic pole. However, this configuration could not represent the different force, field, and direction of vectors of the repulsion and attraction property of two poles of a magnet. Decreased repulsion of N pole at the middle side of magnets could not been explained by increasing effect of S pole, because the synchronic attraction of the two poles on the iron were decreased to minimum at the middle. It can be suggested that, all of the vectors of motion of iron wire are constricted by combination of N and S pole vectors that are not equal in magnitude on the equatorial plane of magnet. Results of this study suggest that continuous appearances of magnetic lines between two poles are actually constricted by two different lines of different poles, which get together at the equator plane of magnet. In addition, if all field lines were to start at N pole and end at the S pole; there should be found a neutralization (interference) state at the middle or continuous energy flux between the poles. Empty dumbbell shape of prevalence of iron wires suggests that most of the magnetic lines should be placed at the surface of the magnets by additional lateral polarizations. Inequality of inverse relation of distance between attraction of S and N pole contradicts the concept of Newtonian force.

Because the axis of rotation and the magnetic poles of the Sun and the Earth are in the same direction; the above results stating that the repulsion between the S poles (at the north) was found to be 21.6 percent greater than the repulsion between N poles could be used to explain why the axis of the earth is tilted (repel) 23.5 degree away from a normal ecliptic plane by the sun. According to information above that the Sun and the Earth have similar inclination of axis of rotation, it can be concluded that; when the Sun is far to the Earth at July, the dominant magnetic interaction is repulsion between S poles, and when the Sun is close to the Earth at January, the dominant magnetic interaction is attraction between S poles of the Earth and N poles of the Sun. Earth's orbital is held on equatorial plane of the Sun, and according to results of this study, magnetic strength of the Sun is weaker here. In addition, constant angle of Earth axial inclination in the Sun's magnetic field could be explained by primary energy of spin (angular momentum) of the Earth. Angular momentum increases by angular velocity, therefore, the required energy for change of rotational axis inclination is increased by angular velocity. Origin of the spin energy of masses is explained by statistical probability that rectilinear kinetic energy of Big Bang was converted to angular momentum via multiple collisions of masses in tangential position while Big Bang [2]. Therefore, it can be concluded that, Earth's spin energy associated with magnetic effect is dominant on the Earth's orbital, and the angles of the Earth's rotational axis is constant around the Sun. According to the information given above, the increase in the angular velocity (spin) of mass may be a unique way "to cluster" masses and the excess (kinetic) energy of the central and primordial universe for the Big Bang, or in other words, spin is the basis of the capacity of mass to store (absorption) energy.

Asymmetric dynamical findings of universe could be explained by application of these findings and suggestions to Physics and Astrodynamics.

REFERENCES

1. Cinar, Y., "Determination of the repulsion between magnetic poles of the same type" (abstract) in *International Conference on Theoretical Physics (TH 2002)*, Edited by D. Iagolnitzer et al., Abstract Book, 190, Paris (2002).
2. Cinar, Y., "Mass and kinetic energy distribution of the universe" (abstract) in *International Conference on Theoretical Physics (TH 2002)*, Edited by D. Iagolnitzer et al., Abstract Book, 191, Paris (2002).

Time and Entropy

Vladimir I. Garaimov

*Astronomy Department, University of Maryland,
College Park, MD 20742, USA*

Abstract. This paper presents alternative ideas on the physics of time that lead to a new interpretation of cosmological redshifts. These ideas are based on the close relationship between the speed of time and entropy processes in our universe. I give numerical estimates and describe laboratory experiments and observational effects that can test the new theory.

POSTULATES

We will define entropy as the irreversible dispersal of energy and will regard time as a physical process.
We postulate that:

1. Time has close relationship with entropy. An increase in entropy will cause a corresponding increase in the speed of time. We can say that the potential energy of matter has been converted into the "kinetic energy"(speed) of time.
2. Any object carries information (as an energy level) about the time when it was created.

The consequences of these postulates are :
1. As the entropy of the Universe increases time is accelerated.

Let's analyze a "stationary" entropy process. During a period of time dT an entropy increase by a factor m results in an acceleration of time by a factor n $(n > 1)$. The next period δT_1 will be $1/n$ times shorter, but during this period the entropy will also increase m times, as the process is "frozen" into the time. This results in time being accelerated exponentially.

Let's analyze this from the position of a cosmological redshift. Assume a photon is emitted with frequency v, and after a certain period of time is registered by a radiation detector. During this period the speed of time is accelerated n times. Therefore the radiation detector registers the photon with a frequency v/n, i.e., the photon is observed with a redshift. It will depend only how long the photon has existed, i.e. the longer the time, the greater redshift effect. In other words, the further from the observer the source of emission is, the greater the redshift, and the principle of symmetry is true for this effect. This means that two objects located some distance from each other both observe one another to have the same redshift. Therefore, cosmological redshift may be explained by the effect of time acceleration.

2. The Hubble constant is the numerical value of the derivative of the speed of time, i.e. it is the acceleration of time (see Fig.1). It has units of acceleration, i.e. a variation

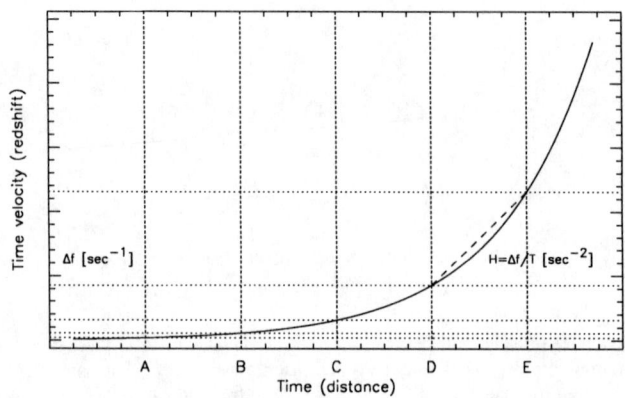

FIGURE 1. Redshift (time velocity) evolution.

of frequency of radiation (sec^{-1}) divided by a time period (sec).

If the universe experiences episodes during which entropy jumps rapidly, such as the first epoch of star formation, we might expect to see a record of this in the variation of the Hubble constant with distance (time).

NUMERICAL ESTIMATES

Let us estimate the time acceleration from the value of the Hubble constant ($H \approx 60 \; km \; sec^{-1} \; MPc^{-1}$). The time it takes light to travel $1 MPc \approx 10^{14} sec$

$$v/c = aT \implies 60/3 \times 10^5 = 10^{14} a$$

where: c - speed of light; a - acceleration of time; T - time period. As result a second decreases by $2 \times 10^{-18} sec$ per second.

We can compare this value with the value calculated from the correction to the equinox after all known precessions are accounted for. Figure 2 shows the Equinox corrections from solar and lunar observations [1].

The Equinox was determined with the help of chronometers. Our standards of time are atomic and are "frozen" into time. From the above discussion it follows that during this time period a second decreased, and resulted in the corrections presented in figure 2.

Let's estimate the time acceleration using a linear approximation and the formula: $dT = at^2/2 \implies a = 2dT/t^2$. The numerical data from [1] ($\dot{E} = +0.00085$ sec per year) give $a \approx 1.7 \times 10^{-18} sec$ per second.

Comparing this result with that calculated from the Hubble constant we see the orders of the values are the same.

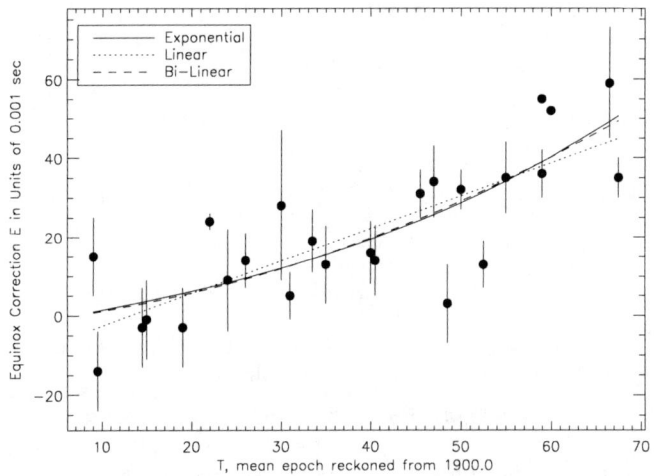

FIGURE 2. Equinox corrections from solar and lunar observations. Three different interpolations are also presented, an exponential interpolation (solid line; sigma =121.87); a linear interpolation (dotted line; sigma=127.56); bi-linear interpolation (dashed line; sigma=122.42).

OBSERVATIONS AND EXPERIMENTS

Observations

From Fig.1 we see that:

- Two equal intervals will show different measured redshifts because it increases non-linearly with the passage of time, i.e for an observation of a single object the redshift will increase non-linearly with time (see Supernovae project result: http://www-supernova.lbl.gov/);
- Hubble constants calculated for objects located at different distances are not equal (the further away an object is the smaller the Hubble constant, i.e. $H_{DE} > H_{CE} > H_{BE} > H_{AE}$).

Laboratory experiments

A localization of entropy processes suggest that the increasing speed of time (its acceleration) is also localized in space and decreases with distance. On the basis of these ideas the following experiment has been devised (see.Fig 3).

The experimental technique is as follows:

a) measure the frequency of radiation from the laser (2)

b) heat the liquid in the vessel (1)

c) measure the frequency of radiation from the laser (2) while the liquid is boiling

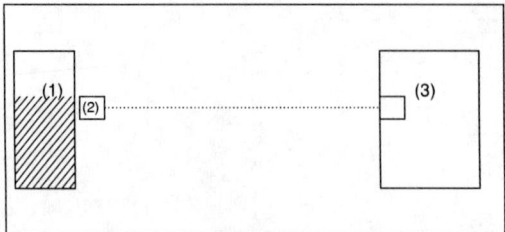

FIGURE 3. An experimental device for measuring entropy blueshift. 1 – entropy source (Ex.: vessel containing a boiling liquid); 2 – laser source(milliwatt); 3 – high resolution spectrometer.

The idea of the experiment consists of the following:

the boiling liquid in the vessel increases the entropy locally, i.e. it accelerates time. The radiation from the laser source is generated at the faster time while the spectrometer measures the radiation at the slower time. As a result the radiation registered from laser source during the boiling of the liquid has a higher frequency than without boiling the liquid. Increasing the intensity of the boil (generating more entropy) will cause the frequency of observed radiation to increase. For this technique a blueshift of the laser radiation will be observed. If the entropy system is located close to the spectrometer a redshift of the laser radiation will be observed.

Also this effect should be observable using spectral absorbtion lines.

The same experiment we can do into the interplanetary space. The Sun is a huge entropy system. The satellite transmitter is a coherent radiation system. Within the bounds of these ideas, unusual redshifts from the radio beacons on the distant spacecraft ("Pioneer" and "Voyager") will be observed.

Also we can observe the cumulative effect of time acceleration. For example, particle collisions (inside a particles accelerator) or nuclear explosions are very powerful entropy processes. If one of two synchronized chronometers is located close to this system and another is located very far away, then after the collision or the explosion the closest chronometer will register more time than the other. It will be because during the collision or explosion the first chronometer will be into the fast speed time zone.

CONCLUSION

The ideas presented here are speculative. However, this theory can be applied on any spatial scale, e.g.: extra-galactic scales (cosmological redshift), planetary scales (equinox correction) and local scales (laboratory experiments). The results of the laboratory experiment will provide a genuine test of the correctness of this theory.

REFERENCES

1. Fricke, W., *Astron.Astrophys.*, **107**, L13-L16 (1982).

The Joint Emergence of Large-scale Structure and Large-scale Magnetic Fields after Combination Time

Howard D. Greyber

10123 Falls Road, Potomac, MD 20854, U.S.A., hgreyber@yahoo.com

Abstract. Two recent discoveries, of surprisingly strong magnetic fields very far from any cluster of galaxies, and of the coherent orientation of quasar polarization vectors over huge distances, are powerful evidence for the validity of the "Strong" Magnetic Field model (SMF) in astrophysics. CalTech's Cosmic Background Imager (CBI) may observe evidence for the beginning of large-scale polarization in the CMB radiation that the SMF model predicts, signifying emerging large-scale magnetic fields and emerging cosmic structure. The important implications of SMF for energy production in the AGN/quasar central engines, for jets in galaxies, quasars and newly forming stars, and galactic structure, were discussed in previous articles.

INTRODUCTION

In my SMF model (1,2) certain important plasma processes occur at and after Combination Time in the Big Bang model of the Universe. The word "Strong" has been used by me since 1961 in precisely the same sense that the pioneering Russian physicist Y. B. Zel'dovich used it in his 1983 remark "A major challenge is to understand strong magnetic fields whose energies greatly exceed those of hydrodynamic motions" (3). Thus, the field may be far, far stronger, or it could be far, far weaker than the earth's magnetic field. Kronberg, Lesch and Hopp have discovered that there exist "large-scale magnetized regions of space outside clusters" (5). Their observations argue for magnetic fields about 10exp-9 gauss averaged over all intergalactic (IG) space. The real question is whether the field is of primordial origin, or is this IG field the result of processes inherent in galactic evolution? Their answer is that very early dwarf galaxies seed the IG medium with magnetic fields. This assumes extremely rapid dynamo amplification will occur in dwarf galsxies at very high redshifts, despite the fact that galactic dynamo models have dubious assumptions. SMF does not agree. A second important discovery is from the observations of D. Hutsemekers and H. Lamy "Confirmation of the Existence of Coherent Orientations of Quasar Polarization Vectors on Cosmological Scales" (6). They found "the polarization vectors in their region A1 are apparently parallel to the plane of the local supercluster, while those of quasars at lower redshifts are not accordingly aligned". A distinct possibility is that the quasar structural axes themselves are coherently oriented and that "the presence of coherent orientation at such large scales seems to indicate the existence of a new, interesting effect of cosmological importance". Obviously, flows surging from individual primeval dwarf galaxies are randomly oriented and cannot possibly explain coherent orientation over cosmological scales. The concept of a constant primordial magnetic field over a prequasar plasma cloud, distorted strongly by gravitational collapse forming a stable storage ring, was assumed as natural by Greyber starting in l961 (7-11). Other implications of SMF are described by Greyber (13-22).

PROCESSES IN THE EARLY UNIVERSE

The Electromagnetic Force, which is the only other long-range force in physics, is 10exp40 times stronger than Gravity. A fact usually ignored is that even in its Dark Ages, the Universe was always a plasma. That electric currents and long range magnetic fields can exist for very long times was pointed out long ago by the Swedish astrophysicist, Nobelist Hannes Alfven. Einasto et al reported "We present evidence for a quasi-regular three-dimensional network of rich superclusters and voids, with regions of high density separated by 120 Mpc. If this reflects the distribution of all matter (luminous and dark), then there must exist some hitherto unknown process that produces regular structure on large scales. - Thus we conclude that our present understanding of structure formation on very large scales needs revision" (23) In a review, De Lapparent wrote "The CfA redshift survey slices show that the galaxy distribution is characterized by sharp, sheet-like structure delineating voids - it is worth emphasizing that so far all existing surveys are consistent with a same discription of the galaxy distribution, namely a cell-like galaxy distribution in which galaxies lie along walls separated by voids with diameters in the range of 10/h to 100/h Mpc" (24). Gregory concluded " The large structures we observe today had to be impressed somehow onto the protogalactic material before it collected into galaxies" (25). In a long article on the "Dynamics of First Order Transitions" , Gunton emphasized that the problem of pattern formation in systems which are attempting to reach equilibrium provides a fascinating example of nonlinear, nonequilibrium phenomena (27). In the classical theory of first order transitions, there are two different types of instability. One is nucleation, the gravitational attraction of matter to a positive density fluctuation. The other is an instability against infinitesmal amplitude, nonlocalized (long wavelength) fluctuations which lead to the initial decay of an unstable state. This instability, first noted by Willard Gibbs in 1906, is called Spinodal Decomposition (SD).

The first successful qualitative theory of SD was initiated by Hillert, and developed by John W. Cahn at NIST, who was awarded the National Medal of Science for this research. Cahn's linearized deterministic theory shows that the initial stages of SD should produce an exponential growth for wavelengths larger than a critical wave number. The real process in our Universe is similar to that in a very quiet lake as the temperature falls extremely slowly through zero degrees. Some patches of the lake will supercool, staying liquid, while other patches will freeze. Similarly, at Combination Time in the Big Bang model, the morphology of the Universe will be determined by SD, i.e. the Universe will consist partly of patches of only slightly ionized plasma, and partly of patches of fully ionized plasma. In those huge patches that have become almost neutral, about 10 eV per atom of photons is released. Because the photon mean free path in fully ionized plasma is far shorter than that in almost neutral plasma, the radiation energy will penetrate only a very short distance into the borders of fully ionized plasma. All those border layers will receive a steady flood of photons from the almost neutral patches, and will heat up. Thus each fully ionized patch will be drastically compressed. SMF makes the reasonable assumption that there are slight differences in matter density and slight differences in the strength of the seed magnetic fields across the interfaces. A huge subset of interfaces are Rayleigh-Taylor stable. Recall that it follows from Maxwell's Equations that there must be an electric current at the boundary between two adjacent volumes of plasma with different magnetic field intensities. This electric current is proportional to the jump across the boundary of the component of the magnetic field (B) parallel to the boundary on each side. Therefore, as the enormous compression of each fully ionized patch proceeds, electric current amplification occurs due to the different equations of state. While magnetic pressure varies as B squared, matter pressure goes as the density to some power like 1, 4/3 or 5/3.

For instance, assuming an isothermal plasma, a twenty-fold doubling, i.e. a compression of 10 exp6 is possible over a million years, increasing v in the expression for current, $j = nev$. (Actually due to the three-dimensional wrinkling of the boundaries of patches, amplification in certain "corner" volumes will be much larger). Simultaneously, the number of current carriers, n, will increase dramatically due to the well-known Pinch Effect that draws in charged particles until they join the current flow, and, also, due to captured random particles. If n has a twenty-fold doubling also, then an increase of current and magnetic field of 10exp12 is expected. Thus a seed magnetic field of 10exp-21 gauss could be amplified to about 10exp-9 gauss. Finally, as the mass density of the current-carrying interfaces is rising, Gravity acts to draw in both neutral and charged particles. Notice that even if the SMF process has increased the matter density at the boundaries by only 0.0001 of the average density in a patch, that is sufficent to imprint the observed structure on the matter. This is because Gravity continues to attract matter into the sheets around voids long after the SMF pressure differences are gone. Eventually the density rises until at some volumes critical density is reached. Gravitational collapse then occurs forming stars, quasars and galaxies on spatially curved sheets of matter around voids, matching the observations. Since the Pinch Effect is known to be unstable (28), the large majority of the matter will instead remain anchored as DARK Matter along the superclusters. It is crucial to note that the SMF model leads to a current along each interface, which becomes the long axis of each supercluster. Recalling the gigantic size of a supercluster, the magnetic induction is essentially constant over the far smaller size of a prequasar plasma cloud. (As described in earlier articles, in the SMF model, quasars have the highest ratio of magnetic energy to rotational energy of all types of galaxies, Seyfert galaxies less, etc.). SMF predicts that each quasar along a supercluster of galaxies will tend to have its structural axis along the field direction. Thus the quasar's electric field vector is coherently oriented with other quasars in that super cluster, and roughly parallel to the long axis of that supercluster - just as Hutsemekers and Lamy discovered. - Galactic dynamo models have been studied for 70 years. The astrophysical dynamo theory came up in an interesting exchange in 1934 when physicist Walter Elsasser related to a friend Thomas G. Cowling's attempt to make a theory for an axially symmetric astrophysical dynamo. "If that simple idea does not work", remarked Albert Einstein, "then dynamo theory will not work". The result was negative, yielding Cowling's well-known Anti-dynamo Theorem. Clearly, my SMF model agrees with Albert Einstein's insight.

REFERENCES

1. H. D. Greyber, 2000, in a Space Telescope Science Institute Report from their 1999 May Symposium "The Largest Explosions Since the Big Bang: Supernovae and Gamma Ray Bursts", Feb. 2000
2. H. D. Greyber, 1994, in A.I.P. Conf. Proc. 336, Dark Matter, Ed. S. Holt, 509
3. Y. Zel'dovich, 1983, in Magnetic Fields in Astrophysics, Gordon and Breach, 9
4. P. P. Kronberg, H. Lesch and U. Hopp, 1999, Ap. J. 511, 56, January 20
5. D. Hutsemekers and H. Lamy, 2001, A. & A. 367, 381
6. H. D. Greyber in Instabilite Gravitationelle - -, 14th Int. Astron. Symp., Memoirs of the Royal Soc. of Liege, XV, 189, 197
7. H. D. Greyber, 1990, in 14th Texas Symp., Ann. New York Acad. Sci. 571, 239
8. J. Einasto, A. Starobinsky et al, 1997, Nature 385, 139
9. V. De Lapparent, 1996, in Mapping the Large-scale Structure in Cosmology, Session LX, Les Houches, Elsevier, eds R. Schaeffer et al
10. S. A. Gregory, 1998, in Mercury, publ. by Astron. Soc. of the Pacific, March-April, 26
11. J. D. Gunton et al, 1983, in Phase Transitions and Critical Phenomena, vol. 8, ed. by C. Domb and J. L. Liebowitz, Chapter 3, 319
12. M. Kruskal & M. Schwarzschild, 1954, Proc. of the Royal Soc. A, 223, 348

List of Attendees

Name	Affiliation	Email
Abel, Tom	Harvard-Smithsonian-CfA	tabel@cfa.harvard.edu
Aguirre, Anthony	Princeton University-IAS	aguirre@ias.eduaguirre@ias.edu
Albrecht, Andreas	U. of California-Davis	albrecht@physics.ucdavis.edu
Arabadjis, John	MIT	jsa@space.mit.edu
Bahcall, Neta A.	Princeton University	neta@astro.princeton.edu
Band, David	NASA/GSFC	dband@lheapop.gsfc.nasa.gov
Barger, Amy	U. of Wisconsin-Madison	barger@astro.wisc.edu
Benford, Dominic	NASA/GSFC	Dominic.J.Benford.1@gsfc.nasa.gov
Bennett, Chuck	NASA/GSFC	Charles.L.Bennett.1@gsfc.nasa.gov
Berrington, Bob	Naval Research Lab	rberring@ssd5.nrl.navy.mil
Bhatia, Anand	NASA/GSFC	bhatia@stars.gsfc.nasa.gov
Bhavsar, Suketu	University of Kentucky	suketu@pa.uky.edu
Boehringer, Hans	MPI for Extraterrestrial Physics	hxb@mpe.mpg.de
Boldt, Elihu A.	NASA/GSFC	boldt@lheavx.gsfc.nasa.gov
Borne, Kirk	IST @ Raytheon	Kirk.Borne@gsfc.nasa.gov
Boughn, Steve	Haverford College	sboughn@haverford.edu
Bullock, James	Harvard-Smithsonian-CfA	james@astronomy.ohio-state.edu
Centrella, Joan	NASA/GSFC	jcentrel@milkyway.gsfc.nasa.gov
Cheng, Edward	NASA/GSFC	ec@cobi.gsfc.nasa.gov
Cheung, Cynthia	NASA/GSFC	ccheung@pop600.gsfc.nasa.gov
Cillis, Analia Nilda	NASA/GSFC	cillis@gamma.gsfc.nasa.gov
Cinar, Yildirim	U. of Abant Izzet Baysal	yildirimcinar@ttnet.net.tr
Cline, Thomas L.	NASA/GSFC	cline@apache.gsfc.nasa.gov
Colbert, James W.	UCLA	colbert@astro.ucla.edu
Cowen, Ron S	cience News	rcowen@sciserv.org
Crannell, Carol Jo	NASA/GSFC	crannell@gsfc.nassa.gov
Drachman, Richard	NASA/GSFC	drachman@stars.gsfc.nasa.gov
Dwek, Eli	NASA/GSFC	Eliahu.Dwek.1@gsfc.nasa.gov
Fahey, Dick	NASA/GSFC	fahey@stars.gsfc.nasa.gov
Felten, James E.	NASA/GSFC	felten@stars.gsfc.nasa.gov
Ferguson, Henry C.	STScI	ferguson@stsci.edu
Figueroa, Enectali	NASA/GSFC	Enectali.Figueroa@gsfc.nasa.gov
Franceschini, Alberto	Astronomical Obs. - Padova	franceschini@pd.astro.it
Frenk, Carlos	University of Durham	c.s.frenk@durham.ac.uk
Garaimov, Valdimir	University of Maryland	gvi@astro.umd.edu
Gardner, Jonathan	NASA/GSFC	gardner@harmony.gsfc.nasa.gov
Gehrels, Neil	NASA/GSFC	Cornelis.A.Gehrels.1@gsfc.nasa.gov
Graber, James	LOC	jgra@loc.gov
Green, James	University of Colorado	jgreen@casa.colorado.edu
Greene, Brian	Columbia University	bg111@columbia.edu
Greenhouse, Matthew	NASA/GSFC	Matthew.A.Greenhouse.1@gsfc.nasa.gov
Greyber, Howard	Greyber Assoc.	hgreyber@yahoo.com
Hammell, Molly	Dartmouth College	molly.hammell@dartmouth.edu
Harding, Alice K.	NASA/GSFC	harding@twinkie.gsfc.nasa.gov
Hasinger, Gunther	MPI for Extraterrestrial Physics	ghasinger@mpe.mpg.de
Heap, Sally	NASA/GSFC	heap@srh.gsfc.nasa.gov

Heckman, Tim	The Johns Hopkins University	heckman@pha.jhu.edu
Hinshaw, Gary	NASA/GSFC	Gary.Hinshaw@gsfc.nasa.gov
Holder, Gil	Princeton University-IAS	holder@ias.edu
Holt, Steve	Olin College	stephen.holt@olin.edu
Horner, Don	STScI	horner@stsci.edu
Hu, Wayne	University of Chicago	whu@background.uchicago.edu
Hudson, Danny	JCA/UMBC	dhudson@jca.umbc.edu
Ikebe, Yasushi,	JCA/UMBC	ikebe@milkyway.gsfc.nasa.gov
Iping, Rosina	Catholic University of America	rosina@taotaomona.gsfc.nasa.gov
Jahoda, Keith M.	NASA/GSFC	Keith.M.Jahoda.1@gsfc.nasa.gov
Jamkhedkar, Priya	University of Maryland	priyaj@glue.umd.edu
Jones, Frank C.	NASA/GSFC	Frank.C.Jones.1@gsfc.nasa.gov
Kainer, Selig	BKG Research	kainer@lepvax.gsfc.nasa.gov
Kallman, Timothy	NASA/GSFC	Timothy.R.Kallman.1@gsfc.nasa.gov
Kazanas, Demosthenes	NASA/GSFC	Demos.Kazanas.1@gsfc.nasa.gov
Kelley, Richard	NASA/GSFC	Richard.L.Kelley.1@gsfc.nasa.gov
Kochanek, Chris	Harvard-Smithsonian-CfA	ckochanek@cfa.harvard.edu
Kogut, Al	NASA/GSFC	Alan.Kogut@gsfc.nasa.gov
Kollmeier, Juna A.	The Ohio State University	jak@astronomy.ohio-state.edu
Kosowsky, Arthur	Rutgers University	kosowsky@physics.rutgers.edu
Kuehn, Frederick	The Ohio State University	kuehn@mps.ohio-state.edu
Kuntz, K. D.	NASA/GSFC/LHEA-UMBC	kuntz@milkyway.gsfc.nasa.gov
Leisawitz, David	NASA/GSFC	David.T.Leisawitz.1@gsfc.nasa.gov
Leventhal, Marv	University of Maryland	ml@astro.umd.edu
Livadiotis, George	University of Athens	
Loeb, Avi	Princeton U.-IAS & Harvard U.	aloeb@cfa.harvard.edu
Majumdar, Subhabrata	U. of Illinois, Urbana-Champaign	subha@astro.uiuc.edu
Malhotra, Sangeeta	STScI	san@stsci.edu
Malkan, Matt	UCLA	malkan@astro.ucla.edu
Maran, Steve	NASA/GSFC	Stephen.P.Maran.1@gsfc.nasa.gov
Marshall, Frank	NASA/GSFC	Francis.E.Marshall.1@gsfc.nasa.gov
Martel, Hugo	University of Texas	hugo@simplicio.as.utexas.edu
Mather, John	NASA/GSFC	john.mather@gsfc.nasa.gov
McConnell, Barbara	National Geographic Magazine	bmcconne@ngs.org
McEnery, Julie	NASA/GSFC – UMBC/JCA	mcenery@milkyway.gsfc.nasa.gov
McKay, Tim	University of Michigan	tamckay@umich.edu
McKee, Maggie	Astronomy Magazine	mmckee@astronomy.com
Miller, Cole	University of Maryland	miller@astro.umd.edu
Mitchell, John W.	NASA/GSFC	John.W.Mitchell.1@gsfc.nasa.gov
Mundy, Lee	University of Maryland	lgm@astro.umd.edu
Mushotzky, Richard F.	NASA/GSFC	richard@xray-5.gsfc.nasa.gov
Nakajima, Reiko	University of Pennsylvania	reiko3@sas.upenn.edu
Natarajan, Priya	Yale University	priyamvada.natarajan@yale.edu
Nelan, Jenica	Dartmouth College	jnelan@dartmouth.edu
Niedner, Mal	NASA/GSFC	Malcolm.B.Niedner.1@gsfc.nasa.gov
Norris, Jay	NASA/GSFC	Jay.P.Norris.1@gsfc.nasa.gov
Oegerle, Bill	NASA/GSFC	William.R.Oegerle.1@gsfc.nasa.gov
Ormes, Jonathan F.	NASA/GSFC	Jonathan.F.Ormes1@gsfc.nasa.gov
Overzier, Roderik	JHU & Leiden Observatory	overzier@strw.leidenuniv.nl
Peacock, John	University of Edinburgh	jap@roe.ac.uk
Petre, Robert	NASA/GSFC	Robert.Petre.1@gsfc.nasa.gov
Porter, F. Scott	NASA/GSFC	Frederick.S.Porter.1@gsfc.nasa.gov
Rees, Martin	Institute of Astronomy-Cambridge	mjr@ast.cam.ac.uk
Reiprich, Thomas	University of Virginia	thomas@reiprich.net
Reynolds, Chris	University of Maryland	chris@astro.umd.edu

Rose, William K.	University of Maryland	wrose@astro.umd.edu
Rosenbaum, Doris	SMU	DRTeplitz@aol.com
Rupke, David	University of Maryland	drupke@astro.umd.edu
Sanders, Wilton	University of Wisconsin-Madison	wtsander@wisc.edu
Schirber, Michael	The Ohio State University	schirber@pacific.mps.ohio-state.edu
Schmid, Christoph	ETH-Zurich	chschmid@itp.phys.ethz.ch
Serlemitsos, Peter	NASA/GSFC	Peter.J.Serlemitsos.1@gsfc.nasa.gov
Shapiro, Paul	University of Texas	shapiro@astro.as.utexas.edu
Silverberg, Bob	NASA/GSFC	Robert.F.Silverberg.1@gsfc.nasa.gov
Sonneborn, George	NASA/GSFC	george.sonneborn@gsfc.nasa.gov
Staggs, Suzanne	rinceton University	staggs@princeton.edu
Stahle, Caroline K.	NASA/GSFC	Caroline.K.Stahle.1@gsfc.nasa.gov
Stebbins, Robin	NASA/GSFC	stebbins@lheapop.gsfc.nasa.gov
Stecker, Floyd W.	NASA/GSFC	Floyd.W.Stecker.1@gsfc.nasa.gov
Steidel, Chuck	Caltech	css@astro.caltech.edu
Steinhardt, Paul	Princeton University	steinh@princeton.edu
Stiller, Bertram		
Swank, Jean	NASA/GSFC	Jean.H.Swank.1@gsfc.nasa.gov
Tan, Jonathan	Princeton University	jt@astro.princeton.edu
Tanaka, Yasuo	MPI for Extraterrestrial Physics	ytanaka@mpe.mpg.de
Teegarden, Bonnard	NASA/GSFC	Bonnard.J.Teegarden.1@gsfc.nasa.gov
Tegmark, Max	University of Pennsylvania	max@physics.upenn.edu
Temkin, Aaron	NASA/GSFC	
Tinker, Jeremy	The Ohio State University	tinker@astronomy.ohio-state.edu
Tittley, Eric	JCA/UMBC	tittley@jca.umbc.edu
Trasco, John	University of Maryland	jtrasco@astro.umd.edu
Trimble, Virginia	UC-Irvine & U. of Maryland	vtrimble@astro.umd.edu
Wang, Xiaomin	University of Pennsylvannia	xiaomin@student.physics.upenn.edu
Wang, JunXian	The Johns Hopkins University	jxw@pha.jhu.edu
Wegner, Gary A.	Dartmouth College	gary.wegner@dartmouth.edu
Weinberg, David	The Ohio State University	dhw@astronomy.ohio-state.edu
Westover, Mike	Harvard-Smithsonian-CfA	mwestover@cfa.harvard.edu
White, Simon	MPI for Astrophysics	swhite@mpa-garching.mpg.de
White, Nicholas E.	NASA/GSFC	Nicholas.E.White.1@gsfc.nasa.gov
Wise, John H.	Penn State University	jwise@astro.psu.edu
Wollack, Ed	NASA/GSFC	Edward.J.Wollack.1@gsfc.nasa.gov
Woodgate, Bruce	NASA/GSFC	woodgate@stars.gsfc.nasa.gov
Xu, Yongzhong	University of Pennsylvania	xuyz@hep.upenn.edu
Yoshida, Naoki	Harvard-Smithsonian-CfA	nyoshida@cfa.harvard.edu
Zentner, Andrew R.	The Ohio State University	zentner@pacific.mps.ohio-state.edu
Zhang, Will	NASA/GSFC	William.W.Zhang.1@gsfc.nasa.gov

Author Index

A

Abel, T., 97
Allen, P. D., 249, 269
Arabadjis, G., 135
Arabadjis, J. S., 135
Arribas, S., 307

B

Barger, A. J., 205
Bautz, M. W., 135
Berrington, R. C., 183
Bharadwaj, S., 303
Bhavsar, S. P., 303
Böhringer, H., 139, 291, 319
Borne, K. D., 307
Boughn, S. P., 67
Bromm, V., 73
Brown, M., 265
Bullock, J. S., 151, 261
Bushouse, H., 307

C

Centrella, J. M., 337
Cinar, Y., 357
Colbert, J. W., 241
Colina, L., 307
Crittenden, R. G., 67

D

Dalal, N., 103
Dalton, G. B., 249, 269
Davé, R., 157, 191
Davies, R. L., 315
Dermer, C. D., 183
Donahue, M., 311

F

Fang, L.-Z., 187
Ferrara, A., 85
Franceschini, A., 215

G

Garaimov, V. I., 361
Gardner, J. P., 245
Greyber, H. D., 365

H

Hammell, M., 249, 269
Hasinger, G., 227
Heckman, T., 265
Henriksen, M., 199
Horner, D. J., 311
Hu, W., 45
Hudson, M. J., 315

I

Ikebe, Y., 139
Iliev, I. T., 85, 89

J

Jamkhedkar, P., 187

K

Kainer, S., 237
Katz, N., 157, 191
Kempner, J. C., 319
Kitayama, T., 139
Kochanek, C. S., 103
Kollmeier, J. A., 157, 191
Kosowsky, A., 325

L

Loeb, A., 73
Lucas, R., 307
Lucey, J. R., 315

M

Malecki, J. J., 315
Malhotra, S., 265
Malkan, M. A., 241, 253
Martel, H., 85, 89
Mather, J. C., 347
McKay, T. A., 123
McKee, C. F., 93
McLean, I. S., 253
Moore, S. A. W., 315
Moustakas, L., 249, 269
Mushotzky, R., 171

N

Natarajan, P., 113
Nelan, J. E., 315
Norman, C., 265

O

Olding, E. J., 249, 269
Overzier, R. A., 257

P

Peacock, J. A., 275

Q

Quinney, S. J., 315

R

Raga, A. C., 89
Reiprich, T. H., 319
Rengelink, R. B., 257
Retzlaff, J., 319
Rhoads, J., 265
Rich, M., 241
Rodighiero, G., 215
Rose, W. K., 143, 237
Röttgering, H. J. A., 257
Rupke, D. S., 195

S

Sanders, D. B., 195
Sarazin, C. L., 319
Scannapieco, E., 85
Schade, D., 315
Schirber, M., 261
Schuecker, P., 291
Shapiro, P. R., 85, 89
Sheth, J. V., 303
Sivakoff, G. R., 319
Skrutskie, M. F., 319
Smith, R. J., 315
Staggs, S. T., 59
Steinhardt, P. J., 33
Suntzeff, N. B., 315

T

Tan, J. C., 93
Tegmark, M., 19
Teplitz, H. I., 253
Tittley, E., 199
Trimble, V., 3
Turok, N., 33

V

Veilleux, S., 195

W

Wang, J., 265
Wegner, G. A., 249, 269, 315
Weinberg, D. H., 157, 191
Wilman, R. J., 257
Wise, J. H., 97

Y

Yoshida, N., 147

Z

Zentner, A. R., 151

Subject Index

2MASS, 248, 308, 319, 320, 322
2dF Galaxy Redshift Survey, 11, 28, 212, 275–279, 281, 283–289

Abell 1644, 319, 322
Abell 1795, 322
Abell 3391, 199–201
Abell 3395, 199–201
Abell 400, 316
Abell clusters, 9–11, 13, 117, 139, 140, 186, 315–318
accretion disks, 110, 143, 239, 337, 340
active galactic nuclei, 143, 145, 191–193, 195, 197, 205, 207–209, 211, 221, 227–235, 238, 239, 241, 243, 257–259, 261–263, 265–268, 270, 293, 308, 309
ALMA, 333, 348
American Astronomical Society, 5
ASCA, 139, 199, 311–314
Atacama Cosmology Telescope, 325, 333, 334
Automated Plate Measuring machine, 276, 278, 283

baryon loading, 52
baryon-photon ratio, 46, 53
bias parameter, 281, 282, 287, 289
big bang, 25, 33, 40, 41, 46, 53, 144
big crunch, 25
BIMA, 57
black holes, 20, 29–31, 73, 81, 82, 99, 100, 143–145, 205, 208, 228, 235, 237–239, 257, 259, 263, 337–344
black holes, primordial, 239
black holes, stellar mass, 29
black holes, supermassive, 29, 73, 81, 143–145, 205, 208, 228, 237–239, 337–343, 345
BOOMerANG, 46, 65
cD-galaxies, 257
Cepheid variables, 9
Chandra Deep Field North, 206, 208–211, 266
Chandra Deep Field South, 206, 228–234, 266
Chandra X-ray Observatory, 135, 205, 227–229, 231, 233–236, 239, 265, 311–314, 319, 322
clusters of galaxies, 3, 29, 135, 139, 183, 199, 249, 269, 291–293, 311, 319, 320, 331, 334, 337
clusters of galaxies, M-L relation, 295, 298
clusters of galaxies, mergers, 183–186
COBE, 45, 46, 51, 67, 69, 151, 216, 217, 325, 350
Coma, 8, 10, 183, 317
concordance model, 23, 26, 28, 298
cooling flow, 135–139
Copernicus, 4
correlation function, redshift-space, 279, 289

correlation function, two-point, 257, 279, 298
Cosmic Background Imager, 46, 57
Cosmic Infrared Background, 215–222, 309, 348
Cosmic Microwave Background, 13, 19–24, 26–29, 31, 33, 39, 40, 45–48, 50, 52, 56–65, 67–69, 73, 85, 87, 161, 275, 285–289, 291, 315, 325–334, 365
Cosmic Microwave Background, acoustic oscillations, 49, 50, 53, 55, 56, 59, 61–63, 326
Cosmic Microwave Background, acoustic peaks, 46, 54, 56, 57, 328
Cosmic Microwave Background, anisotropies, 45, 67, 285, 286
Cosmic Microwave Background, dipole anisotropy, 13
Cosmic Microwave Background, first peak, 22, 45, 52
Cosmic Microwave Background, polarization, 47
Cosmic Microwave Background, power spectrum, 22, 31, 63, 327
Cosmic Microwave Background, Sachs-Wolfe effect, 45, 46, 50, 54, 55, 57, 67, 69, 70
Cosmic Microwave Background, second peak, 46
Cosmic Microwave Background, surface of last scattering, 63, 325, 329, 332
cosmic variance, 69, 270, 279, 282, 286

cosmological constant, 11, 23, 24, 67, 69, 151, 291, 333
cosmological defects, 50
cyclic model, 33–39, 41

D/H ratio, 285
dark energy, 19, 23–26, 29–31, 34–36, 45, 52, 54, 55, 57, 157, 163, 164, 168, 291, 300, 301
dark matter, 19, 24, 26, 29–31, 46, 54, 55, 57, 73, 81, 86, 93, 97, 98, 103, 105, 113–117, 121, 123, 124, 129, 132, 135, 136, 139–144, 148, 151, 157, 158, 160, 186, 187, 238, 258, 288, 295, 326, 330, 343
dark matter, cold, 24, 29, 30, 46, 73, 97, 117, 121, 123, 135, 144, 158
dark matter, halos, 30, 73, 113, 123, 124, 132, 135, 148, 186, 343
dark matter, self-interacting, 30, 103
dark matter, warm, 30, 135, 188–190
DASI, 46, 56, 64, 65
data mining, 307–309
decoupling, epoch of, 61
DUET, 301
dynamical friction, 238, 239, 340

EGRET, 183, 186
Einstein field equations, 21
ekpyrotic scenario, 25
entropy, 35, 93, 123, 361–364
excitation temperature, 85
extremely red objects, 224, 243, 244, 257–259, 269, 270, 307

Faber-Jackson relations, 317, 318
filaments, 10, 11, 75, 158, 199, 288, 303–305, 311
Finger of God, 279
first stars, 73, 75, 78, 79, 85, 97, 98, 147
fluctuations, curvature, 45, 48–50, 56, 57
fluctuations, metric, 27, 48, 54
fluctuations, quantum, 35, 41, 123
fluctuations, temperature, 31, 47, 54, 56, 87, 325, 326, 328–330
Friedmann equation, 24, 25, 329
Friedmann-Robertson-Walker metric, 22, 35

Galactic halo, 100
galaxies, dwarf, 81, 238, 365
galaxies, elliptical, 13, 81, 143, 145, 237, 270, 282
galaxies, evolution, 215, 228, 241, 249, 251
galaxies, formation, 13, 26, 28, 73, 93, 143, 164, 195, 249, 257, 307, 308
galaxies, harassment, 116
galaxies, luminosity function, 160, 208, 220, 221, 227, 228, 234, 235, 245–247, 261, 278, 289, 294, 295, 298
galaxies, Lyman Break, 164, 193, 253, 254, 271
galaxies, spiral, 237, 238, 347
galaxies, starburst, 193, 218, 220, 230, 231, 239, 262, 270
GALEX, 245, 248

gamma-ray bursts, 73, 79, 80, 100, 307
general relativity, 19, 21, 24, 30, 338
GEO600, 339
globular clusters, 8
gravitational lensing, 26, 27, 29, 30, 57, 63, 103, 253, 329, 332
gravitational microlensing, 105, 106, 110
gravitational waves, 21, 39–41, 56, 57, 59, 63, 65, 82, 238, 330, 337–343, 345
gravitational waves, primordial, 41
gravitational waves, spectrum, 39
Greeks, 4
GUT, 329

HEAO1 A2, 67, 68
Hercules Deep Field, 309, 351
Herschel observatory (FIRST), 243, 244
Hubble Deep Field North, 228, 229, 232–234
Hubble Deep Field South, 217, 223
Hubble expansion, 12
Hubble Space Telescope, 216, 217, 224, 241, 242, 255, 284, 287, 309, 311, 312, 348, 349

inflationary cosmology, 25, 26, 33, 34, 36–41, 46, 49, 50, 56, 57, 59, 60, 63, 151, 168, 237, 239, 328, 329
Infrared Imaging Surveyor, 350
Infrared Space Observatory, 216–222

initial mass function, 73, 74, 100, 222
intergalactic medium, 73, 80, 86, 89, 91, 97, 98, 158, 163, 167, 168, 197, 198, 211, 263
intermittency, 187–189
interstellar medium, 103–106, 108, 109, 265
interstellar obscuration, 6
intracluster medium, 139–142, 183, 185, 291
intracluster medium, density profile, 140
intracluster medium, temperature profile, 140, 141
ionization fronts, 89
IRAS, 220, 221, 308, 309, 350
IRAS, Faint Source Catalog, 309
island universe, 5, 10

James Webb Space Telescope (NGST), 348, 349, 351, 353, 354
jets, 79, 143, 337, 341, 365

Kaiser distortion, 275

LALA survey, 265, 266
large-scale structure, 19, 20, 26, 28, 45, 67, 68, 70, 135, 259, 275, 289, 291, 292, 294
Las Campanas Redshift Survey, 303–306
LIGO, 40, 339, 345
LISA, 40, 337, 340–346
Local Group, 8, 9, 13
Lockman Hole, 219, 220
LOFAR, 85, 87, 88
Lyman-Werner bands, 81, 98, 149

M31, 3, 7, 144
M87, 145
MAP, 20, 57, 64, 326, 329, 330, 348
MAT, 333
Maxima, 45, 46, 65
Mesopotamians, 3
Milky Way, 5–7, 9, 12, 15, 29, 144, 151, 152, 292, 347
minihalos, 85–87, 89–92
modified Newtonian Dynamics, 30
MS0302, 311–313
MS1512-cB58, 253

National Virtual Observatory, 307–309
Next Generation Sky Survey, 349–351
NFW models, 141, 153
nucleosynthesis, 46, 60
numerical relativity, 341
NVSS, 257, 258

Olbers' paradox, 4
Ostriker-Vishniac effect, 331
Oxford-Dartmouth Survey, 249, 252, 271, 272

pair instability, 97–100
Palomar Observatory Sky Survey, 10
photoevaporation, 89–92
Planck density, 25
power spectrum, CMB, 22, 31, 63, 327
power spectrum, galaxy, 29, 275, 283, 287, 289
power spectrum, primordial, 19, 28, 29, 87, 151, 326, 328, 330

power spectrum, tilted, 151–153
Press-Schechter, 87, 90, 97
primordial gas, 73, 75, 81, 147, 149, 150, 238
primordial stars, 97
protogalaxies, 238
proximity effect, 191, 261–264

QSOs, 187, 188, 228, 231–233, 262, 285
QSOs, type-2, 228, 232

radiation-dominated universe, 34, 52
radio galaxies, 11, 145, 257–259, 342
radio relics, 183
recombination, 47, 51, 52, 54–56, 85, 123, 158
redshift-space distortions, 275, 281, 288, 289
REFLEX survey, 291–293, 295, 298–301
reionization, 28, 48, 57, 62, 65, 82, 85, 87, 89, 90, 92, 168, 275, 288, 337, 338, 341–343, 345, 351
ringdown radiation, 183, 186, 227, 234, 291–293, 311, 312, 320
ROSAT, 139
ROSAT All Sky Survey, 200, 202, 291–293, 301
ROSITA, 26, 65, 89
Rubin-Ford effect, 12

Saskatoon, 333
scalar perturbations, 63, 326–329
scale invariant spectrum, 35, 36

scale invariant spectrum, nearly, 33, 34, 36, 39, 46, 49
SCUBA, 221, 307
Seyfert nuclei, 144
SIRCE, 349, 352
SIRTF, 224, 245, 248, 347–349
Sloan Digital Sky Survey, 11, 123, 125–127, 129, 131, 132, 157, 161, 167, 208, 209, 211, 212, 245–248, 276, 278, 279
SOFIA, 349, 350
sound horizon, 51, 53, 327
SPECS, 347, 349, 353
SPICA, 349, 352
spin temperature, 85
SPIRIT, 349, 353
square-kilometer array, 85–88
SS433, 145
star formation, rate, 164, 191, 193, 196, 222, 224, 241, 249
Sunyaev-Zeldovich effect, 331
Sunyaev-Zeldovich effect, thermal, 331
superclusters, 10, 311, 366
supernovae, 19, 25, 45, 67, 76, 81, 97, 98, 100, 105, 164, 193, 195, 291, 300, 301, 352
supernovae, type-1a, 19, 23–26, 29, 299
superstring theory, branes, 34, 35
superstring theory, M-theory, 34, 37

tensor perturbations, 328, 329
tidal stripping, 116
Toco, 333
topology, 19, 20, 22, 23, 26, 329

ultraluminous infrared galaxies, 195–198, 222, 308, 309, 351
ultraviolet background, 98, 157

Very Large Telescope, 162, 167, 229, 230
very luminous infrared galaxies, 307–309
VIRGO, 339

X-ray background, 67–70, 78, 205, 227
XMM-Newton observatory, 139, 140, 227, 228, 231, 319, 322